NANOTECHNOLOGY SCIENCE AND TECHNOLOGY

GRAPHENE

PROPERTIES, SYNTHESIS AND APPLICATIONS

NANOTECHNOLOGY SCIENCE AND TECHNOLOGY

Additional books in this series can be found on Nova's website under the Series tab.

Additional E-books in this series can be found on Nova's website under the E-book tab.

NANOTECHNOLOGY SCIENCE AND TECHNOLOGY

GRAPHENE

PROPERTIES, SYNTHESIS AND APPLICATIONS

ZHIPING XU
EDITOR

Nova Science Publishers, Inc.
New York

For permission to use material from this book please contact us:
Telephone 631-231-7269; Fax 631-231-8175
Web Site: http://www.novapublishers.com

LIBRARY OF CONGRESS CATALOGING-IN-PUBLICATION DATA

Graphene : properties, synthesis, and applications / editor, Zhiping Xu.
p. cm.
Includes bibliographical references and index.
ISBN 978-1-61470-949-7 (hardcover : alk. paper) 1. Graphene. I. Xu, Zhiping, 1981-
QD341.H9G694 2011
549'.27--dc23
2011029443

Published by Nova Science Publishers, Inc. † New York

CONTENTS

PREFACE

Graphene is an ultimately monolayer material with single-atom thickness. It possesses many fascinating properties including massless Dirac electronic structure, extraordinary high electron mobility, thermal conductivity, stiffness and strength, and large surface volume ratio etc. This edited book is organized into two parts. The first part focuses on the electronic and mechanical properties of graphene materials. The second part of this book covers the application aspect of graphene and related materials. Graphene is a simple material, but also a building block for very complicated applications. The approach to understand the role of graphene and related materials in these fields relies not only on individual research by scientists in their own fields but also, on interdisciplinary cross-talk, as the authors present here in this book.

Chapter 1 - Since the advent of graphene, the two-dimension allotrope of carbon, there has been a growing interest in calculation of quantum mechanical properties of this crystal and its derivatives. To this end, the tight-binding model has been in frequent use for a relatively long period of time. This review, presents an in-depth survey on this method as possible to graphene and two of the most important related structures, graphene anti-dot lattice, and graphane. The two latter derivatives of graphene are semiconducting, while graphane is semimetallic. The authors present band structure and quantum mechanical orbital calculations and justify the results with the density function theory.

Chapter 2 - In this chapter only ballistic transport was considered. In principle, disorder and electron-electron interactions could be incorporated within the dynamical approach in the way the Boltzmann equation approach can be extended beyond the linear response. In fact, using a phenomenological methodology, disorder has recently been incorporated for the ac field. Similarly Coulomb interactions and the pair annihilations into photons, phonons etc. can be taken into account. Generally though, when the system has a large number of electron-hole pairs, the screening by the neutral plasma is more effective and the influence of impurities and interactions diminishes. Understanding these effects is crucial for investigating the (nonlinear) plasma waves and their damping.

Chapter 3 - Tuning properties of graphene system is a key issue for its actual application in nanoelectronic devices. The authors' first-principles calculations reveal surprisingly high sensitivity of the field-induced energy gap of bilayer graphene to changes in its interlayer spacing. Small adjustments in the interlayer spacing near its equilibrium value produce large modulations in the gap over a wide range of field strength. For epitaxial bilayer graphene grown on SiC substrate, the first C layer or the buffer layer is electronically active to the second C layers in the presence of electric and mechanical tuning. The second graphene on C-terminated SiC exhibits an *n*-to-*p*-type transition under electric field, while the Fermi level of the second graphene on Si-terminated SiC is significantly shifted by compressing interlayer distance with respect to its buffer layer. These results provide insights for fundamental understanding and innovative device applications of bilayer graphene materials.

Chapter 4 - The authors have presented here an extensive study on electrical spectroscopy using conducting tip atomic force microscopy (CTAFM) of graphene sheets and its edges. The authors have tried to address in the present study two main issues: (1) How does the electrical property of the graphene sheet changes when the graphene sheet is displaced by shear forces over highly oriented pyrolytic graphite (HOPG) substrate? and (2) How does the electrical property of the graphene sheet change across it edges? While addressing these two issues the authors observed that the conductivity of graphene sheets is higher whenever they are loosely bound to the underlying bulk HOPG substrate. The authors also observe that certain edges of the graphene layers show sharp rise in current when biased, indicating enhanced electronic density of states spatially localized near those edges. In certain edges, the authors do not observe this phenomenon. These two observations are substantiated by a theoretical study of these edges (passivated with hydrogen) using molecular orbital calculations based on density functional theory (DFT) with B3LYP functional. Structure of edges of 2D graphene sheets are either in zigzag (trans) or armchair (cis) forms. In their theoretical investigation the authors observed high electron density along the zigzag edge and low density at the armchair edge. The authors also found out that the electron density at the central region of the sheet increases as the interlayer separation among the graphene layers are increased which supports their experimental data. The authors also find that the HOMO-LUMO gap of the graphene sheet varies as a function of its size . The authors have shown here that the electron density around a mono-vacancy in a graphene sheet is high, which further leads to bending of the graphene sheet.

Chapter 5 - Graphene, a two-dimensional sp^2 hybridized lattice of carbon atoms, has attracted intense interest from both theoretical and experimental scientists due to its unique electronic, mechanical, chemical, and catalytic properties. Recent synthetic efforts have been paid in diverse directions. For instance, chemical vapor deposition on catalytic metal surfaces, carbon segregation from silicon carbide or carbon-doped metal substrates, micromechanical cleavage of graphite, exfoliation of graphite in solvents, and chemical conversion from graphite have been developed to produce single-layer or few-layer graphene sheets. Among them, chemical conversion from graphite and exfoliation of graphite in solvents appears to be efficient approaches to produce graphene or chemically modified graphene (CMG) on large scale and at low cost. Although chemical modification partly destroys the sp^2 crystalline structure of graphene, the functional groups or defects of CMGs may provide them with new properties or functions. Compared to graphene films produced from physical methods, graphene from this chemical way is readily dispersible in solvents and easily processable. Thus, the further manipulation, modification, or applications is possible. As a result, CMG have a variety of technological applications for including nanoelectronic and photoeletronic devices, energy storage materials, polymer composites, and electrocatalysis. However, many limitations exist in the chemical synthesis of graphene layers. It is still insurmountable challenges to synthesize graphene with controlled thickness, high crystalline structure, large size, or high concentration suspension. Therefore, it is of significance to understand the influence of synthesis parameters on the resulting graphene files in order to control the synthesis process to realize the desired properties. The content of this chapter mainly focuses on recent progress in chemical methods for graphene synthesis (*i.e.*, chemical conversion from graphite and exfoliation of graphite in solvents), the unique properties and functions of graphene/CMG generated from different chemical approaches, and the potential applications. Moreover, future developments are also addressed.

Chapter 6 - Graphene,a single atomic layer of two-dimensional carbon nanomaterials, owning excellent electrical, thermal and mechanical properties, has attracted intensive attentions since it was first separated by mechanical cleavage. However, this method can't provide large area graphene for research and application. To find a low cost preparation method that yields large scale and high qualitygraphene seems significantly important. Rapid progress has been made recently.Large area graphene films with thickness varying from single layer to few layers havebeen synthesized. These films, which show high conductance, transmittance and flexibility, have potentialapplicationin electronics and energy industry. However, some challenges must be overcomebefore industrial production. For example, there exists a gap between theoreticalproperties and practical properties due to thedefects present in the film.Furthermore, some synthesis methods will raise pollution. Similarly, there are both opportunities and challenges for practical application.Solar cells composed of graphene show potential application prospects but required furtherimprovement before practical in industry. Besides, graphene based composites display distinct properties, thanks to the unique structure of graphene.In this chapter, current status of graphene synthesis, including achievement and challenge, will be reviewed comprehensively. Oxidation reduction, liquid-phase exfoliation and chemical vapor deposition, which are promising for future research and application, will be introduced in details. Besides, research on graphene based solar cells and composites will be reviewed as well.

Chapter 7 - The electrochemical performance and utilisation of graphene is explored and overviewed. Starting with graphene's use as a sensor substrate through to its utilisation in energy production and storage devices, this chapter details the reported enhancements that graphene has imposed upon various electrochemical applications. A comprehensive approach is employed in which the possible limitations of graphene's exploitation within electrochemistry are highlighted and insight into the future prospects of graphene electrochemistry is sought at each stage.

Chapter 8 - Capacitive deionisation (CDI) is a promising alternative technology in desalination. It targets to remove the salt ions which are only a small percentage of the feed solution, as compared to most other technologies that aim to shift water which accounts of 90% of the feed solution. As a result, CDI requires less energy to operate and the electrodes are easily regenerated. Porous carbon materials are the most important component in the CDI processes, as they are used as the electrodes that play a significant role in the efficiency of the desalting process. The ideal electrode materials for CDI should be both highly conductive and of high surface area and suitable pore structures. The currently available carbon electrodes limit the desalination efficiency of CDI due to their low conductivity and non-ideal pore structure and pattern. Previous research has demonstrated that the efficiency of CDI strongly depends upon the surface properties of the carbon electrodes, such as surface area and pore microstructure. Many kinds of carbon materials have been investigated as CDI electrodes such as carbon aerogel, carbon cloth, carbon nanotubes and mesoporous carbons. Recently, graphene-like nano-flakes (GNFs) with relatively high specific surface area have been prepared and used as electrodes for capacitive deionization. The GNFs were synthesized by a modified Hummers' method using hydrazine for reduction. The GNFs prepared by this process had the specific surface area as high as 222.01 m^2/g. It was found that the ratio of nitric acid and sulfuric acid plays a vital role in determining the specific surface area of GNFs. They were characterized by atomic force microscopy, N_2 adsorption at 77 K and electrochemical workstation. Its electrosorption performance was much better than commercial activated carbon (AC), suggesting a great potential in capacitive deionisation application. Further, the electrosorptive performance of GNFs electrodes at various operational conditions, include different electrical potentials, flow rates and ionic strengths were measured and the electrosorption isotherm and kinetics were investigated. It is found that the specific electrosorptive capacity of the GNFs was 23.18 μmol/g for sodium ions (Na^+) when the initial concentration was at 25 mg/l, which was higher than that of previously reported data using graphene and AC under the same experimental condition. In addition, the equilibrium electrosorption capacity was determined as 73.47 μmol/g at 2.0 V by fitting data through the Langmuir isotherm, and the rate constant was found to be 1.01 min^{-1} by fitting data through pseudo-first-order adsorption. The results suggested that GNFs synthesized by a modified chemical oxidation and reduction method can be used as effective electrode materials in CDI process for brackish water desalination. It was found that highly charged multivalent ions with smaller hydrated radius would be strongly attracted to the electrodes. For cations with same charge, the one with smaller hydrated radius would be more effectively removed.

Chapter 9 - Graphene, a recently discovered carbon nanomaterial with carbon atoms tightly packed into a two dimensional honeycomb lattice, possesses many novel and unique physical and chemical properties because of its unusual monolayer atomic structure. It has found potential applications in many fields such as nanocomposites, batteries, supercapacitors, nanoelectronic devices, and sensors, etc. This chapter describes the application of graphene in the field of electrocatalyst preparation, and presents the fabrication, characterization and electrocatalytic characteristics of the metal–graphene nanocomposites. The electrochemically deposition of Au and Pt–Au nanostructures on the surface of graphene sheets with the formation of the Au–graphene and Pt–Au–graphene nanocomposites, respectively, and characterization of these nanocomposites with scanning electron microscopy (SEM), energy–dispersive spectroscopy (EDS), X–ray powder diffraction (XRD), and electrochemical methods are depicted. The dependence of the morphology, size, and composition of the Au and Pt–Au nanostructures on the experimental parameters, such as electrodeposition time, the concentration of the metal precursors, and the molar ratio of these precursors is also described. The electrocatalytic activities of the Au–graphene and Pt–Au–graphene nanocomposites toward oxygen reduction and some small organic molecules (glucose, methanol) oxidation were investigated by cyclic voltammetry. For comparison, the Au and Pt–Au nanostructures were also electrodeposited on the surface of the conventional carbon materials, such as glassy carbon and Vulcan XC–72 carbon, and the electrocatalytic activities of these nanocatalysts were also studied. These studies presented in this chapter can provide a general route for fabrication of graphene–based metal nanocatalysts, which could be potential utility to fuel cells and the related fields.

Chapter 10 - Graphene, a flat monolayer of carbon atoms tightly packed into a two–dimensional (2D) honeycomb lattice, holds great promise for application in electrochemistry due to its large surface area, outstanding electrical conductivity, inert electrochemical activity, wide potential window, and so forth. In particular, graphene has high potential application in redox protein immobilization for fabrication of protein–graphene nanohybrid. This chapter describes the preparation and characterization of glucose oxidase/graphene (GOx–graphene) and laccase/graphene (Lac–graphene) nanohybrids. The direct electron transfer characteristics of the GOx on the surface of graphene, the electrocatalytic activities of the GOx–graphene nanohybrid toward the reduction of oxygen, and application in the detection of glucose in human blood serum are presented. The electrocatalytic reduction of oxygen by the Laccase–graphene (Lac–graphene) nanohybrid in the presence of 2,2–azino–bis(3–ethylbenzothiazoline–6–sulfonic acid) (ABTS) as electrochemical mediator is depicted. The detection of the extracellular oxygen released from human erythrocytes with use of the Lac–graphene nanohybrid is also presented in this chapter. This chapter establishes a general route for fabricating redox proteins–based graphene nanohybrids, which has potential utility to bioelectroanalytical chemistry, cellular biology, pathophysiology, and so on.

Chapter 11 - Graphene, a true 2-dimensional nanomaterial, has received increasing attention due to its unique physicochemical properties like large surface area, excellent conductivity, high mechanical strength and ease of bio-functionalization. Based on its unique optical, physical and electrical properties, graphene has found applications in a wide spectrum of biomedical utilities like imaging, sensing and therapeutics. Attempts are also being made to synthesize highly stable sheets of graphene, which can be easily ingested or injected into the body. Graphene can be characterized with various characterization tools like transmission electron microscopy (TEM) and atomic force microscopy (AFM). In a recent report, we have characterized graphene oxide (GO) sheets by flow cytometry, which stands out as a remarkable and versatile means for study of physical properties of graphene and graphene-cell interactions. Flow cytometry also provides rapid access to high-dimension statistical information on individual GO sheets that is not achievable with existing characterization tools. This review discusses latest advancements in graphene research highlighting its unique characteristics, potential biomedical applications, future prospective, as well as toxicity considerations.

In: Graphene: Properties, Synthesis and Applications
Editor: Zhiping Xu
ISBN: 978-1-61470-949-7

Chapter 1

AN OVERVIEW OF TIGHT-BINDING METHOD FOR TWO-DIMENSIONAL CARBON STRUCTURES

Behnaz Gharekhanlou and Sina Khorasani*
School of Electrical Engineering, Sharif University
of Technology, Tehran, Iran

Abstract

Since the advent of graphene, the two-dimension allotrope of carbon, there has been a growing interest in calculation of quantum mechanical properties of this crystal and its derivatives. To this end, the tight-binding model has been in frequent use for a relatively long period of time. This review, presents an in-depth survey on this method as possible to graphene and two of the most important related structures, graphene anti-dot lattice, and graphane. The two latter derivatives of graphene are semiconducting, while graphane is semimetallic. We present band structure and quantum mechanical orbital calculations and justify the results with the density function theory.

Keywords: Graphene, Graphane, Carbon Nanostructures, Tight-binding Method, Quantum Mechanics.

I. TIGHT-BINDING METHOD

I.1. Introduction

The tight-binding model (or TB model) is an approach to the calculation of electronic band structure using an approximate set of wave functions based upon superposition of wave functions for isolated atoms located at each atomic site. This model describes the properties of tightly bound electrons in solids. The electrons in this model should be tightly bound to the

* E-mail address: khorasani@sina.sharif.edu

atom to which they belong and they should have limited interaction with states and potentials on surrounding atoms of the solid. As a result the wave function of the electron will be rather similar to the atomic orbital of the free atom where it belongs to. This method was developed by Bloch [1] in 1928 considered only the *s* atomic orbital. In 1934 Jones, Mott, and Skinner [2] considered different atomic orbitals.

Tight-binding models are applied to a wide variety of solids. The model gives good qualitative results in many cases and can be combined with other models that give better results where the tight-binding model fails. Though the tight-binding approach is a one-electron model, it also provides a basis for more advanced calculations like the calculation of surface states and application to various kinds of many-body problem and quasiparticle calculations. It is also a common practice to use optimized tight-binding methods, in which the values of the matrix elements are derived approximately or fitted to experiment or other theories [3].

I.2. Secular Equation

In a crystalline solid,the translational symmetry along the directions of thelattice vectors $\mathbf{a}_i$, enforces the Bloch's theorem to be satisfied by any wave function of the lattice. Mathematically, this can be written as

$$\mathbb{T}\psi_k(r) = e^{i\mathbf{k}\cdot\mathbf{a}_i}\psi_k(r) \qquad (i = 1,2,3), \tag{1}$$

where $\mathbb{T}$ is the translation operator along the lattice vector $\mathbf{a}_i$, and $\mathbf{k}$ is the Bloch wave vector [4, 5]. In the tight-binding model of electronic structures, single electron wave functions are normally expanded in terms of atomic orbitals [6, 7]

$$\varphi_{nlm}\left(r,\theta,\Phi\right) = R_{nl}(r)Y_{lm}\left(\theta,\Phi\right), \tag{2}$$

where R_{nl} and Y_{lm} are radial and spherical-harmonic functions in polar coordinates and n, l, and m are the principal, angular-momentum, and magnetic quantum numbers, respectively.

The wave function can be defined in variety of models, but the most commonly used form is a linear combination of atomic orbitals. Thus the wave function $\phi(\mathbf{k},\mathbf{r})$ is defined as a summation on the atomic wavefunction $\varphi(\mathbf{r}-\mathbf{R}_j)$ on the site j

$$\phi_j(\mathbf{k},\mathbf{r}) = \frac{1}{\sqrt{N}}\sum_{\mathbf{R}_j}^{N} e^{i\mathbf{k}\cdot\mathbf{R}_j}\varphi(\mathbf{r}-\mathbf{R}_j) \qquad j = 1,\ldots,n, \tag{3}$$

where $\mathbf{R}_j$ is the position of the atom and N is the number of atomic wavefunctions in the unit cell. It is easy to verify that this function satisfies Bloch condition

$$\begin{aligned}\phi_j(\mathbf{k},\mathbf{r}+\mathbf{a}) &= \frac{1}{\sqrt{N}}\sum_{\mathbf{R}_j}^{N} e^{i\mathbf{k}\cdot\mathbf{R}_j}\varphi(\mathbf{r}+\mathbf{a}-\mathbf{R}_j) \\ &= e^{i\mathbf{k}\cdot\mathbf{a}_j}\frac{1}{\sqrt{N}}\sum_{\mathbf{R}_j-\mathbf{a}}^{N} e^{i\mathbf{k}\cdot(\mathbf{R}_j-\mathbf{a})}\varphi\left[\mathbf{r}-(\mathbf{R}_j-\mathbf{a})\right] \\ &= e^{i\mathbf{k}\cdot\mathbf{a}_j}\phi_j(\mathbf{k},\mathbf{r}).\end{aligned} \tag{4}$$

In a solid, eigenfunctions $\psi_j(\mathbf{k},\mathbf{r})$ $(j=1,\ldots,n)$ are defined as a linear recombination of Bloch functions

$$\psi_j(\mathbf{k},\mathbf{r}) = \sum_{j'=1}^{n} C_{jj'}(\mathbf{k})\,\phi_j(\mathbf{k},\mathbf{r}). \tag{5}$$

Here, $C_{jj'}(\mathbf{k})$ are coefficients which have to be determined.

The eigenvalues of the system described by the Hamiltonian $\mathbb{H}$ is given by [8]

$$E_j(\mathbf{k}) = \frac{\langle\psi_j|\mathbb{H}|\psi_j\rangle}{\langle\psi_j|\psi_j\rangle} = \frac{\int \psi_j^*(\mathbf{k},\mathbf{r})\mathbb{H}\,\psi_j(\mathbf{k},\mathbf{r})\,dr}{\int \psi_j^*(\mathbf{k},\mathbf{r})\,\psi_j(\mathbf{k},\mathbf{r})\,dr}. \tag{6}$$

Now, substituting $\psi_j(\mathbf{k},\mathbf{r})$ as defined in (5) leads to

$$E_j(\mathbf{k}) = \frac{\sum_{j,j'=1}^{n} C_{ij}^* C_{ij'}\langle\phi_j|\mathbb{H}|\phi_{j'}\rangle}{\sum_{j,j'=1}^{n} C_{ij}^* C_{ij'}\langle\phi_j|\phi_{j'}\rangle} = \frac{\sum_{j,j'=1}^{n} H_{jj'}(\mathbf{k})C_{ij}^*C_{ij'}}{\sum_{j,j'=1}^{n} S_{jj'}(\mathbf{k})C_{ij}^*C_{ij'}}, \tag{7}$$

where $H_{jj'}(\mathbf{k})$ and $S_{jj'}(\mathbf{k})$ are transfer and overlap matrices, respectively, and are defined by

$$H_{jj'}(\mathbf{k}) = \langle\phi_j|\mathbb{H}|\phi_{j'}\rangle, \tag{8}$$

$$S_{jj'}(\mathbf{k}) = \langle\phi_j|\phi_{j'}\rangle. \tag{9}$$

For a given $\mathbf{k}$ value, the coefficient $C_{jj'}^*(\mathbf{k})$ is optimized so as to minimize $E_i(\mathbf{k})$

$$\frac{\partial E_i(\mathbf{k})}{\partial C_{ij}^*(\mathbf{k})} = \frac{\sum_{j'=1}^{n} H_{jj'}(\mathbf{k})C_{ij'}(\mathbf{k})}{\sum_{j,j'=1}^{n} S_{jj'}(\mathbf{k})C_{ij}^*(\mathbf{k})C_{ij'}(\mathbf{k})}$$

$$-\frac{\sum_{j,j'=1}^{n} H_{jj'}(\mathbf{k})C_{ij}^{*}(\mathbf{k})C_{ij'}(\mathbf{k})}{\left[\sum_{j,j'=1}^{n} S_{jj'}(\mathbf{k})C_{ij}^{*}(\mathbf{k})C_{ij'}(\mathbf{k})\right]^2}\sum_{j'=1}^{n} S_{jj'}(\mathbf{k})C_{ij'}(\mathbf{k}) = 0. \tag{10}$$

This further can be rewritten as

$$\frac{\sum_{j'=1}^{n} H_{jj'}(\mathbf{k})C_{ij'}(\mathbf{k})}{\sum_{j,j'=1}^{n} S_{jj'}(\mathbf{k})C_{ij}^{*}(\mathbf{k})C_{ij'}(\mathbf{k})}$$

$$-\left[\frac{\sum_{j,j'=1}^{n} H_{jj'}(\mathbf{k})C_{ij}^{*}(\mathbf{k})C_{ij'}(\mathbf{k})}{\sum_{j,j'=1}^{n} S_{jj'}(\mathbf{k})C_{ij}^{*}(\mathbf{k})C_{ij'}(\mathbf{k})}\right]\frac{\sum_{j'=1}^{n} S_{jj'}(\mathbf{k})C_{ij'}(\mathbf{k})}{\sum_{j,j'=1}^{n} S_{jj'}(\mathbf{k})C_{ij}^{*}(\mathbf{k})C_{ij'}(\mathbf{k})} = 0. \tag{11}$$

This in turn can be readily simplified as

$$\sum_{j'=1}^{n} H_{jj'}(\mathbf{k})C_{ij'}(\mathbf{k}) - E_i(\mathbf{k})\sum_{j'=1}^{n} S_{jj'}(\mathbf{k})C_{ij'}(\mathbf{k}) = 0. \tag{12}$$

Putting in the matrix form we get

$$\mathbf{H} - E_i(\mathbf{k})\,\mathbf{S} \quad C_i \ \mathbf{k} \quad = 0. \tag{13}$$

If the matrix $\mathbf{H} - E_i(\mathbf{k})\,\mathbf{S}$ has an inverse, the vector $C_i \ \mathbf{k}$ will be identically zero and this leads to the trivial solution. Thus,non-trivial solutions require

$$\left|\mathbf{H} - E_i(\mathbf{k})\,\mathbf{S}\right| = 0. \tag{14}$$

This equation is called the secular equation, whose eigenvalues $E_i(\mathbf{k})$ give the energy bandstructure.

As an example, suppose that the primitive basis contains one atom with the atomic orbital function $\varphi(\mathbf{r})$. We get

$$\phi_j(\mathbf{k},\mathbf{r}) = \frac{1}{\sqrt{N}}\sum_{\mathbf{R}_j}^{N} e^{i\mathbf{k}\cdot\mathbf{R}_j}\,\varphi(\mathbf{r}-\mathbf{R}_j) \quad (j = 1,\ldots,n). \tag{15}$$

Here we use the nearest-neighbor Hamiltonian of the system; it means that only the effect of the nearest-neighbors of the atom in the unit cell is considered.The transfer and overlap matrices are now defined respectively as follows

$$\begin{aligned}
H_{jj}(\mathbf{k}) &= \left\langle \phi_j \left| \mathbb{H} \right| \phi_j \right\rangle = \frac{1}{N} \sum_{\mathbf{R}_j}^{N} \sum_{\mathbf{R}'_j} e^{i\mathbf{k}\cdot(\mathbf{R}_j - \mathbf{R}'_j)} \left\langle \varphi(\mathbf{r}-\mathbf{R}_j) \left| \mathbb{H} \right| \varphi(\mathbf{r}-\mathbf{R}'_j) \right\rangle \\
&= \frac{1}{N} \sum_{\mathbf{R}_j = \mathbf{R}'_j}^{N} \left\langle \varphi(\mathbf{r}-\mathbf{R}_j) \left| \mathbb{H} \right| \varphi(\mathbf{r}-\mathbf{R}_j) \right\rangle \\
&+ \frac{1}{N} \sum_{\mathbf{R}'_j = \mathbf{R}_j + \mathbf{a}}^{N} e^{i\mathbf{k}\cdot\mathbf{a}} \left\langle \varphi(\mathbf{r}-\mathbf{R}_j) \left| \mathbb{H} \right| \varphi\left[\mathbf{r}-(\mathbf{R}_j+\mathbf{a})\right] \right\rangle + \ldots \\
&= -\alpha - \gamma_0 \sum_{\mathbf{R}'_j = \mathbf{R}_j + \mathbf{a}}^{N} e^{i\mathbf{k}\cdot\mathbf{a}}
\end{aligned} \tag{16}$$

$$\begin{aligned}
S_{jj}(\mathbf{k}) &= \left\langle \phi_j \middle| \phi_j \right\rangle = \frac{1}{N} \sum_{\mathbf{R}_j}^{N} \sum_{\mathbf{R}'_{jj}} e^{i\mathbf{k}\cdot(\mathbf{R}_j - \mathbf{R}'_j)} \left\langle \varphi(\mathbf{r}-\mathbf{R}_j) \middle| \varphi(\mathbf{r}-\mathbf{R}'_j) \right\rangle \\
&= \frac{1}{N} \sum_{\mathbf{R}_j = \mathbf{R}'_j}^{N} \left\langle \varphi(\mathbf{r}-\mathbf{R}_j) \middle| \varphi(\mathbf{r}-\mathbf{R}_j) \right\rangle \\
&+ \frac{1}{N} \sum_{\mathbf{R}'_j = \mathbf{R}_j + \mathbf{a}}^{N} e^{i\mathbf{k}\cdot\mathbf{a}} \left\langle \varphi(\mathbf{r}-\mathbf{R}_j) \middle| \varphi\left[\mathbf{r}-(\mathbf{R}_j+\mathbf{a})\right] \right\rangle + \ldots = 1.
\end{aligned} \tag{17}$$

Here, we assume that the atomic wavefunction is normalized.Then, the secular equation reduces to

$$E(\mathbf{k}) = -\alpha - \gamma_0 \sum_{\mathbf{R}'_j = \mathbf{R}_j + \mathbf{a}}^{N} e^{i\mathbf{k}\cdot\mathbf{a}}. \tag{18}$$

The constant parameters α and γ_0 are usually found by fitting this equation to high symmetric points of experimental data.

We can use an important approximation in the Tight-binding method, when considering the orbital number l, as well. One can rewrite (3) as

$$\phi_{li}(\mathbf{k},\mathbf{r}) = \frac{1}{\sqrt{N}} \sum_{\mathbf{R}}^{N} e^{i\mathbf{k}\cdot\mathbf{R}} \varphi(\mathbf{r}-\mathbf{t}_i-\mathbf{R}). \tag{19}$$

The corresponding eigenfunction would be given by

$$\psi(\mathbf{k},\mathbf{r}) = \sum_{l,i}^{n} C_{li}(\mathbf{k}) \phi_{li}(\mathbf{k},\mathbf{r}). \tag{20}$$

Hence, the Schrödinger Hamiltonian equation takes the form

$$\mathbf{H}\psi = E\psi,$$
$$\sum_{l,i}\Big[\big\langle\phi_{mj}(\mathbf{k},\mathbf{r})\big|\mathbb{H}\big|\phi_{li}(\mathbf{k},\mathbf{r})\big\rangle - E(\mathbf{k})\big\langle\phi_{mj}(\mathbf{k},\mathbf{r})\big|\phi_{li}(\mathbf{k},\mathbf{r})\big\rangle\Big]C_{li}\ \mathbf{k}\ = 0. \tag{21}$$

In order to solve this linear system we need to be able toevaluate the following parameters:

$$\begin{aligned}\big\langle\phi_{mj}(\mathbf{k},\mathbf{r})\big|\phi_{li}(\mathbf{k},\mathbf{r})\big\rangle &= \frac{1}{N}\sum_{\mathbf{R}} e^{i\mathbf{k}\cdot\mathbf{R}}\big\langle\varphi_m(\mathbf{r}-\mathbf{t}_j)\big|\varphi_l(\mathbf{r}-\mathbf{t}_i-\mathbf{R})\big\rangle\\ &= \sum_{\mathbf{R}}^{N} e^{i\mathbf{k}\cdot\mathbf{R}}\big\langle\varphi_m(\mathbf{r}-\mathbf{t}_j)\big|\varphi_l(\mathbf{r}-\mathbf{t}_i-\mathbf{R})\big\rangle.\end{aligned} \tag{22}$$

In a similar way, we have

$$\begin{aligned}\big\langle\phi_{mj}(\mathbf{k},\mathbf{r})\big|\mathbb{H}\big|\phi_{li}(\mathbf{k},\mathbf{r})\big\rangle &= \frac{1}{N}\sum_{\mathbf{R}} e^{i\mathbf{k}\cdot\mathbf{R}}\big\langle\varphi_m(\mathbf{r}-\mathbf{t}_j)\big|\mathbb{H}\big|\varphi_l(\mathbf{r}-\mathbf{t}_i-\mathbf{R})\big\rangle\\ &= \sum_{\mathbf{R}}^{N} e^{i\mathbf{k}\cdot\mathbf{R}}\big\langle\varphi_m(\mathbf{r}-\mathbf{t}_j)\big|\mathbb{H}\big|\varphi_l(\mathbf{r}-\mathbf{t}_i-\mathbf{R})\big\rangle.\end{aligned} \tag{23}$$

At this point we can use an important approximation [11]: the overlap matrix elements in (9) is non-zero only for the same orbitals on the same atom, i.e.

$$\big\langle\varphi_m\ \mathbf{r}-\mathbf{t}_j\ \big|\varphi_l\ \mathbf{r}-\mathbf{t}_i-\mathbf{R}\ \big\rangle = \delta_{lm}\delta_{ij}\delta(\mathbf{R})\,. \tag{24}$$

This is referred to as an "orthogonal basis", since any overlap between different orbitals on the same atom or orbitals on different atoms is taken to be zero. So it will be a unit matrix.

Similarly, we will take the Hamiltonian matrix elements in (8) to be non-zero only if the orbitals are on the same atom which are referred to as the "on-site energies":

$$\big\langle\varphi_m(\mathbf{r}-\mathbf{t}_j)|\mathbb{H}|\varphi_l(\mathbf{r}-\mathbf{t}_i-\mathbf{R})\big\rangle = \delta_{lm}\delta_{ij}\delta(\mathbf{R})\varepsilon_l\,. \tag{25}$$

The above relation implies that if the orbitals belong to different atoms, but situated at nearest neighbor sites, denoted in general as $\mathbf{d}$, one would get the following

$$\big\langle\varphi_m(\mathbf{r}-\mathbf{t}_j)|\mathbb{H}|\varphi_l(\mathbf{r}-\mathbf{t}_i-\mathbf{R})\big\rangle = \delta\ \mathbf{t}_j-\mathbf{t}_i-\mathbf{R}-\mathbf{d}\ V_{lm,ij}\,. \tag{26}$$

The $V_{lm,ij}$ are also referred to as "hopping" matrix elements.

Now, one needs only to find the on-site and hopping matrix elements, from which the energy eigenvalues of the related Hamiltonian matrix are found.

II. CARBON ATOM

II.1. Introduction

Carbon is present in nature in many allotropic forms likes graphite, diamond, fullerenes, nanotubes, etc.This is because of the versatility of its bonding.Carbon is the sixth element of the periodic table. Each carbon atom has six electrons in the configuration$1s^2$, $2s^2$, and $2p^2$ atomic orbitals, i.e. 2 electrons fill the inner shell $1s$, which is close to the nucleus and which is irrelevant for chemical reactions, whereas 4 electrons occupy the outer shell of 2s and 2p orbitals. Because of low difference in energy of $2s$ and $2p$ levels in carbon, the electronic wave functions for these four electrons can be mixed with each other, changing the occupation of the 2s and three 2p atomic orbitals so as to enhance the binding energy of the C atom with its neighboring atoms [8, 9]. These orbitals are called hybrid orbitals. In carbon, mixing of 2s and 2p orbitals leads to three possible hybridizations sp, sp^2 and sp^3, generally called sp^n hybridization with $n = 1,2,3$. Other elements of this group such as Si, Ge exhibit sp^3 hybridization. Carbon differsfrom Si and Ge because it only has the spherical 1s orbitals as inner atomic orbitals so that the absence of nearby inner orbitals facilitates hybridizations involving only valence s and p orbitals for carbon. The lack of sp and sp^2 hybridization in Si and Ge might be related to the absence of "organic materials" made of Si and Ge [8].

II.2. Hybridization in Carbon Allotropes

II.2.1. sp Hybrid Orbitals

The sp hybrid atomic orbitals are one of the possible states of electron in an atom, especially when it is bonded to others. These electron states have half 2s and half 2p characters. We can see this kind of hybridization in chemical compounds such as ethyne. This type of hybridization has been also recently observed in free carbon nanoparticles [10].

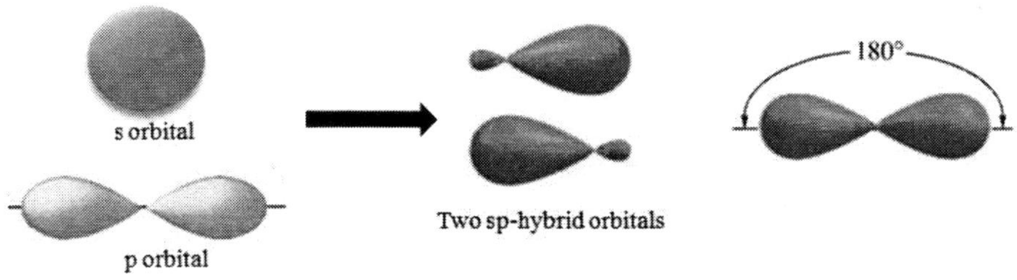

Figure II.2.1. *sp* hybridization.

For illustration, we choose the $|2p_x\rangle$ state. From a mathematical view point, there are two ways to combine the 2s and 2p atomic orbitals, obtained by the symmetric and anti-symmetric combinations

$$\begin{aligned} \left|sp_+\right\rangle &= C_1\left|2s\right\rangle + C_2\left|2p_x\right\rangle, \\ \left|sp_-\right\rangle &= C_3\left|2s\right\rangle + C_4\left|2p_x\right\rangle. \end{aligned} \tag{27}$$

Using the ortho-normality conditions $\left\langle sp_+ \middle| sp_- \right\rangle = 0$, $\left\langle sp_+ \middle| sp_+ \right\rangle = 1$, and $\left\langle sp_- \middle| sp_- \right\rangle = 1$, the relationships between the coefficients are obtained as follow

$$\begin{aligned} &C_1C_3 + C_2C_4 = 0, \qquad && C_1^2 + C_2^2 = 1, \\ &C_3^2 + C_4^2 = 1, && C_1^2 + C_3^2 = 1. \end{aligned} \tag{28}$$

The result is

$$\begin{aligned} \left|sp_+\right\rangle &= \frac{1}{\sqrt{2}}\ \left|2s\right\rangle + \left|2p_x\right\rangle\ , \\ \left|sp_-\right\rangle &= \frac{1}{\sqrt{2}}\ \left|2s\right\rangle - \left|2p_x\right\rangle\ . \end{aligned} \tag{29}$$

These energy states have a region of high electron probability each, and the two atomic orbitals are located opposite to each other, centered on the atom.

II.2.2. sp^2 Hybrid Orbitals

If we mix two of the 2p orbitals with a 2s orbital, we end up with three sp^2 hybridized orbitals. In the case of a superposition of the 2s and two 2p orbitals, which we maychoose to be the $\left|2p_x\right\rangle$ and the $\left|2p_y\right\rangle$ states, one obtains the planar sp^2 hybridization. These three orbitals lie on a plane and make equal angles of 120°, pointing to the vertices of an equilateral triangle. When the central atom makes use of sp^2 hybridized orbitals, the compound so formed has a trigonal shape as shown in Figure II.2.2.

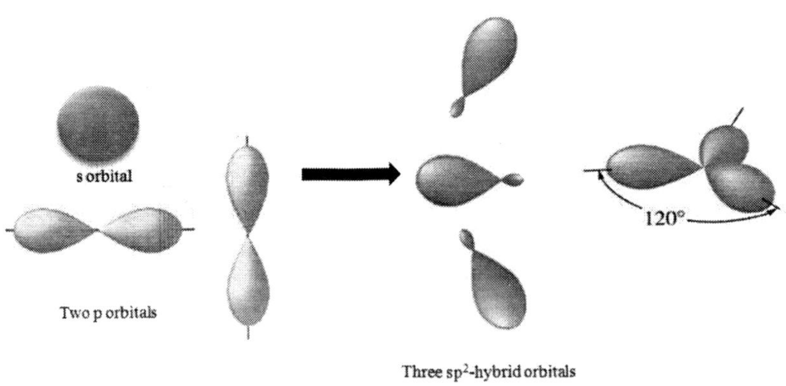

Figure II.2.2. *sp*^{2} hybridization.

From a mathematical view point we have

$$\begin{aligned}
\left|sp_1^2\right\rangle &= C_1\left|2s\right\rangle - \sqrt{1-C_1^2}\left|2p_y\right\rangle, \\
\left|sp_2^2\right\rangle &= C_2\left|2s\right\rangle + \sqrt{1-C_2^2}\left\{\frac{\sqrt{3}}{2}\left|2p_x\right\rangle + \frac{1}{2}\left|2p_y\right\rangle\right\}, \\
\left|sp_3^2\right\rangle &= C_3\left|2s\right\rangle + \sqrt{1-C_3^2}\left\{-\frac{\sqrt{3}}{2}\left|2p_x\right\rangle + \frac{1}{2}\left|2p_y\right\rangle\right\}.
\end{aligned} \tag{30}$$

Using ortho-normality conditions, we obtain

$$\begin{aligned}
&C_1^2 + C_2^2 + C_3^2 = 1 \\
&C_1C_2 - \frac{1}{2}\sqrt{1-C_1^2}\sqrt{1-C_2^2} = 0 \quad \Rightarrow \quad C_1 = C_2 = \frac{1}{\sqrt{3}},\ C_3 = -\frac{1}{\sqrt{3}}. \\
&C_1C_3 - \frac{1}{2}\sqrt{1-C_1^2}\sqrt{1-C_3^2} = 0
\end{aligned} \tag{31}$$

Hence, the three possible quantum-mechanical states are given by

$$\begin{aligned}
\left|sp_1^2\right\rangle &= \frac{1}{\sqrt{3}}\left|2s\right\rangle - \sqrt{\frac{2}{3}}\left|2p_y\right\rangle, \\
\left|sp_2^2\right\rangle &= \frac{1}{\sqrt{3}}\left|2s\right\rangle + \sqrt{\frac{2}{3}}\left\{\frac{\sqrt{3}}{2}\left|2p_x\right\rangle + \frac{1}{2}\left|2p_y\right\rangle\right\}, \\
\left|sp_3^2\right\rangle &= -\frac{1}{\sqrt{3}}\left|2s\right\rangle + \sqrt{\frac{2}{3}}\left\{-\frac{\sqrt{3}}{2}\left|2p_x\right\rangle + \frac{1}{2}\left|2p_y\right\rangle\right\}.
\end{aligned} \tag{32}$$

When carbon atoms make use of sp^2 hybrid orbitals for σ bonding, the three bonds lie on the same plane. Such carbon allotropes include the two-dimensional graphite andgraphene, the one-dimensional carbon nanotube, and the zero-dimensional fullerene, in which, every pair of carbon atoms make use of sp^2 hybrid orbitals. The remaining p orbitals of carbon atoms overlap to form a π bond.

π bonds result from the overlap of atomic orbitals that are in contact through two areas of overlap. π bonds are more diffuse bonds than the σ bonds. Electrons in π bonds are sometimes referred to as π electrons.

In the case of sp^2 hybridization, the carbon atom is a special case. Because the only orbital bonded to the nucleus is 1s, the size of atoms is small and the resultant band is considerably strong. Other elements of group IVnormally appear in sp^3 hybridization. Going down the table of periodic element of this group, and with the physical size of the elements increasing, the bond energy is reduced,and eventually the last element of this group, that is Pb, becomes a metal rather than being a semiconductor. Since the π bonds are much weaker than σ bonds, forming of π bonds in the other elements of this group would be highly unstable. While the bonding energy of π orbitals in Si is only about 25Kcalmol^{-1}, this value is about 60Kcalmol^{-1} for carbon.

II.2.3. sp³ Hybrid Orbitals

Carbon atoms in diamond provide a simple example of sp^3 hybrid orbitals. Mixing one s and all three p atomic orbitals produces a set of four equivalent sp^3 hybrid atomic orbitals. When carbon atoms make use of sp^3 hybrid orbitals, the four bonds around each carbon atom point toward the vertices of a regular tetrahedron and make angles of 109.5°.

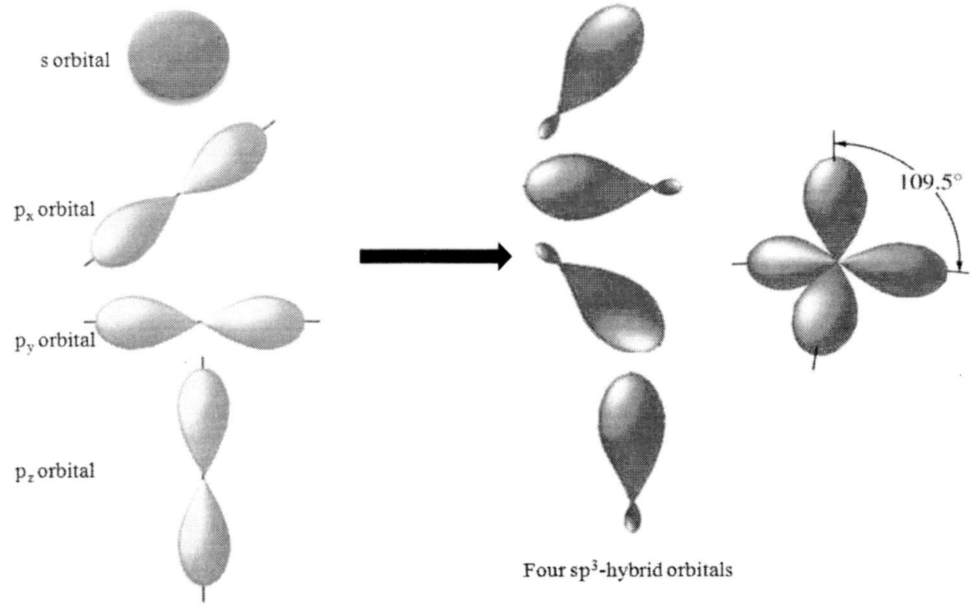

Figure II.2.3. sp^3 hybridization.

Using calculations similar to previous sections, four sp^3 hybrid orbitals are given by

$$\begin{aligned}
\left|sp_1^3\right\rangle &= \frac{1}{2}\ \left|2s\right\rangle + \left|2p_x\right\rangle + \left|2p_y\right\rangle + \left|2p_z\right\rangle \ , \\
\left|sp_2^3\right\rangle &= \frac{1}{2}\ \left|2s\right\rangle - \left|2p_x\right\rangle - \left|2p_y\right\rangle + \left|2p_z\right\rangle \ , \\
\left|sp_3^3\right\rangle &= \frac{1}{2}\ \left|2s\right\rangle - \left|2p_x\right\rangle + \left|2p_y\right\rangle - \left|2p_z\right\rangle \ , \\
\left|sp_4^3\right\rangle &= \frac{1}{2}\ \left|2s\right\rangle + \left|2p_x\right\rangle - \left|2p_y\right\rangle - \left|2p_z\right\rangle \ .
\end{aligned} \tag{33}$$

When sp^3 hybrid orbitals are used for the central atom in the formation of molecule, the molecule is said to have the shape of a tetrahedron.

In general for sp^n hybridization, $n+1$ electrons belong to the carbon atom occupied in the hybridized σ orbital and $4- \ n+1$ electrons remain in the π orbital.

III.3. Vanishing and Non-Vanishing Orbital Interactions based on TB model

To study carbon allotrops, there are four atomic orbitals per atom: one s-type and three p-type p_x, p_y, p_z . We want to use Tight-binding method to obtain the electronic band structure. Therefore we work within the orthogonal basis of orbitals and nearest neighbor interactions only, as described by the equations of (14).To solve this eigenvalue problem we need the Hamiltonian matrix elements between these atomic orbitals at different interatomic distance. For the sp-bonding, there are only four non-zero overlapping integrals as shown in Figure II.3.1 because of spherical orientation of these atomic orbitals. The remaining cases like those shown in Figure II.3.2 correspond to the matrix elements which vanish because of symmetry constraints. σ and π bondings are defined such that the axes of the involved p orbitals are parallel and normal to the interatomic vector, respectively.

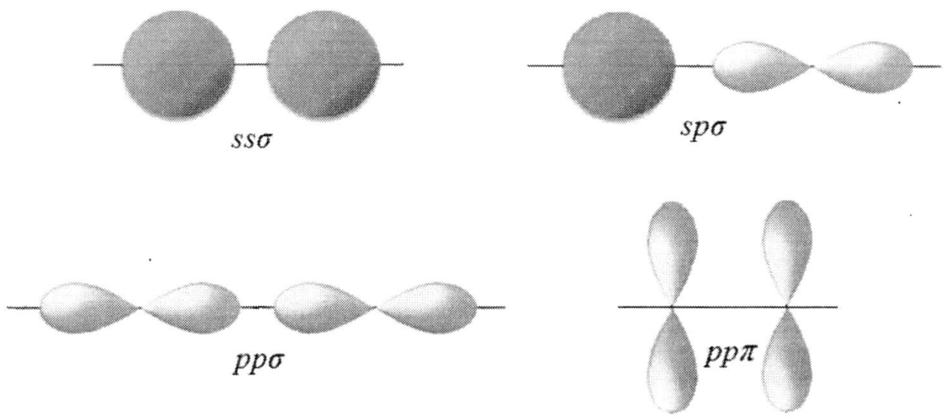

Figure II.3.1. *sp*-bonding correspond to non-vanishing matrix elements.

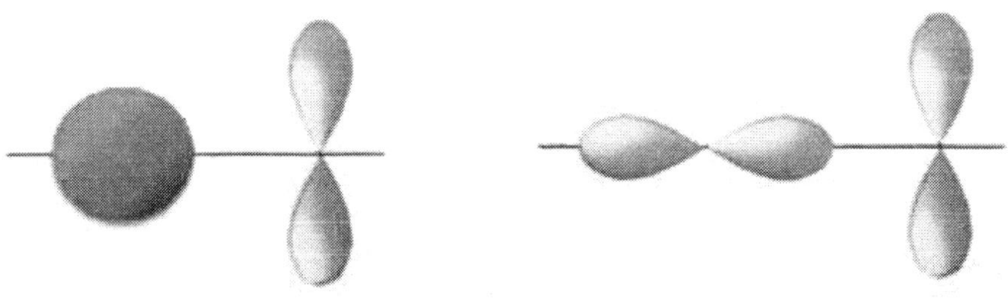

Figure II.3.2. *sp*-bonding correspond to vanishing matrix elements.

In general the p orbitals are notjust parallel or lie on the line that joins the atomic positions. We can always project each p orbital into two components, one of them along the bonding line, the other one perpendicular to it. These are called σ. and π components, respectively. This then leads to the general description of the interaction between two porbitals,or one s and one p orbital [11]. Consider the Hamiltonian matrix element between the s orbital on one atom and one of the p orbitals, which is here denoted by p_a, on another atom in random directions θ relative to the joining line of their centers. Let $\mathbf{d}$ be the unit vector along the bond and $\mathbf{a}$ is the unit vector along one of the Cartesian axes x, y, z as shown in Figure II.3.3.

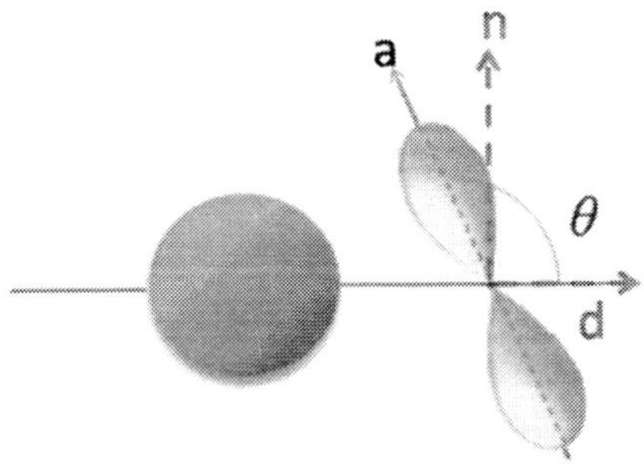

Figure II.3.3. The p orbital is in random directions θ relative to the joining line to the s orbital.

The p orbitals can be decomposed into two components that are parallel and normal to $\mathbf{d}$ as illustrated below

$$|p_a\rangle = \mathbf{a}\cdot\mathbf{d}|p_d\rangle + \mathbf{a}\cdot\mathbf{n}|p_n\rangle = \cos\theta|p_d\rangle + \sin\theta|p_n\rangle. \tag{34}$$

Now the Hamiltonian matrix element $|s\rangle$ and $|p_a\rangle$ is given by

$$\langle s|\mathbb{H}|p_a\rangle = \langle s|\mathbb{H}|p_d\rangle\cos\theta + \langle s|\mathbb{H}|p_n\rangle\sin\theta = H_{sp\sigma}\cos\theta. \tag{35}$$

Note that the overlap between $|s\rangle$ and $|p_n\rangle$ orbitals is zero by symmetry.

Consider the Hamiltonian matrix element between $|p_1\rangle$ and $|p_2\rangle$ orbitals bonded along the unit vector $\mathbf{d}$ in random directions θ_1 and θ_2, respectively. This situation is shown in Figure II.3.4.

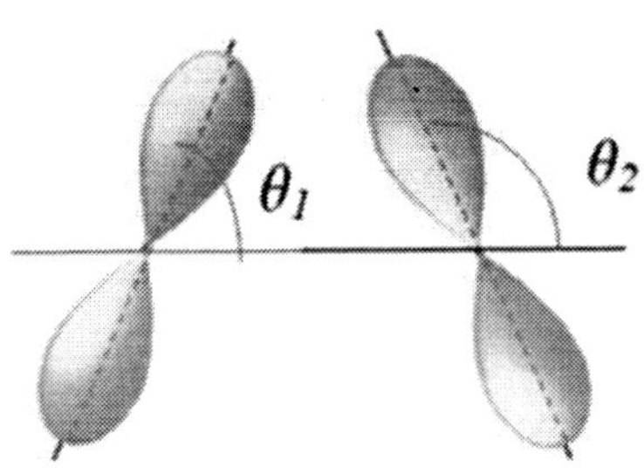

Figure II.3.4. Two p orbitals bonding in random directions.

The p orbitals can be decomposed into two components that are parallel and normal to **d** as we have done below in (36) and demonstrated in Figure II.3.5.

$$\begin{aligned} |p_1\rangle &= \mathbf{a}_1 \cdot \mathbf{d} |p_{d1}\rangle + \mathbf{a}_1 \cdot \mathbf{n} |p_{n1}\rangle = \cos\theta_1 |p_{d1}\rangle + \sin\theta_1 |p_{n1}\rangle, \\ |p_2\rangle &= \mathbf{a}_2 \cdot \mathbf{d} |p_{d2}\rangle + \mathbf{a}_2 \cdot \mathbf{n} |p_{n2}\rangle = \cos\theta_2 |p_{d2}\rangle + \sin\theta_2 |p_{n2}\rangle. \end{aligned} \tag{36}$$

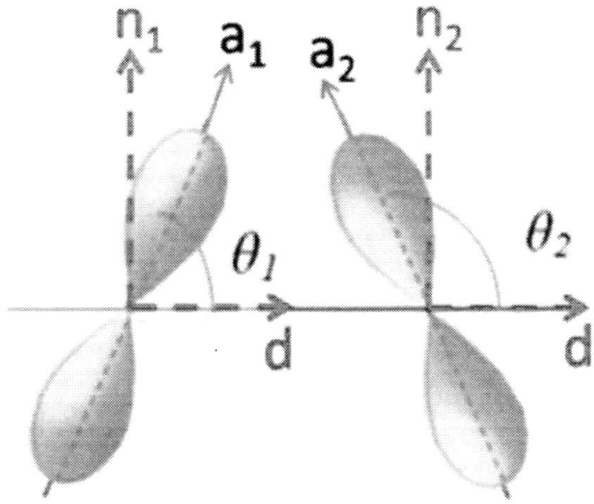

Figure II.3.5. Decomposition of p orbitals into parallel and normal components with respect to **d** .

The Hamiltonian matrix element is then given by

$$\begin{aligned} \langle p_1 |\mathbb{H}| p_2\rangle &= \langle p_{d1} |\mathbb{H}| p_{d2}\rangle \cos\theta_1 \cos\theta_2 + \langle p_{n1} |\mathbb{H}| p_{n2}\rangle \sin\theta_1 \sin\theta_2 \\ &= H_{pp\sigma} \cos\theta_1 \cos\theta_2 + H_{pp\pi} \sin\theta_1 \sin\theta_2 , \end{aligned} \tag{37}$$

where we have used the fact that overlap between orthogonal p orbitals are zero by symmetry.

III. Energy Band Calculation of Graphene

III.1. Introduction

Graphene is an allotrope of carbon. Its structure is one-atom-thick planar sheets of sp^2 bonded carbon atoms that are densely packed in a honeycomb crystal lattice.The honeycomb lattice can be described in terms of two triangular sub-lattices, A and B (Figure III.1.1) so there are two atoms in each unit cell.

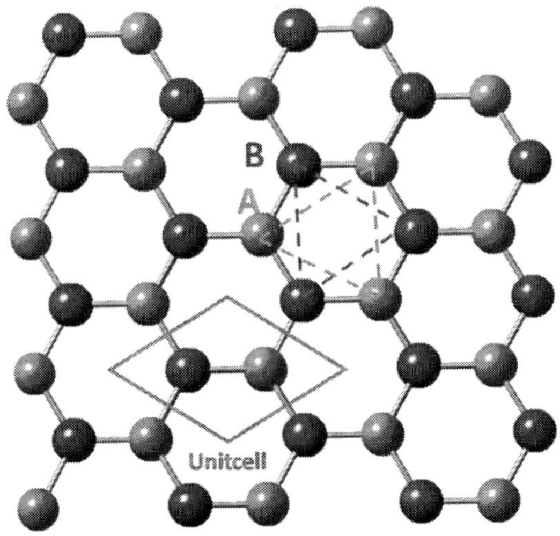

Figure III.1.1. Graphene lattice is composed of two triangular sub-lattices. The lattice constant is a .

The carbon-carbon bond length in graphene is about 1.42Å and its lattice constant a is about 3.49Å. The unit vectors of the underlying triangular sub-lattices are given by

$$\begin{aligned} \mathbf{a}_1 &= \frac{a}{2}\ \sqrt{3},+1\ , \\ \mathbf{a}_2 &= \frac{a}{2}\ \sqrt{3},-1\ . \end{aligned} \tag{38}$$

Graphene is a new material, and known to be not only the thinnest ever semimetal, but also the strongest material, too. As a conductor of electricity, it performs as well as copper. As a conductor of heat it outperforms all other known materials. It is almost completely transparent, yet so dense that not even helium, the smallest atomic gas, can pass through it. Furthermore, graphene is an excellent support of spin-current, and provides an ideal support bed for the future spintronic devices [12].

Graphene is a basic building block for graphitic materials of all other dimensionalities. It can be wrapped up into 0D fullerenes, rolled into 1D nanotubes or stacked into 3D graphite [13]. Graphite is obtained by stacking of graphene layers that is stabilized by weak van der Waalse interactions [14]. Diamond also can be obtained from graphene under extreme pressure and temperatures. Carbon nanotubes are formed by graphene wrapping, while fullerenes are obtainable from graphene by modifying the hexagons into pentagons and heptagons in a systematic way [15]. It is obvious that there has been enormous interest in understanding physical properties of graphene. Graphene has been the subject of many recent investigations due to its peculiar transport properties [16, 17]. The 2010 Nobel Prize has been awarded to Andre Geim and Konstantin Novoselov for groundbreaking experiments regarding the two-dimensional material graphene [18] (Figure III.1.3).

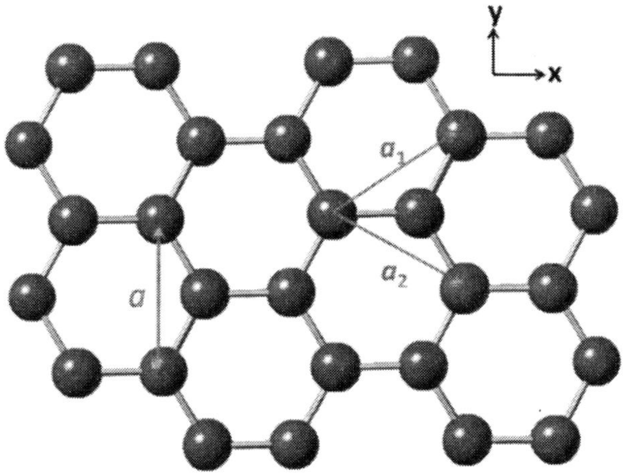

Figure III.1.2. The unit vectors and unit cell in graphene lattice.

Figure III.1.3. Konstantin Novoselov (on the left) accepts the Nobel Prize in Physics in December 10, 2010, “for groundbreaking experiments regarding the two-dimensional material graphene” [18].

III.2. Orbital Hybridization in Graphene

Graphene is a flat monolayer of carbon atoms tightly packed into a two-dimensional (2D) honeycomb lattice. Because of this,carbon atoms have the sp^2 hybridization. In this case, three σ bonds lie on the same plane with bond angles 120°. The remaining $2p_z$ orbital for each carbon overlaps to form a π bond.

III.3. Calculation of σ Bands Using TB

The σ bands are related to σ bonds between carbon atoms. It is should be pointed out that each unit cell of graphene has two carbon atoms with sp^2 hybridization, and every atom participates with three 2s, $2p_x$ and $2p_y$ orbitals in the bonding. Thus we deal with six orbitals in each unit cell yielding six σ bands, three of which appear below the Fermi energy called σ, and the other three above the Fermi energy being referred to as σ^*.

We have to use the secular equation (14). Here the transfer and overlap matrices are represented by 6 × 6 matrices.

We start with the transfer matrix

$$H = \begin{array}{c} \\ 2s_A \\ 2p_{xA} \\ 2p_{yA} \\ 2s_B \\ 2p_{xB} \\ 2p_{yB} \end{array} \begin{array}{c} \begin{array}{cccccc} 2s_A & 2p_{xA} & 2p_{yA} & 2s_B & 2p_{xB} & 2p_{yB} \end{array} \\ \left(\begin{array}{ccc|ccc} H_{11} & \cdots & \cdots & \cdots & \cdots & H_{16} \\ \vdots & \ddots & & & & \vdots \\ \vdots & & \ddots & & & \vdots \\ \hline \vdots & & & \ddots & & \vdots \\ \vdots & & & & \ddots & \vdots \\ H_{61} & \cdots & \cdots & \cdots & \cdots & H_{66} \end{array}\right)_{6\times 6} \end{array}$$

$$= \left(\begin{array}{c|c} H_{AA} & H_{AB} \\ \hline H_{BA} & H_{BB} \end{array}\right). \tag{39}$$

Here, AA and AB terms belong to integrals between orbitals of atom A in the unit cell, and between A and B atomic orbitals, respectively.

Written out in component form, the H_{AA} is

$$\begin{aligned} H_{AA}(\mathbf{k}) &= \langle \phi_A | \mathbb{H} | \phi_A \rangle \\ &= \frac{1}{N} \sum_{\mathbf{R}_A}^{N} \sum_{\mathbf{R}'_A} e^{i\mathbf{k}\cdot(\mathbf{R}_A - \mathbf{R}'_A)} \langle \varphi(\mathbf{r} - \mathbf{R}_A) | \mathbb{H} | \varphi(\mathbf{r} - \mathbf{R}'_A) \rangle \\ &= \frac{1}{N} \sum_{\mathbf{R}_A = \mathbf{R}'_A}^{N} \langle \varphi(\mathbf{r} - \mathbf{R}_A) | \mathbb{H} | \varphi(\mathbf{r} - \mathbf{R}_A) \rangle. \end{aligned} \tag{40}$$

Each matrix element is determined using below relations

$$h_{11}=\left\langle \phi_{2s}\middle|\mathbb{H}\middle|\phi_{2s}\right\rangle=\frac{1}{N}\sum_{\mathbf{R}_A=\mathbf{R}'_A}^{N}\left\langle \varphi_{2s}(\mathbf{r}-\mathbf{R}_A)\middle|\mathbb{H}\middle|\varphi_{2s}(\mathbf{r}-\mathbf{R}_A)\right\rangle=\varepsilon_{2s},$$
$$h_{22}=\left\langle \phi_{2p_x}\middle|\mathbb{H}\middle|\phi_{2p_x}\right\rangle=\frac{1}{N}\sum_{\mathbf{R}_A=\mathbf{R}'_A}^{N}\left\langle \varphi_{2p_x}(\mathbf{r}-\mathbf{R}_A)\middle|\mathbb{H}\middle|\varphi_{2p_x}(\mathbf{r}-\mathbf{R}_A)\right\rangle=\varepsilon_{2p}, \quad (41)$$
$$h_{33}=\left\langle \phi_{2p_y}\middle|\mathbb{H}\middle|\phi_{2p_y}\right\rangle=h_{22}=\varepsilon_{2p}.$$

The remaining terms are zero because of the spherical harmonics $Y_{lm}\ \theta,\varphi$ of p_x and p_y orbitals. In the tabular form, H_{AA} is given by

$$H_{AA}=\begin{pmatrix}\varepsilon_{2s} & 0 & 0\\ 0 & \varepsilon_{2p} & 0\\ 0 & 0 & \varepsilon_{2p}\end{pmatrix}. \quad (42)$$

Suggested model parameters values fitted tothe electronic band structure along the high-symmetry lines are $\varepsilon_{2s}=-8.7\,\text{eV}$ and $\varepsilon_{2p}=0$ eV.

We define tabular form of H_{AB} as

$$H_{AB}=\begin{pmatrix}h_{11} & h_{12} & h_{13}\\ h_{21} & h_{22} & h_{23}\\ h_{31} & h_{32} & h_{33}\end{pmatrix} \quad (43)$$

which its components have the general form

$$H_{AB}(\mathbf{k})=\left\langle \phi_A\middle|\mathbb{H}\middle|\phi_B\right\rangle=\frac{1}{N}\sum_{\mathbf{R}_A}\sum_{\mathbf{R}_B}e^{i\mathbf{k}\cdot(\mathbf{R}_A-\mathbf{R}_B)}\left\langle \varphi(\mathbf{r}-\mathbf{R}_A)\middle|\mathbb{H}\middle|\varphi(\mathbf{r}-\mathbf{R}_B)\right\rangle. \quad (44)$$

Here we consider only the nearest-neighbor interaction. There are three such nearest neighbors, and hence the vectors $\mathbf{R}_{Ai}$ are given by

$$\mathbf{R}_{A1}=\frac{a}{\sqrt{3}}\ -1,0\ ,$$
$$\mathbf{R}_{A2}=\frac{a}{2\sqrt{3}}\ 1,+\sqrt{3}\ , \quad (45)$$
$$\mathbf{R}_{A3}=\frac{a}{2\sqrt{3}}\ 1,-\sqrt{3}\ .$$

As discussed before, a is the lattice constant of the graphene. Figure III.3.1 shows these vectors.

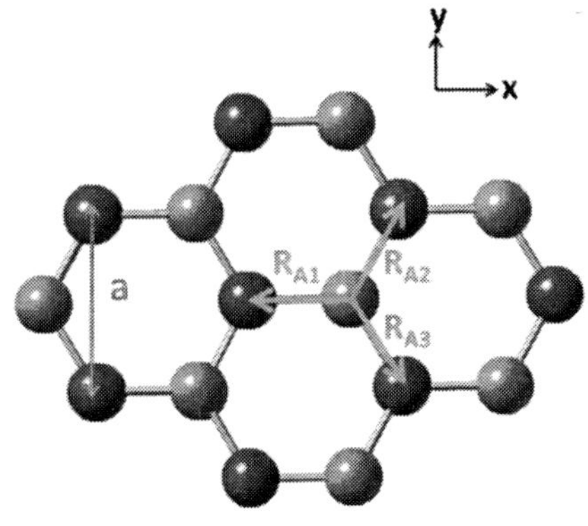

Figure III.3.1. Three nearest neighbors of the carbon atom shown by $\mathbf{R}_{Ai}$ vectors.

Because of the different orientations of the nearest-neighbor atoms B_i correspond to the atom A, it is necessary to consider general forms for (40) and (42). Hence, one can write H_{AB} components as

$$\begin{aligned} h_{11} = \langle \phi_{A2s} | \mathbb{H} | \phi_{B2s} \rangle &= \frac{1}{N} \sum_{\mathbf{R}_A}^{N} \sum_{\mathbf{R}_B} e^{i\mathbf{k}\cdot(\mathbf{R}_A - \mathbf{R}_B)} \langle \varphi_{A2s}(\mathbf{r} - \mathbf{R}_A) | \mathbb{H} | \varphi_{B2s}(\mathbf{r} - \mathbf{R}_B) \rangle \\ &= H_{ss\sigma} \sum_{i=1}^{3} e^{i\mathbf{k}\cdot\mathbf{R}_{Ai}}. \end{aligned} \tag{46}$$

$\mathbf{R}_{Ai}$ are the vectors pointing from A to B_i atoms. As it was discussed above, these vectors show the nearest neighbors of A atom. Therefore, we get the following

$$\begin{aligned} h_{12} &= \langle \phi_{A2s} | \mathbb{H} | \phi_{B2p_x} \rangle \\ &= \frac{1}{N} \sum_{\mathbf{R}_A}^{N} \sum_{\mathbf{R}_B} e^{i\mathbf{k}\cdot(\mathbf{R}_A - \mathbf{R}_B)} \langle \varphi_{A2s}(\mathbf{r} - \mathbf{R}_A) | \mathbb{H} | \varphi_{B2p_x}(\mathbf{r} - \mathbf{R}_B) \rangle \\ &= H_{sp\sigma} \sum_{i=1}^{3} e^{i\mathbf{k}\cdot\mathbf{R}_{Ai}} \cos\theta, \end{aligned} \tag{47}$$

$$\begin{aligned} h_{13} &= \langle \phi_{A2s} | \mathbb{H} | \phi_{B2p_y} \rangle \\ &= \frac{1}{N} \sum_{\mathbf{R}_A}^{N} \sum_{\mathbf{R}_B} e^{i\mathbf{k}\cdot(\mathbf{R}_A - \mathbf{R}_B)} \langle \varphi_{A2s}(\mathbf{r} - \mathbf{R}_A) | \mathbb{H} | \varphi_{B2p_y}(\mathbf{r} - \mathbf{R}_B) \rangle \\ &= H_{sp\sigma} \sum_{i=1}^{3} e^{i\mathbf{k}\cdot\mathbf{R}_{Ai}} \sin\theta, \end{aligned} \tag{48}$$

$$\begin{aligned} h_{22} &= \left\langle \phi_{A2p_x} \middle| \mathbb{H} \middle| \phi_{B2p_x} \right\rangle \\ &= \frac{1}{N} \sum_{\mathbf{R}_A}^{N} \sum_{\mathbf{R}_B} e^{i\mathbf{k}\cdot(\mathbf{R}_A - \mathbf{R}_B)} \left\langle \varphi_{A2p_x}(\mathbf{r}-\mathbf{R}_A) \middle| \mathbb{H} \middle| \varphi_{B2p_x}(\mathbf{r}-\mathbf{R}_B) \right\rangle \\ &= \sum_{i=1}^{3} \left[H_{pp\sigma} \cos^2\theta + H_{pp\pi} \quad 1-\cos^2\theta \right] e^{i\mathbf{k}\cdot\mathbf{R}_{Ai}}, \end{aligned} \tag{49}$$

$$\begin{aligned} h_{23} &= \left\langle \phi_{A2p_x} \middle| \mathbb{H} \middle| \phi_{B2p_y} \right\rangle \\ &= \frac{1}{N} \sum_{\mathbf{R}_A}^{N} \sum_{\mathbf{R}_B} e^{i\mathbf{k}\cdot(\mathbf{R}_A - \mathbf{R}_B)} \left\langle \varphi_{A2p_x}(\mathbf{r}-\mathbf{R}_A) \middle| \mathbb{H} \middle| \varphi_{B2p_y}(\mathbf{r}-\mathbf{R}_B) \right\rangle \\ &= \sum_{i=1}^{3} \quad H_{pp\sigma} - H_{pp\pi} \quad \cos\theta \sin\theta \; e^{i\mathbf{k}\cdot\mathbf{R}_{Ai}}, \end{aligned} \tag{50}$$

$$\begin{aligned} h_{33} &= \left\langle \phi_{A2p_y} \middle| \mathbb{H} \middle| \phi_{B2p_y} \right\rangle \\ &= \frac{1}{N} \sum_{\mathbf{R}_A}^{N} \sum_{\mathbf{R}_B} e^{i\mathbf{k}\cdot(\mathbf{R}_A - \mathbf{R}_B)} \left\langle \varphi_{A2p_y}(\mathbf{r}-\mathbf{R}_A) \middle| \mathbb{H} \middle| \varphi_{B2py}(\mathbf{r}-\mathbf{R}_B) \right\rangle \\ &= \sum_{i=1}^{3} \left[H_{pp\sigma} \sin^2\theta + H_{pp\pi} \quad 1-\sin^2\theta \right] e^{i\mathbf{k}\cdot\mathbf{R}_{Ai}}. \end{aligned} \tag{51}$$

Other parameters can be written based on the facts that $\langle s|\mathbb{H}|p\rangle = -\langle p|\mathbb{H}|s\rangle$, $\langle p_1|\mathbb{H}|p_2\rangle = \langle p_2|\mathbb{H}|p_1\rangle$ and $\langle s_1|\mathbb{H}|s_2\rangle = \langle s_2|\mathbb{H}|s_1\rangle$. We choose the model parameters value to be $H_{sp\sigma} = 5.5\,\text{eV}$, $H_{ss\sigma} = -6.7\ \text{eV}$, $H_{pp\sigma} = 5.1\ \text{eV}$, and $H_{pp\pi} = -3.1\ \text{eV}$, which are close to reported values in [8]. These parameters are obtained from a fit to the experimental electronic band structure along the high-symmetry lines.

The same treatment yields the overlap matrix elements. We assume that the atomic wavefunctions are normalized as shown in (17), so the S_{AA} matrix is given by

$$S_{AA} = \begin{pmatrix} 1 & 0 & 0 \\ 0 & 1 & 0 \\ 0 & 0 & 1 \end{pmatrix}. \tag{52}$$

Similarly to H_{AB},the elements of S_{AB} is described as below

$$s_{11} = \left\langle \phi_{A2s} \middle| \phi_{B2s} \right\rangle = S_{ss\sigma} \sum_{i=1}^{3} e^{i\mathbf{k}\cdot\mathbf{R}_{Ai}}, \tag{53}$$

$$s_{12} = \left\langle \phi_{A2s} \middle| \phi_{B2p_x} \right\rangle = S_{sp\sigma} \sum_{i=1}^{3} e^{i\mathbf{k}\cdot\mathbf{R}_{Ai}} \cos\theta, \tag{54}$$

$$s_{13}=\left\langle\phi_{A2s}\middle|\phi_{B2p_y}\right\rangle=S_{sp\sigma}\sum_{i=1}^{3}e^{i\mathbf{k}\cdot\mathbf{R}_{Ai}}\sin\theta, \quad (55)$$

$$s_{22}=\left\langle\phi_{A2p_x}\middle|\phi_{B2p_x}\right\rangle=\sum_{i=1}^{3}\left[S_{pp\sigma}\cos^2\theta+S_{pp\pi}\left(1-\cos^2\theta\right)\right]e^{i\mathbf{k}\cdot\mathbf{R}_{Ai}}, \quad (56)$$

$$s_{23}=\left\langle\phi_{A2p_x}\middle|\phi_{B2p_y}\right\rangle=\sum_{i=1}^{3}e^{i\mathbf{k}\cdot\mathbf{R}_{Ai}}\left(S_{pp\sigma}-S_{pp\pi}\right)\cos\theta\sin\theta, \quad (57)$$

$$s_{33}=\left\langle\phi_{A2p_y}\middle|\phi_{B2p_y}\right\rangle=\sum_{i=1}^{3}\left[S_{pp\sigma}\sin^2\theta+S_{pp\pi}\left(1-\sin^2\theta\right)\right]e^{i\mathbf{k}\cdot\mathbf{R}_{Ai}}. \quad (58)$$

Other parameters can be written based on the facts that $\langle s|p\rangle=-\langle p|s\rangle$, $\langle p_1|p_2\rangle=\langle p_2|p_1\rangle$, and $\langle s_1|s_2\rangle=\langle s_2|s_1\rangle$. Suggested tight-binding parameters are $S_{sp\sigma}=-0.10\,\text{eV}$, $S_{ss\sigma}=0.20\text{ eV}$, $S_{pp\sigma}=-0.15\,\text{eV}$, and $S_{pp\pi}=0.12\text{ eV}$.

The two carbon atoms in the unit cell of graphene are identical. Hence the transfer matrix element H_{AA}, which is given by the interaction of an atom at site A with itself an all other A atoms in the crystal, is exactly the same as H_{BB}. Similarly, H_{AB} is simply the complex conjugate of H_{BA}. This condition is also satisfied for S_{AA} and S_{BB}, and S_{AB} and S_{BA}. So we have

$$\begin{aligned} H_{BB}&=H_{AA}, \quad H_{BA}=H_{AB}^{*}, \\ S_{BB}&=S_{AA}, \quad S_{BA}=S_{AB}^{*}. \end{aligned} \quad (59)$$

Finally, the secular equation is expressed in the most general form of

$$\begin{bmatrix} H_{AA(3\times3)} & H_{AB(3\times3)} \\ H_{BA(3\times3)} & H_{BB(3\times3)} \end{bmatrix}-E(\mathbf{k})\begin{bmatrix} S_{AA(3\times3)} & S_{AB(3\times3)} \\ S_{BA(3\times3)} & S_{BB(3\times3)} \end{bmatrix}=0. \quad (60)$$

Solving this eigenvalue problem yields the electronic band structure of graphene. Since all possible eigenstates are specified by the wavevector **k** within any one primitive cell of the periodic lattice in reciprocal space, the first Brillouin zone is the uniquely defined cell that is the most compact possible cell. So it is enough to calculate the band structure in this region. For graphene, the first Brillouin zone is a hexagon. In this case, six symmetry axes exist which can be deduced from one another by rotations of $\pi/6$ with an invariance under rotation equal to $\pi/3$. Figure III.3.2 shows this region.

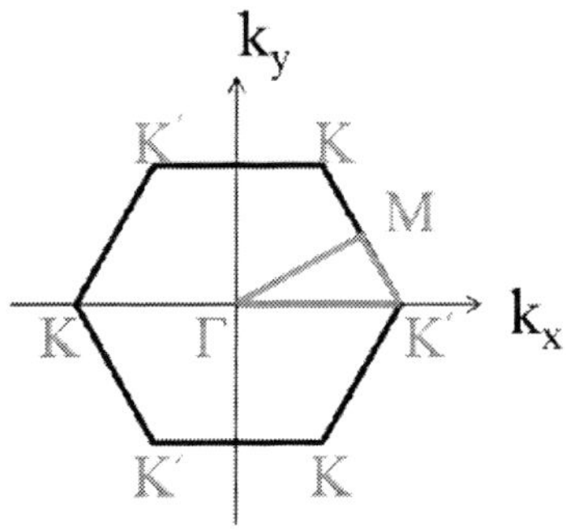

Figure III.3.2. First Brillouin zone of the hexagonal lattice. The irreducible zone is shown in red.

The range over which **k** is studied can be further reduced by considering the other symmetries which constitutes the irreducible Brillouin zone of the hexagonal lattice. It can be demonstrated that the irreducible Brillouin zone is the half of an equilateral triangle with apexes Γ, K and M as shown in Figure III.3.2. It is generally sufficient to let the wave vector **k** describes the edge of the irreducible Brillouin zone, while the inner region is ignored.

Figure III.3.3 shows the σ band structure of the graphene lattice calculated using Tight-binding model [19].

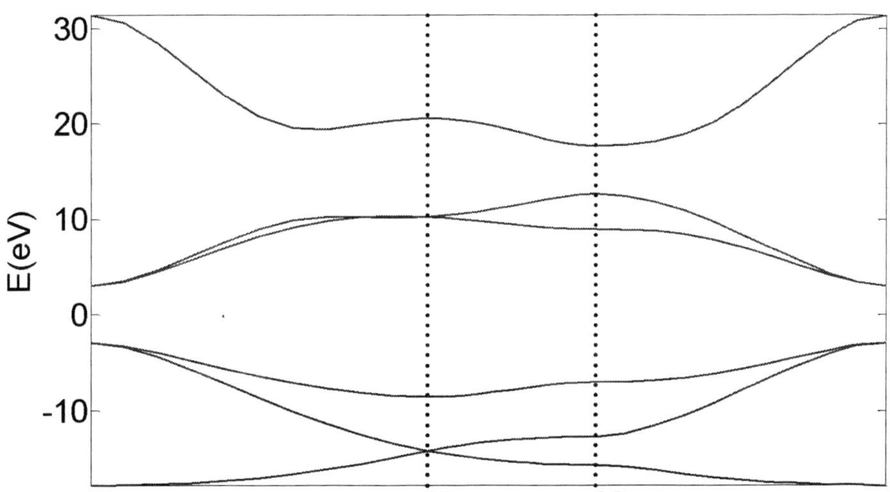

Figure III.3.3. σ electronic band structure of graphene calculated using tight-binding method.

III.4. Calculation of π Bands Using TB

Band structure of graphene around the Fermi level and for transition energies well above the optical range is determined by π orbitals; the Fermi level is normally placed at the Dirac point for unbiased and undoped graphene. A common approximation for the low energy electronic properties is hence a tight-binding Hamiltonian including only the carbon $2p_z$ states. Thep orbitals are all parallel to one another and each contains one electron (Figure III.4.1).

As mentioned before, each unit cell of graphene has two carbon atoms which construct π bonds using p_z orbitals. Using the secular equation (14), the transfer and overlap matrices are obtained as 2×2 matrices.

We start with transfer matrix

$$\mathbf{H} = \begin{array}{c} \\ 2p_{zA} \\ 2p_{zB} \end{array} \begin{array}{c} \begin{array}{cc} 2p_{zA} & 2p_{zB} \end{array} \\ \left(\begin{array}{c|c} H_{AA} & H_{AB} \\ \hline H_{BA} & H_{BB} \end{array} \right)_{2\times 2} \end{array} . \tag{61}$$

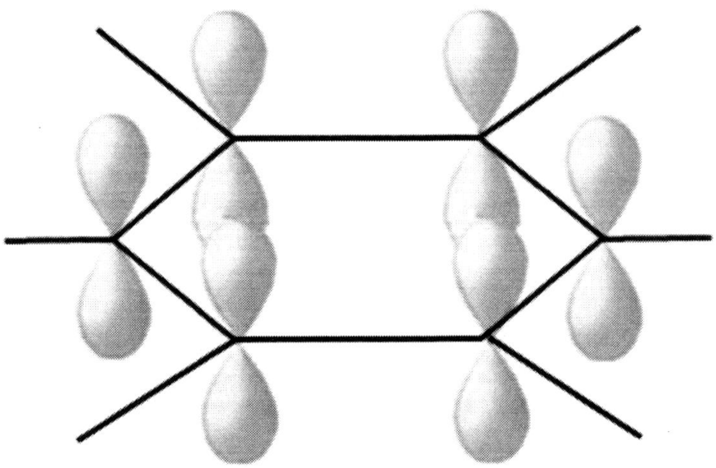

Figure III.4.1. p orbitals of carbon atoms in the graphene structure.

Written out in component form, the $\mathbf{H}$ is

$$\begin{aligned} H_{AA} &= \left\langle \phi_{2p_z} \middle| \mathbb{H} \middle| \phi_{2p_z} \right\rangle \\ &= \frac{1}{N} \sum_{\mathbf{R}_A = \mathbf{R}'_A}^{N} \left\langle \varphi_{2p_z}(\mathbf{r} - \mathbf{R}_A) \middle| \mathbb{H} \middle| \varphi_{2p_z}(\mathbf{r} - \mathbf{R}_A) \right\rangle = \varepsilon_{2p}, \end{aligned} \tag{62}$$

$$\begin{aligned} H_{AB} &= \left\langle \phi_{A2p_z} \middle| \mathbb{H} \middle| \phi_{B2p_z} \right\rangle \\ &= \frac{1}{N} \sum_{\mathbf{R}_A}^{N} \sum_{\mathbf{R}_B} e^{i\mathbf{k}\cdot(\mathbf{R}_A - \mathbf{R}_B)} \left\langle \varphi_{A2p_z}(\mathbf{r} - \mathbf{R}_A) \middle| \mathbb{H} \middle| \varphi_{B2p_z}(\mathbf{r} - \mathbf{R}_B) \right\rangle \\ &= \sum_{i=1}^{3} H_{pp\pi} e^{i\mathbf{k}\cdot\mathbf{R}_{Ai}}. \end{aligned} \tag{63}$$

The same treatment yields the overlap matrix elements as follows

$$\mathbf{S} = \begin{array}{c|c|c} & 2p_{zA} & 2p_{zB} \\ 2p_{zA} & S_{AA} & S_{AB} \\ \hline 2p_{zB} & S_{BA} & S_{BB} \end{array}_{2\times 2} \tag{64}$$

which can be written out in the component form

$$S_{AA} = \left\langle \phi_{2p_z} \middle| \phi_{2p_z} \right\rangle = \frac{1}{N} \sum_{\mathbf{R}_A = \mathbf{R}'_A}^{N} \left\langle \varphi_{2p_z}(\mathbf{r} - \mathbf{R}_A) \middle| \varphi_{2p_z}(\mathbf{r} - \mathbf{R}_A) \right\rangle = 1, \tag{65}$$

$$\begin{aligned} S_{AB} &= \left\langle \phi_{A2p_z} \middle| \phi_{B2p_z} \right\rangle = \\ &= \frac{1}{N} \sum_{\mathbf{R}_A}^{N} \sum_{\mathbf{R}_B} e^{i\mathbf{k}\cdot(\mathbf{R}_A - \mathbf{R}_B)} \left\langle \varphi_{A2p_z}(\mathbf{r} - \mathbf{R}_A) \middle| \varphi_{B2p_z}(\mathbf{r} - \mathbf{R}_B) \right\rangle \\ &= \sum_{i=1}^{3} S_{pp\pi} e^{i\mathbf{k}\cdot\mathbf{R}_{Ai}}. \end{aligned} \tag{66}$$

Again, in (65) we assume that the atomic wave functions are normalized.

As mentioned before, the transfer matrix element H_{AA}, given by the interaction of an atom at site A with itself an all other A atoms in the crystal, is exactly the same as H_{BB}. Similarly, H_{AB} is simply the complex conjugate of H_{BA}. This condition is also satisfied for S_{AA}, S_{BB}, and S_{AB} and S_{BA}.

The secular equation is constructed as below

$$\begin{vmatrix} H_{AA(1\times1)} & H_{AB(1\times1)} \\ H_{BA(1\times1)} & H_{BB(1\times1)} \end{vmatrix} - E(\mathbf{k}) \begin{vmatrix} S_{AA(1\times1)} & S_{AB(1\times1)} \\ S_{BA(1\times1)} & S_{BB(1\times1)} \end{vmatrix} = 0. \tag{67}$$

Solving this eigenvalue problem yields to the π electronic bands of graphene shown in Figure III.4.2 [19].

Figure III.4.3 shows σ and π band structures of graphene together. There is a very good agreement between these results and those reported in [20] obtained from first-principles calculations shown in Figure III.4.4.

It is remarkable that graphene is a gapless semiconductor, with no gap at the Dirac point [16]. Many interesting ways have been proposed and investigated to tune the band gap of graphene. However, it is well known that the presence of defects such as vacancies or impurities [19, 21, 22], the interaction with a substrate [23], or the edge structure in graphene nanoribbons [24, 25, 26] can lead to a bandgap opening in graphene.

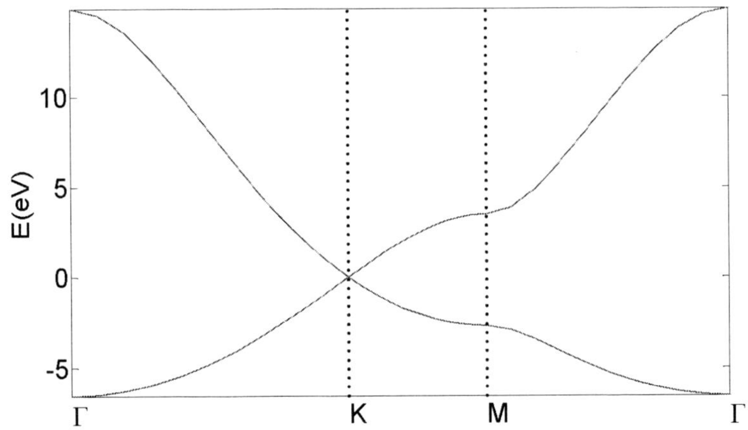

Figure III.4.2. π electronic band structurecalculated using tight-binding method.

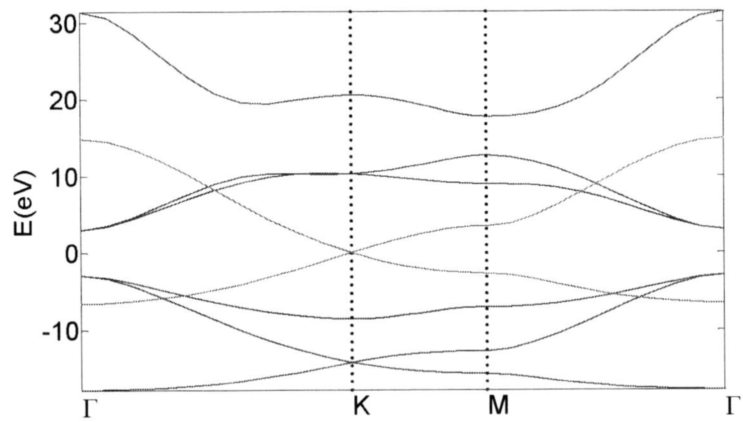

Figure III.4.3. σ (blue) and π (red) band structures calculated using Tight-binding method.

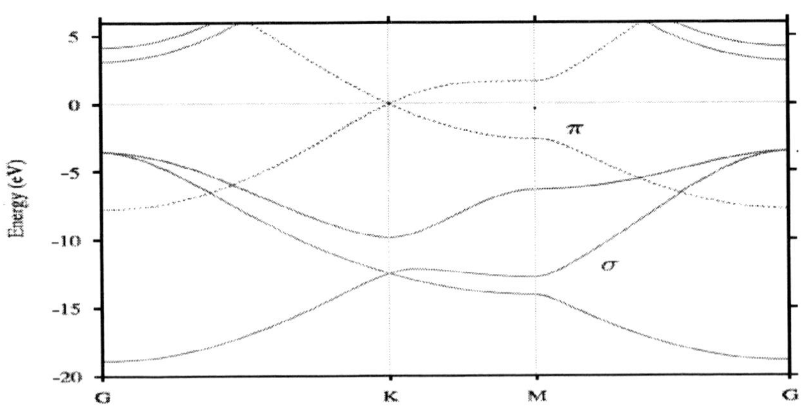

Figure III.4.4. σ (red) and π (dashed blue) band structures from first-principles calculations [20].

The idea of introducing a periodic array of vacancies into the graphene lattice was first mentioned in [19]. Many works have been done with the same idea, since then, on the so-called anti-dot graphene lattices [27, 28, 29, 30, 31].

Electronic structures of coupled semiconductor quantum dots arranged as graphene hexagonal lattice are studied theoretically using the tight-binding method up to the third-nearest-neighbors. The novel two-dimensional Dirac-like electronic excitations in graphene are found in these artificial planar quantum dot structures which provide the theoretical basis for searching Dirac fermions in quantum dot materials and have great significance for investigating and making semiconductor quantum dot devices [32].

Band gap engineering in hexagonal h-BN antidot lattices is another work done in this field. Although graphene is a gapless semimetal due to the symmetry between two sublattices of the honeycomb structure, h-BN is a semiconductor with a wide bandgaparound 5.5 eV [33]. It is found that when the anti-dot lattice is along the zigzag direction, the band gap opening can be related to the intervalley scattering and does not follow the simple scaling rule previously proposed in the literature for the anti-dot lattice along the armchair direction. For h-BN, the calculations show that the anti-dot lattice results in reducing of band gaps. Coupled with doping of carbon atoms, the bandgap of h-BN antidot lattice can be reduced to below 2 eV, which might have implications in light-emitting devices or photo-electrochemistry [34].

Regular hydrogen adsorption on graphene is another way to open a bandgap in graphene [35] to be discussed briefly later in the next section.

IV. Energy Band Calculation of Graphane

IV.1. Introduction

The chemical modification of graphene to create derivatives with different structures and properties has so far been restricted to graphene oxide [36, 37, 38], a disordered structure that bears an assortment of functional groups.

Another possible modification, which has been predicted theoretically [39], is the addition of hydrogen atoms to graphene, altering the sp^2 hybridization of carbon atoms to sp^3, and thus changing the structure and electronic properties, to form a new two-dimensional hydrocarbon called graphane. Recently, the Nobel Laureates Andre Geim and Konstantin Novoselov from the University of Manchester and co-workers from Russia and the Netherlands have hydrogenated mechanically exfoliated graphene by exposing it to hydrogen plasma [40]. The electronic properties of the new material change markedly: the highly conductive graphene is converted from a semimetal into a semiconductor. Because of its structure and lowdimensionality, it provides a fertile playground for fundamental science andtechnological applications.

It is reported that hydrogenation of mono, bi, and trilayer graphene samples via exposure to H_2 plasma occurs as a result of electron irradiation of H_2O adsorbates on the samples, rather than H species in the plasma as reported [41]. It is proposed that the hydrogenation mechanism is electron-impact fragmentation of H_2O adsorbates into H^+ ions. At the incident electron energies of 60eV, hydrogenation is observed which is significantly more stable at temperatures 200°C than previously reported [42].

This hexagonal arrangement of carbons in the sp^3 hybridization with C: H = 1 forms two constructions: if the H atoms connect at opposite sides of the plane of C-C bond, it is called *chairlike*, as illustrated in Figure IV.1.1, and if on the same side of the C-C plane, it is called *boatlike*. The typical C-C bond length for chair like conformer is 1.52Å which is similar to the sp^3 bond length of 1.53Å in diamond and that for boatlike conformer it is 1.56Å [39].

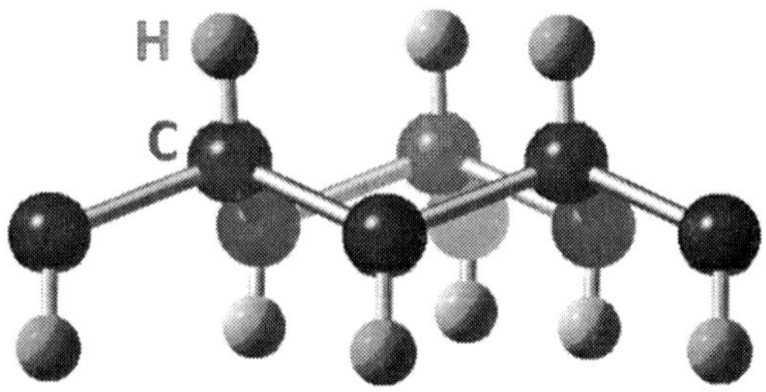

Figure IV.1.1. The chairlike structure of graphane.

Graphane was first predicted to exist only in the boat and chair configurations [37]. But a new report has just pointed out that the existence of a new conformer, being referred to as the stirrup, is also plausible [42]. Nevertheless, all these three conformers are expected to exhibit very close chemical and physical properties. The structural and phonon properties of these structures, as well as their relative stability can be established by first-principles calculations. It is shown that all these three graphane conformers respond to any arbitrarily oriented extension with a much smaller lateral contraction than the one calculated for graphene. Furthermore, the boatlike-graphane has a small and negative Poisson ratio along the armchair and zigzag principal directions of the carbon honeycomb lattice. Moreover, the chairlike-graphane admits both softening and hardening hyperelasticity, depending on the direction of applied load [43]. Lattice thermal properties of graphane lattice have been also calculated [44]. Some further properties of graphene/graphane interfaces are discussed in [45, 46]. It is also anticipated that doped graphane could serve as a high temperature superconductor with a transition temperature well exceeding 90K [47, 48, 49].

IV.2. Orbital Hybridization in Graphene

Graphane is again an arrangement of carbon atoms with sp^3 hybridization saturated by hydrogen forming a honeycomb shape. Transformation of the sp^2 hybridization in graphene to the sp^3 hybridization in graphane results in a change of the bond lengths and angles (Figure IV.2.1).

A typical bond length for sp^2 C-C bondand the standard bond angle for graphene and graphite is 1.42Å and 120°, respectively. For sp^3 hybridization, the standard value of C-C bond length is 1.54Å, and the corresponding angle is 109.5°. A typical value for the single C-H bond length is about 1.14Å. But in graphane, depending on the number of the added hydrogen, the length of C-C bond is between 1.42Å and 1.54Å. Also C-C-H and C-C-C angles are intermediate between 90° and 109.5° and 120° and 109.5°, respectively. This implies an intermediate character of the hybridization between sp^2 and sp^3 [50].

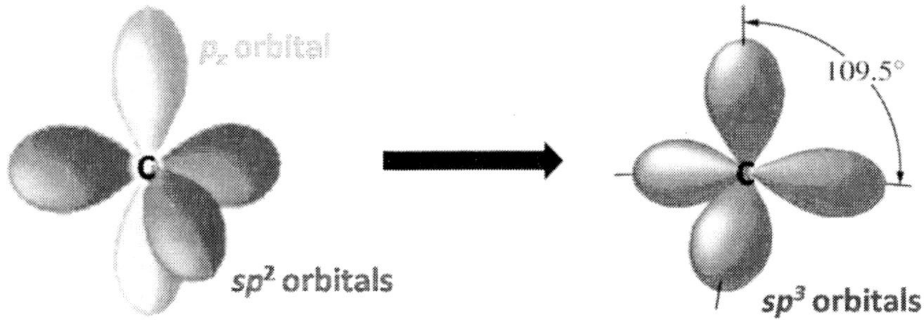

Figure IV.2.1. Transformation of the sp^2 hybridization of carbon to the sp^3.

IV.3. Electronic Band Structure Calculated Using TB

As mentioned before, graphane is a honeycomb structure of carbon atoms which their p_z orbitals are saturated with hydrogen atoms. Here we consider chairlike structure. The top view of this structure is a hexagonal shape, like in graphene. Thus, its unit cell has a rhombus cross section and there are four atoms in each unit cell: two carbon and two hydrogen atoms as shown in Figure IV.3.1.

Each carbon atom participates in the bonding in sp^3 hybridization using 2s, $2p_x$, $2p_y$ and $2p_z$ orbitals. Each hydrogen atom also participates through its 1s orbital. Hence, the Hamiltonian matrix related to (25) and (26) is sized originally as 10×10.

Therefore, Hamiltonian matrix has a general form given as

$$
\mathbf{H} = \begin{array}{c} \\ 2s_A \\ 2p_{xA} \\ 2p_{yA} \\ 2p_{zA} \\ 1s_A \\ 2s_B \\ 2p_{xB} \\ 2p_{yB} \\ 2p_{zB} \\ 1s_B \end{array}
\begin{array}{c} 2s_A \quad 2p_{xA} \quad 2p_{yA} \quad 2p_{zA} \quad 1s_A \quad 2s_B \quad 2p_{xB} \quad 2p_{yB} \quad 2p_{zB} \quad 1s_B \\
\left(\begin{array}{ccccc|ccccc}
H_{11} & & & & & & & & & H_{1,10} \\
& \ddots & & & & & & & & \\
& & \ddots & & & & & & & \\
& & & \ddots & & & & & & \\
& & & & \ddots & & & & & \\
\hline
& & & & & \ddots & & & & \\
& & & & & & \ddots & & & \\
& & & & & & & \ddots & & \\
& & & & & & & & \ddots & \\
H_{10,1} & & & & & & & & & H_{10,10}
\end{array} \right)_{10\times 10} \end{array}
$$

$$
= \left(\begin{array}{c|c} H_{AA} & H_{AB} \\ \hline H_{BA} & H_{BB} \end{array} \right). \tag{68}
$$

A and B terms belong to each carbon and hydrogen connected to it in the unit cell as labeled in Figure IV.3.1.

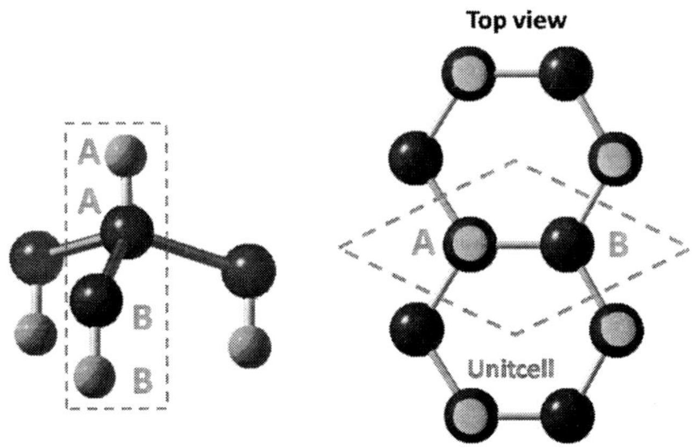

Figure IV.3.1. Unit cell of chairlike graphane.

We define tabular form of H_{AA} as

$$H_{AA} = \begin{pmatrix} h_{11} & h_{12} & h_{13} & h_{14} & h_{15} \\ h_{21} & h_{22} & h_{23} & h_{24} & h_{25} \\ h_{31} & h_{32} & h_{33} & h_{34} & h_{35} \\ h_{41} & h_{42} & h_{43} & h_{44} & h_{45} \\ h_{51} & h_{52} & h_{53} & h_{54} & h_{55} \end{pmatrix}. \tag{69}$$

This matrix is referred to on-site energies as defined in (25). Writing out H_{AA} in its components gives

$$h_{ji} = \left\langle \varphi_m \;\; \mathbf{r} - \mathbf{t}_j \;\, |\mathbb{H}| \varphi_l \;\; \mathbf{r} - \mathbf{t}_i - \mathbf{R} \;\right\rangle = \delta_{lm}\delta_{ij}\delta(\mathbf{R})\varepsilon_l \,. \tag{70}$$

Now, one can find the matrix elements as

$$\begin{aligned} h_{11} &= \left\langle \varphi_{2s} \;\; \mathbf{r} - \mathbf{t}_j \; \middle|\mathbb{H}\middle| \varphi_{2s} \;\; \mathbf{r} - \mathbf{t}_j \;\right\rangle = \varepsilon_{2s}, \\ h_{ii} &= \left\langle \varphi_{2p_n} \;\; \mathbf{r} - \mathbf{t}_j \; \middle|\mathbb{H}\middle| \varphi_{2p_n} \;\; \mathbf{r} - \mathbf{t}_j \;\right\rangle = \varepsilon_{2p}, \\ i &= 2,3,4 \qquad i = 2,\, n = x \\ &\qquad\qquad\quad\;\; i = 3, n = y \\ &\qquad\qquad\quad\;\; i = 4, n = z \\ h_{55} &= \left\langle \varphi_{1s} \;\; \mathbf{r} - \mathbf{t}_j \; \middle|\mathbb{H}\middle| \varphi_{1s} \;\; \mathbf{r} - \mathbf{t}_j \;\right\rangle = \varepsilon_{1s}. \end{aligned} \tag{71}$$

The remaining terms h_{ij} with $i \neq j$ are zero. In tabular form H_{AA} is given by

$$H_{AA} = \begin{pmatrix} \varepsilon_{2s} & 0 & 0 & 0 & 0 \\ 0 & \varepsilon_{2p} & 0 & 0 & 0 \\ 0 & 0 & \varepsilon_{2p} & 0 & 0 \\ 0 & 0 & 0 & \varepsilon_{2p} & 0 \\ 0 & 0 & 0 & 0 & \varepsilon_{1s} \end{pmatrix}. \tag{72}$$

Suggested model parameters values fitted to the electronic band structure along the high-symmetry lines are $\varepsilon_{2s} = -5.2\text{eV}$, $\varepsilon_{2p} = 2.3\text{eV}$, and $\varepsilon_{1s} = -4\text{eV}$.

Similarly, H_{AB} has the tabular form as

$$H_{AB} = \begin{pmatrix} h_{11} & h_{12} & h_{13} & h_{14} & h_{15} \\ h_{21} & h_{22} & h_{23} & h_{24} & h_{25} \\ h_{31} & h_{32} & h_{33} & h_{34} & h_{35} \\ h_{41} & h_{42} & h_{43} & h_{44} & h_{45} \\ h_{51} & h_{52} & h_{53} & h_{54} & h_{55} \end{pmatrix}. \tag{73}$$

The components of the H_{AB} have the general form of

$$h_{ji} = \left\langle \varphi_m \ \mathbf{r} - \mathbf{t}_j \ |\mathbb{H}| \varphi_l \ \mathbf{r} - \mathbf{t}_i - \mathbf{R} \ \right\rangle = \delta \ \mathbf{t}_j - \mathbf{t}_i - \mathbf{R} - \mathbf{d} \ V_{lm,ij}. \tag{74}$$

Here, $\mathbf{d}$ denotes the nearest neighbor atoms position.Thus,what remains is to determine the rest of four nearest neighbors vectors $\mathbf{R}_{Ai}$ of each carbon atom shown in Figure IV.3.2.

We need to project these vectors to Cartesian coordinate. Figure IV.3.3 shows the tetrahedral structure constructed by the center carbon atom, denoted by the letter A, and its nearest neighbors.

The bonding length of two carbon atoms $a_{\text{C–C}}$ is about 1.52Å and the bonding length of carbon and hydrogen atoms $a_{\text{C–H}}$ is about 1.1Å. Also C-C-H angle θ is about 109.5°. Figure IV.3.4 shows the triangles constructed on A, B and C points and A, C and D points. Here β is the complementary of the angle θ, given by $\beta = \pi - \theta$. Using the triangular relations, one can get the distances h and w as well.

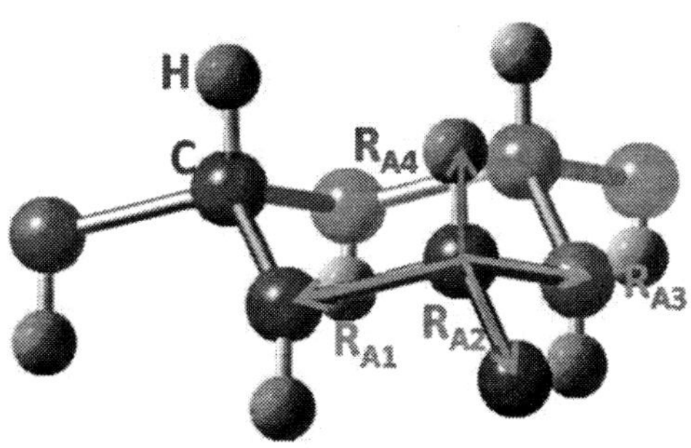

Figure IV.3.2. Four nearest neighbors of the carbon atom: three carbon and one hydrogen atoms.

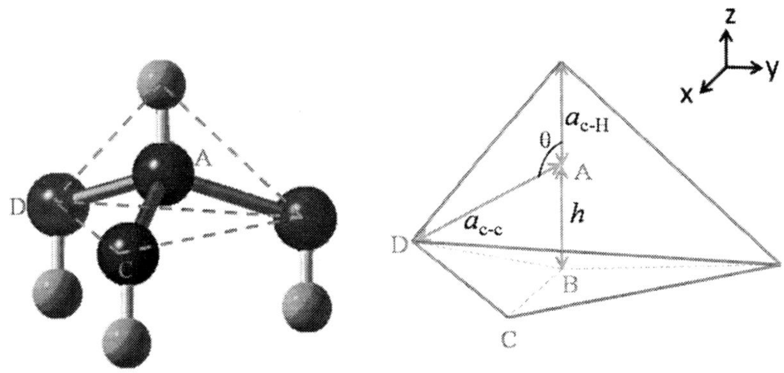

Figure IV.3.3. The tetrahedral structure constructed by atoms in graphane.

$$w = 2a_{cc}\sin(\theta/2) \qquad h = a_{cc}\cos\beta = a_{cc}\cos\theta$$

Figure IV.3.4. The triangles structure constructed to calculate *h* and *w*.

Now, suppose that the central carbon atom C_0 is located at the origin of the Cartesian coordinate. Using the recent calculated parameters, one can get the location of the nearest neighbors of the carbon atom A as

$$\begin{aligned} &C_0 : \ 0,0,0 \ , \\ &C_1 : \left(\frac{w}{\sqrt{3}}, 0, -h\right), \\ &C_2 : \left(-\frac{w}{2\sqrt{3}}, \frac{w}{2}, -h\right), \\ &C_3 : \left(-\frac{w}{2\sqrt{3}}, -\frac{w}{2}, -h\right), \\ &H : \ 0,0,a_{C-H} \ . \end{aligned} \tag{75}$$

Figure IV.3.5 shows the projected location of these atoms on xy-plane.

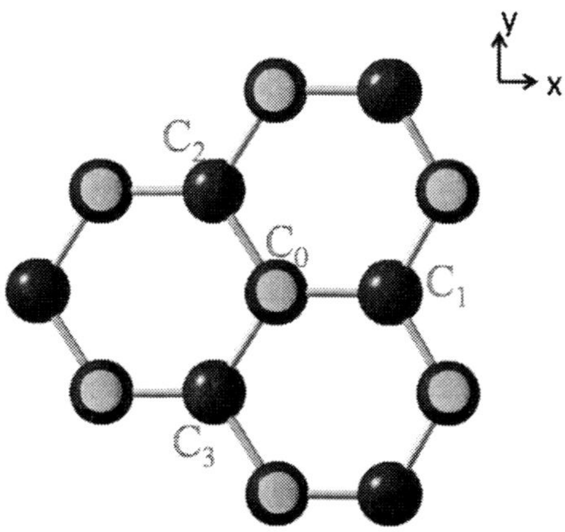

Figure IV.3.5. Top view of the atoms specifiedwith (75).

Now, one can write $\mathbf{R}_{Ai}$ vectors needed to calculate Hamiltonian parameters H_{AB} as

$$\begin{aligned}
&\mathbf{R}_{A1}:\left(\frac{w}{\sqrt{3}},0,-h\right),\\
&\mathbf{R}_{A2}:\left(-\frac{w}{2\sqrt{3}},\frac{w}{2},-h\right),\\
&\mathbf{R}_{A3}:\left(-\frac{w}{2\sqrt{3}},-\frac{w}{2},-h\right),\\
&\mathbf{R}_{A4}:\ 0,0,a_{C-H}\ .
\end{aligned} \tag{76}$$

For each of the $\mathbf{R}_{Ai}$ vectors, one can define the corresponding angles made with respect to the x, y and z axes as θ_x, θ_y and θ_z. Considering these vectors, the components of H_{AB} can be written out as

$$h_{11}=\left\langle\varphi_{A2s}\left|\mathbb{H}\right|\varphi_{B2s}\right\rangle=V_{ss\sigma}\sum_{i=1}^{3}e^{i\mathbf{k}\cdot\mathbf{R}_{Ai}}, \tag{77}$$

$$h_{12}=\left\langle\varphi_{A2s}\left|\mathbb{H}\right|\varphi_{B2p_x}\right\rangle=V_{sp\sigma}\sum_{i=1}^{3}\cos\theta_x\ e^{i\mathbf{k}\cdot\mathbf{R}_{Ai}}, \tag{78}$$

$$h_{13}=\left\langle\varphi_{A2s}\left|\mathbb{H}\right|\varphi_{B2p_y}\right\rangle=V_{sp\sigma}\sum_{i=1}^{3}\cos\theta_y\ e^{i\mathbf{k}\cdot\mathbf{R}_{Ai}}, \tag{79}$$

$$h_{14}=\left\langle\varphi_{A2s}\left|\mathbb{H}\right|\varphi_{B2pz}\right\rangle=V_{sp\sigma}\sum_{i=1}^{3}\cos\theta_z\ e^{i\mathbf{k}\cdot\mathbf{R}_{Ai}}, \tag{80}$$

$$h_{15}=\left\langle\varphi_{A2s}\left|\mathbb{H}\right|\varphi_{H1s}\right\rangle=V_{1s1s}e^{i\mathbf{k}\cdot\mathbf{R}_{A4}}, \tag{81}$$

$$h_{22}=\left\langle\varphi_{A2p_x}\left|\mathbb{H}\right|\varphi_{B2p_x}\right\rangle=\sum_{i=1}^{3}\left[V_{pp\sigma}\cos^2\theta_x+V_{pp\pi}\ 1-\cos^2\theta_x\ \right]e^{i\mathbf{k}\cdot\mathbf{R}_{Ai}}, \tag{82}$$

$$h_{23}=\left\langle\varphi_{A2p_x}\left|\mathbb{H}\right|\varphi_{B2p_y}\right\rangle=\sum_{i=1}^{3} V_{pp\sigma}-V_{pp\pi} \cos\theta_x \cos\theta_y \, e^{i\mathbf{k}\cdot\mathbf{R}_{Ai}}, \tag{83}$$

$$h_{24}=\left\langle\varphi_{A2p_x}\left|\mathbb{H}\right|\varphi_{B2p_z}\right\rangle=\sum_{i=1}^{3} V_{pp\sigma}-V_{pp\pi} \cos\theta_x \cos\theta_z \, e^{i\mathbf{k}\cdot\mathbf{R}_{Ai}}, \tag{84}$$

$$h_{33}=\left\langle\varphi_{A2p_y}\left|\mathbb{H}\right|\varphi_{B2p_y}\right\rangle=\sum_{i=1}^{3}\left[V_{pp\sigma}\cos^2\theta_y+V_{pp\pi} \; 1-\cos^2\theta_y\right]e^{i\mathbf{k}\cdot\mathbf{R}_{Ai}}, \tag{85}$$

$$h_{34}=\left\langle\varphi_{A2p_y}\left|\mathbb{H}\right|\varphi_{B2p_z}\right\rangle=\sum_{i=1}^{3} V_{pp\sigma}-V_{pp\pi} \cos\theta_y \cos\theta_z e^{i\mathbf{k}\cdot\mathbf{R}_{Ai}}, \tag{86}$$

$$h_{35}=\left\langle\varphi_{A2py}\left|\mathbb{H}\right|\varphi_{1s}\right\rangle=-V_{1s2p}\cos\theta_y \, e^{i\mathbf{k}\cdot\mathbf{R}_{A4}}, \tag{87}$$

$$h_{44}=\left\langle\varphi_{A2p_z}\left|\mathbb{H}\right|\varphi_{B2p_z}\right\rangle=\sum_{i=1}^{3}\left[V_{pp\sigma}\cos^2\theta_z+V_{pp\pi} \; 1-\cos^2\theta_z\right]e^{i\mathbf{k}\cdot\mathbf{R}_{Ai}}, \tag{88}$$

$$h_{45}=\left\langle\varphi_{A2pz}\left|\mathbb{H}\right|\varphi_{1s}\right\rangle=-V_{1s2p}\cos\theta_z \, e^{i\mathbf{k}\cdot\mathbf{R}_{A4}}, \tag{89}$$

$$h_{55}=\left\langle\varphi_{H1s}\left|\mathbb{H}\right|\varphi_{H1s}\right\rangle=V_{1ss}e^{i\mathbf{k}\cdot\mathbf{R}_{A4}}. \tag{90}$$

Other parameters can be written based on the facts that $\left\langle s\left|\mathbb{H}\right|p\right\rangle=-\left\langle p\left|\mathbb{H}\right|s\right\rangle$, $\left\langle p_1\left|\mathbb{H}\right|p_2\right\rangle=\left\langle p_2\left|\mathbb{H}\right|p_1\right\rangle$ and $\left\langle s_1\left|\mathbb{H}\right|s_2\right\rangle=\left\langle s_2\left|\mathbb{H}\right|s_1\right\rangle$. We choose the model parameter values, fitted tothe electronic band structure along the high-symmetry lines, to be $V_{sp\sigma}=3.8\text{eV}$, $V_{ss\sigma}=-4.4\text{eV}$, $V_{pp\sigma}=5.6\text{eV}$, $V_{pp\pi}=-1.8\text{eV}$, $V_{1s1s}=0\text{eV}$, $V_{1s2p}=4\text{eV}$, and $V_{1ss}=-3.5\text{eV}$ .As mentioned before, the matrix element H_{AA} is exactly the same as H_{BB} and H_{AB} is simply the complex conjugate of H_{BA}.

Figure IV.3.6 shows the resultant electronic band structure of graphane obtained from direct Tight-binding calculations. The irreducible Brillouin zone is again the half of an equilateral triangle with apexes $\wp$, K and M, exactly like in graphene, as shown in Figure III.3.2.

There is an excellent agreement compared to the reported result in [39], which are obtained from density-functional theory calculations, as shown in Figure IV.3.7. This establishes the usefulness and high accuracy of the Tight-binding method as applied to the planar graphene and graphane structures. We here notice that the former structure is a semi-metal with a zero-gap at its Dirac points, while the latter is a semiconductor with a direct energy gap.

Chemical functionalization is a viable route toward bandgap engineering of graphene-based materials, as first demonstrated in [40], where exposure to a stream of hydrogen atoms has led to 100% hydrogenation of a graphene sheet,the so-called graphane.

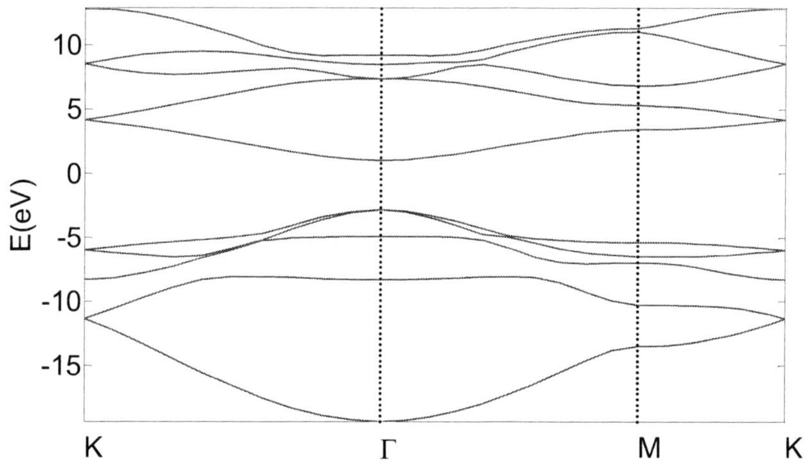

Figure IV.3.6. The top view of the atoms specified

The band gap of periodically doped graphene with hydrogen is investigated. It is found through a Tight-binding model that for certain periodicities, called nongap periodicities (NGPs), no gap is opened at the Dirac point. Density-functional-theory calculations show that a tiny gap is opened for NGPs due to exchange effects, not taken into account in the TB model. However, this tiny gap is one order of magnitude smaller than the gap opened for other periodicities different from NGPs. This remarkable reduction in the band-gap opening for NGPs provokes a crossing of the midgap and the conduction bands and, consequently, the metallization of the system. This result is also valid for other adsorbates different from atomic hydrogen [51].

Although graphene is a gapless semimetal due to the symmetry between two sublattices of the honeycomb structure, graphane is a semiconductor with a wide bandgap around 3.5eV [39]. This gives good idea for opening gap in Dirac points of graphene. Furthermore, it is possible to control the gap in graphane by removing some of its hydrogen atoms periodically [52, 53]. The role of hydrogen vacancies is to decrease the energy gap, and at the same time, to increase the thermoelectric figure of merit [54], as well as significant lattice contraction. It has been also noticed thatsuch hydrogen vacancy clusters in graphane may appear as quantum dots and induction of stable ferromagnetism [55]. Furthermore, it has been shown that single-sided hydrogenation also causes reduction of energy gap to about 1.1eV [52].

Obviously, research on functionalized graphane devices is at an embryonic stage. Several issues must be addressed to introduce graphane in future generations of electron devices. From this perspective, theoretical simulations have been done to explore possible solutions for device fabrication and design, in order to provide an early assessment of the opportunities of functionalized graphane in nanoscale electronics.

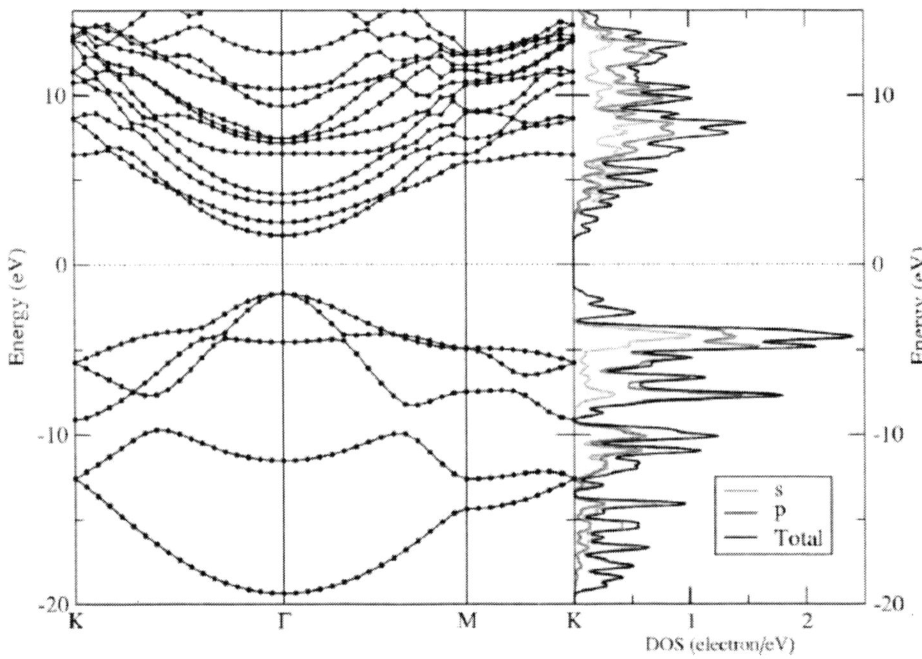

Figure IV.3.7. Band structure and density of states of graphane [40].

with the 2-D geometry and unique transport features of graphene, the possibility of doping graphane makes it possible to define p and n regions so that 2-D p-n junctions become feasible with small reverse currents. This junction is designed based on patterned graphane and its current-voltage characteristics is also predicted [52]. Experimental characteristics of such a junction is observed in [56]. Recent analysis also shows possibilities of designing bipolar junction transistors based on patterned graphane [57]. For this reason, donor and acceptor impurities should be introduced to the graphane with controlled concentrations, the effects of which have been also studied [58].

A performance analysis of field-effect transistors (FETs) based on recently fabricated 100% hydrogenated graphene (graphane) [55] and theoretically predicted semihydrogenated graphene (graphone) [59] has been also reported. The approach is based on accurate calculations of the energy bands by means of GW approximation, subsequently fitted with a three-nearest neighbor sp^3 tight-binding Hamiltonian, and finally used to compute ballistic transport in transistors based on functionalized graphene. Due to the large energy gap, the proposed devices have many of the advantages provided by one-dimensional graphene nanoribbon FETs, such as large I_{on} and I_{on}/I_{off} ratios, reduced band-to-band tunneling, without the corresponding disadvantages in terms of prohibitive lithography and patterning requirements for circuit integration [60]. Recent advaces in the fabrication of graphene integrated circuits paves the way toward the expected birth of two-dimensional nanoscale planar electronics based on the graphene and graphane [61].

REFERENCES

[1] Bloch, F. *Z. Phys.* 1928, vol. 52.

[2] Slater, JC; Koster, GF. *Phys. Rev.* 1954, vol. 94, 1498.

[3] Martin, RM. *Electronic Structure: Basic Theory and Practice Methods*, Cambridge University Press, 2004.

[4] Kittel, C. *Introduction to Solid State Physics*, 7th ed., Wiley, 1996.

[5] Ashcroft, NW; Mermin, ND. *InSolid State Physics,* Holt, Rine-hart and Winston, New York, NY,1976.

[6] Sutton, AP. *Electronic Structure of Materials,* Oxford University Press, Oxford, UK, 1993.

[7] Harrison, AW. *Structure and the Properties of Solids,* Dover, New York, 1989.

[8] Saito, R; Dresselhaus, G; Dresselhaus, MS. *Physical Properties of Carbon Nanotubes*, Imperial, London, 1998.

[9] Dresselhaus, MS; Dresselhaus, G; Sugihara, K; Spain, IL; Goldberg, HA. *Springer Series in Materials Science,* 1988, vol. 5.

[10] Ravagnan, L; Mazza, T; Bongiorno, G; Devetta, M; Amati, M; Milani, P; Piseri, P; Coreno, M; Lenardi, C; Evangelista, F; Rudolf, P. *Chem. Commun.*, 2011, vol. 47, 2952.

[11] Kaxiras, E; *Atomic and Electronic Structure of Solids*, Cambridge University Press, 2003.

[12] Zhang, Q.;.Chan, K. S.; Lin, Z.; *Appl. Phys. Lett.*, 2011, vol. 98, 032106.

[13] Geim, AK; Novoselov, KS. *Nature Materials,* 2007, vol. 6, 183.

[14] Pauling, L; *The Nature of the Chemical Bond,* Cornell U. P., Ithaca, NY, 1972.

[15] Peres, NMR; Guinea, F; Castro Neto, AH. *Phys. Rev. B*, 2006, vol. 73, 125411.

[16] Novoselov, KS. *Nature*, 2005, vol. 438, 197.

[17] Abanin, DA; Lee, PA; and Levitov, LS. *Phys. Rev. Lett*, 2006, vol. 96, 176803.

[18] http://nobelprize.org/nobel_prizes/physics

[19] Gharekhanlou, B.; Alavi, M; Khorasani, S. *Semicond. Sci. Technol.*, 2008, vol. 23, no. 7, 075026.

[20] Boukhvalov, DW; Katsnelson, MI.; Lichtenstein, A. I. *Phys. Rev. B*, 2008, vol. 77, 035427.

[21] Peres, NMR.; Guinea, F; Castro Neto, AH. *Phys. Rev. B*, 2006, vol. 73, 125411.

[22] Pereira, VM.; Guinea, F; Lopes dos Santos, JMB; Peres, NMR; Castro Neto, AH. *Phys. Rev. Lett*, 2006, vol. 96, 036801.

[23] Vanin, M; Mortensen, JJ; Kelkkanen, AK; García-Lastra, JM; Thygesen, KS; Jacobsen, KW. *Phys. Rev. B*, 2010, vol. 81, 081408.

[24] Lu, YH; Feng,YP. *J. Phys. Chem.* , 2009, C 113, 20841.

[25] Han, MY; Ozyilmaz, B; Zhang, Y; Kim, P. *Phys. Rev. Lett*, 2007, vol. 98, 206805.

[26] Son, YW; Cohen, ML; Louie, SG. *Phys. Rev. Lett*, 2006, vol. 97, 216803.

[27] Pedersen, TG; Flindt, C; Pedersen, J; Mortensen, NA; Jauho, A-P; Pedersen, K. *Phys. Rev. Lett,* 2008, vol. 100, 136804.

[28] Tiwari, RP; Stroud, D.*Phys. Rev. B*, 2009, vol. 79, 205435.

[29] Liu, W; Wang, ZF; Shi, QW; Yang, J; Liu, F. *Phys. Rev. B*, 2009, vol. 80, 233405.

[30] Bai, J; Zhong, X; Jiang, S; Huang, Y; Duan, X. *Nature Nanotech.,* 2010, vol. 5, 190.

[31] Schultz, MH; Jauho, AP; Pedersen, TG. *Phys. Rev. B*, 2011, vol. 84, 045428.
[32] Penga, J; Li, S. *App. Phys. Lett*, 2010, vol. 97, 242105.
[33] Song, L; Ci, LJ; Lu, H; Sorokin, PB; Jin, CH; Ni, J; Kvashnin, AG; Kvashnin, DG; Lou, J; Yakobson, BI; Ajayan, PM. *Nano Lett.*, 2010, vol. 10, 3209.
[34] Zhang, A; Teoh, HF; Dai, Z; Feng, YP; Zhang, C. *App. Phys. Lett*, 2011, vol. 98, 023105.
[35] Balog, R; Jørgensen, B; Nilsson, L; Andersen, M; Rienks, E; Bianchi, M; Fanetti, M; Lægsgaard, E; Baraldi, A; Lizzit, S; Sljivancanin, Z; Besenbacher, F; Hammer, B; Pedersen, TG; Hofmann P; Hornekær, L.*Nature Materials*, 2010, vol. 9, 315.
[36] Dreyer, RD; Park, S; WB, Christopher ; Ruoff, S. R. *Chem. Soc. Rev.*, 2010, vol. 39, 228.
[37] Kovtyukhova, N I; Karpenko, GA; Chuiko, AA. *Russ. J. Inorg. Chem.* 1992, vol. 37, 566.
[38] Liu, P; Gong, K; Xiao, P. *Carbon*, 1999, vol. 37, 2073.
[39] Sofo, JO; Chaudhari, AS; Barber; GD. *Phys. Rev. B*, 2007, vol. 75, 153401.
[40] Elias, DC; Nair, RR; Mohiuddin, TMG; Morozov, SV; Blake, P; Halsall, MP; Ferrari, AC; Boukhvalov, DW; Katsnelson, MI; Geim, AK; Novoselov, KS.*Science*, 2009, vol. 323, 610.
[41] Jones, JD; Hoffmann, WD; Jesseph, AV; Morris, CJ; Verbeck, GF; Perez, J. M. *App. Phys. Lett*, 2010, vol. 97, 233104.
[42] Bhattacharya, A.; Bhattacharya, S.; Majumder, C.; Das, G. P. *Phys. Rev. B*, 2011, vol. 83, 033404.
[43] Cadelano, E; Palla, PL; Giordano, S; Colombo, L. *Phys. Rev. B*, 2010, vol. 82, 235414.
[44] Neek-Amal, M; Peeters, FM. *Phys. Rev. B*, 2011, vol. 83, 235437.
[45] Ijäs, M; Havu, P; Harju, A; Pasanen, P. *Phys. Rev. B*, 2011, vol. 84, 041403(R).
[46] Averill, FW; Morris, JR; *Phys. Rev. B*, 2011, vol. 84, 035411.
[47] Savini, G; Ferrari, AC; Giustino, F. *Phys. Rev. Lett.*, 2011, vol. 105, 037002.
[48] Kristoffel, N; Rägo, K. *Phys. Lett. A*, 2011, vol. 375, 2246.
[49] Loktev, VM; Turkowski, V; *J. Low Temp. Phys.*, 2011, vol. 164, 264.
[50] Boukhvalov, DW; Katsnelson, MI; Lichtenstein, A. I. *Phys. Rev. B*, 2008, vol. 77, 035427.
[51] García-Lastra, J. M. *Phys. Rev. B*, 2010, vol. 82, 235418.
[52] Gharekhanlou, B; Khorasani, S. IEEE *Trans. Electron Dev.*, 2010, vol. 57, 209.
[53] Savchenko, A.*Science*, 2009, vol. 323, 589.
[54] Ni, X; Liang, G; Wang, J-S; Li, B. *App. Phys. Lett.*, 2009, vol. 95, 192114.
[55] Singh, AK; Penev ES; Yakobson, BI. *ACS Nano*, 2010, vol. 4, 3510.
[56] Courtesy of N. Ranjan-Ray (Saha Institute, India) private communication.
[57] Gharekhanlou, B; Tousaki, SB; Khorasani, S. *J. Phys.: Conf. Series*, 2010, vol. 248, 012061.
[58] Wang, Y; Ding, Y; Shi, S; Tang, W. *Appl. Phys. Lett.*, 2011, vol. 98, 163104.
[59] Pujari, BS; Gusarov, S; Brett, M; Kovalenko, A; *Phys. Rev. B*, 2011, vol. 84, 041402(R).
[60] Fiori, G; Lebègue, S; Betti, A; Michetti, P; Klintenberg, M; Eriksson, O; Iannaccone, G. *Phys. Rev. B*, 2010, vol. 82, 153404.
[61] Lin, YM; Valdes-Garcia, A; Han, SJ; Farmer, DB; Meric, I; Sun, Y; Wu, Y; Dimitrakopoulos, C; Grill, A; Avouris, P; Jenkins, KA. *Science*, 2011, vol. 332, 1294.

In: Graphene: Properties, Synthesis and Applications ISBN: 978-1-61470-949-7
Editor: Zhiping Xu

Chapter 2

BALLISTIC TRANSPORT IN GRAPHENE: A WINDOW INTO RELATIVISTIC PHYSICS OF MASSLESS FERMIONS

***H. C. Kao*[1], *M. Lewkowicz*[2] *and B. Rosenstein*[3,4]**
[1]Physics Department, National Taiwan Normal University
Taipei 11677, Taiwan, R. O. C.
[2]Physics Department, Ariel University Center of Samaria
Ariel 40700, Israel
[3]Electrophysics Department, National Chiao Tung University
Hsinchu 30050, Taiwan, R. O. C.
[4]National Center for Theoretical Sciences
Hsinchu 30043, Taiwan, R. O. C.

Abstract

Excitations near Dirac points in graphene are $2 + 1$ dimensional massless (pseudo) relativistic fermions with Fermi velocity $v_g = 10^6 m/s$ playing a role of velocity of light. Electron-hole pairs are copuously created by an DC or AC applied electric field, a process well described using a dynamic "first quantized" approach. Evolution of current density, number of pairs and energy in ballistic regime exhibits host of novel relativistic phenomena. We show how the physics far from the two Dirac points enters various physical quantities in linear response via the chiral anomaly. The series in electric field strength E up to third order indicate that at certain time scale $t_{nl} \propto E^{-1/2}$ the physical behaviour dramatically changes and the perturbation theory breaks down. Beyond the linear response physics is explored using an exact solution of the first quantized equations. While for small electric fields the I–V curve is linear characterized by the universal minimal conductivity $\sigma = \pi/2(e^2/h)$, at $t > t_{nl}$ the current grows linearly with time due to copious Schwinger's pair creation (with rate $E^{3/2}$). Eventually it leads to creation of the electron-hole plasma, a process terminated by a relaxational recombination. Since there is no energy gap, the radiation due to the pairs annihilation is enhanced. The spectrum of radiation exhibits a maximum at $\omega = \sqrt{eEv_g/\hbar}$ and has a distinctive angular and polarization dependence.

PASC numbers: 72.80.Vp 73.20.Mf 12.20.-m.

*E-mail address: vortexbar@yahoo.com

I. Introduction

It has been demonstrated recently that a graphene sheet, especially one suspended on leads, is one of the purest electronic systems: electronic mobility in graphene exceeding that in best semiconductor 2D systems [1]. The scattering of charge carriers in suspended graphene samples of submicron length is so negligible that the transport is ballistic[2, 3]. The ballistic flight time can be estimated as $t_L = L/v_g$, where v_g is the graphene velocity characterizing the "relativistic" spectrum of graphene near the Dirac points and L is the length of the sample. During the ballistic flight one therefore can neglect interactions with phonons, ripplons, disorder, etc.[4], as well as electron-electron interactions (for a reason, not completely understood yet) [5].

In a simplified model of a single graphene sheet (namely neglecting scattering processes and electron-electron interactions) the chemical potential is located right between the valence and conductance bands and the Fermi "surface" consists of two Dirac points of the Brillouin zone [4]. The effective model near this chemical potential is identical to the two-dimensional relativistic and massless Dirac model with the velocity of light c replaced by v_g, and therefore possesses pseudo-Lorentz invariance that can be effectively utilized. Consequently graphene can be considered a condensed matter bench-top laboratory to study purely relativistic effects like (Schwinger's) particle-antiparticle pair creation by an electric field and their annihilation into either photons or other excitations, electron-hole plasma, Klein paradox, chiral symmetry and its spontaneous breaking, chiral anomaly, etc. These phenomena constitute the main subject of the present chapter. We start however with a brief recount of a controversy (both theoretical and experimental) over the value of the minimal conductivity of graphene. As will be demonstrated later, this question is intimately connected with the above purely relativistic phenomena.

A lot of effort has been devoted theoretically to the question of dc and ac transport in pure graphene due to the surprising fact that the conductivity is finite without any dissipation process present. A widely accepted "standard" value of the minimal dc conductivity" at zero temperature,

$$\sigma_1 = \frac{4}{\pi}\frac{e^2}{h}, \tag{1}$$

was obtained very early on using a simplified Dirac model by several methods. Direct application of the Kubo formula at zero frequency, disorder strength, temperature and chemical potential [6, 7] makes use of certain "regularizations" like the $i\eta$ regulator which is interpreted as an "infinitesimal disorder". The regulator is removed at the end of the calculation. A more customary application of the Kubo formula starts with finite frequency. As noted by Ziegler and others [8], the order of limits is important and several other values different from σ_1 were provided for the *same* system. The standard value σ_1 is obtained using a rather unorthodox procedure when the dc limit $\omega \to 0$ is made *before* the zero disorder strength limit $\eta \to 0$ is taken. If the order of limits is reversed, one obtains:

$$\sigma_2 = \frac{\pi}{2}\frac{e^2}{h}. \tag{2}$$

When the limit is taken holding $\omega = \eta$, one can even obtain a value of $\sigma_3 = \pi\frac{e^2}{h}$ [8], thus solving the experimental so called "missing π" problem (the same value was obtained more recently using yet another regularization in [9]). Indeed, at least early experiments on graphene sheets on Si substrates provided values roughly three times larger than σ_1 [10]. Recent experiments on suspended graphene [2] demonstrated that the dc conductivity is lower, $1.7\sigma_1$, as temperature is reduced to $4K$, and even more recently it is consistent with σ_2 [11]. Hence, while σ_3 seems to be inconsistent with experiment, one still faces the question of what is the proper theoretical value. Also resumming the series in disorder by a self consistent Born approximation gives σ_1 [12], while other resummations can lead to a different result [13].

In contrast to the controversy with respect to the dc conductivity, both the experimental and the theoretical situation for the ac conductivity at the high frequency is quite different. The theoretically predicted value in the Dirac model is σ_2 independent of frequency under condition $\omega \gg T/\hbar$ [12, 14]. The Dirac model becomes inapplicable when ω is of order of $\gamma/\hbar \approx 4 \cdot 10^{15} Hz$ or larger, where γ is the hopping energy of graphene. It was shown theoretically using the tight binding model and experimentally in [15] that the optical conductivity at frequencies higher or of order $\gamma/\hbar$ becomes slightly larger than σ_2. Moreover, in light transmittance measurements at frequencies down to $2.5 \cdot 10^{15} Hz$ it was found equal to σ_2 within 4%. The tight binding model *does not contain any other time* scales capable of changing the limiting value of the ac conductivity all the way to $\omega \rightarrow 0$. Therefore one would expect that the dc conductivity is σ_2 rather than σ_1.

The physical nature of the quasi-Ohmic "resistivity" without either charge carriers or dissipation in infinite samples (namely neglecting leads or "reservoir") is as follows [16, 17]. The electric field creates electron-hole excitations in the vicinity of the Dirac points similar to the Zener tunneling effect in narrow gap semiconductors or electron-positron pair creation in Quantum Electrodynamics. Importantly, in graphene the energy gap is zero, thus the pair creation is possible at zero temperature and arbitrary small $\mathbf{E}$, even within linear response. Although the absolute value of the quasiparticle velocity v_g cannot be altered by the electric field due to the "ultra-relativistic" dispersion relation, the orientation of the velocity can be influenced by the applied field. The electric current, $e\mathbf{v}$, proportional to the projection of the velocity $\mathbf{v}$ onto the direction of the electric field is increased by the field. These two sources of current, namely creation of moving charges by the electric field (polarization) and their reorientation (acceleration) are responsible for the creation of a stable current. The result within linear response is that the current settles very fast, on the microscopic time scale of $t_\gamma = \hbar/\gamma \simeq 0.24\ fs$, on the value of σ_2.

A deeper analysis of the "quasi-Ohmic" graphene system beyond the leading order in perturbation theory in the electric field reveals [18] that on the time scale

$$t_{nl} = \sqrt{\frac{\hbar}{eEv_g}}, \tag{3}$$

the linear response breaks down due to intensive Schwinger's pair creation and leads to a linear increase with time[19, 20]:

$$J\left(t\right) = \sigma_2 \left(\frac{\sqrt{3}}{2}E\right)^{3/2} \left(\frac{ev_g}{\hbar}\right)^{1/2} t. \tag{4}$$

This is in contrast to dissipative systems, in which the linear response limit can be taken directly at infinite time. This perhaps is the origin of the "regularization" ambiguities in graphene, since large time and small field limits are different. In most transport experiments the ratio $t_L/t_{nl} = L\sqrt{eE/\hbar v_g}$ is smaller than 1. The time scale on which nonlinear effects become dominant is not always very large; for example, for experiments not necessarily dedicated to large current transport measurements [21] in which $E = 10^4 V/m$ and $L = 0.3\mu m$, nonlinearity sets in at $t_{nl} = 0.3ps$, which is of order of the ballistic time. Moreover, graphene flakes under large fields of order $2{\cdot}10^6 V/m$ have been recently studied in specially designed high current density experiments [22]. For a sample length $L = 2\mu m$ this results in a very large ratio $t_L/t_{nl} = 100$.

The physics of pair creation is highly non-perturbative and non-linear in nature and therefore, instead of the linear response, Schwinger found an exact formula using functional methods. The rate can be intuitively understood using the much simpler instanton approach originally proposed in the context of particle physics [23] (extended later to low dimensions [24]); in condensed matter physics it is known as the Landau–Zener tunneling probability [20, 25]. In particle physics it is extremely difficult to observe Schwinger's creation rate and it would be interesting to establish experimentally this dynamical phase in low dimensional condensed matter physics featuring the massless Dirac quasiparticle spectrum like graphene or novel materials (sharing with graphene the massless Dirac spectrum) like topological insulators or tuned semiconductor heterojunctions [26].

In this chapter we use the dynamical approach to study rigorously transport in ballistic samples and show that indeed the physics behind the two values is different despite the fact that numerically $\sigma_2 = 1.57e^2/h$ is just 24% higher than $\sigma_1 = 1.27e^2/h$ (this is quite accidental due to nonuniversality of the mesoscopic value). An alternative and independent approach to ballistic transport in mesoscopic graphene samples with finite length L, was pioneered in[27] following the idea to consider "Zitterbewegung" contributions to the current [28]. They applied the Landauer-Büttiker formula to the interband transitions within the Weyl model with an infinite potential barrier and obtained σ_1 in the limit of a large aspect ratio W/L. Indeed, it was shown recently, using the dynamical approach [29], that in mesoscopic samples the intraband mechanism dominates the conductivity for small and intermediate times only. At large times, however, the interband mechanism takes over leading to σ_1.

As argued above, the basic pseudo-relativistic physical effect of the electric field is a coherent creation of electron-hole pairs, mostly, but not exclusively, near the two Dirac points. To describe this process effectively we develop a dynamical approach to charge transport in clean graphene using a "first quantized" approach to pair creation physics similar to that developed in relativistic physics for the description of electron-positron pairs creation[30]. To the end of visualizing the phenomenon of dissipationless resistivity at zero temperature, doping, frequency and disorder, we follow the experimental situation as closely as possible by a straight forward calculation of the electric current's time evolution after switching on an electric field. This is presented in section II within the tight binding model. In section III we show in detail that some aspects of the linear response physics are not dominated by the two Dirac points of the Brillouin zone at which the spectrum is gapless. For example, large contributions (infinite, when the size of the Brillouin zone $2\pi/a$ is being considered infinite; $a = 3\mathring{A}$ is the lattice spacing) to the conductivity from the vicinity of the Dirac points are cancelled by contributions from the region between them. This phenomenon is related to the chiral anomaly in field theory [31]. We analyze the use of the massless Dirac (Weyl) model approximation to the tight binding model and propose an effective chiral invariant regularization for it.

In section IV the range of applicability of the linear response approximation is determined by the calculation of higher orders in the (dc and ac) electric field and also by an exact solution of the model in the nonlinear electromagnetic response regime. We first obtain perturbative corrections up to the third order in the electric field E. At this order qualitatively new phenomena like the third harmonic generation and the imaginary part of the conductivity appear. The calculation of the corrections allows to estimate the time scale at which perturbation theory breaks down, $t_{nl} \propto E^{-1/2}$. At this scale the physical behavior is expected to change qualitatively, but one has to resort to nonperturbative methods. Long ballistic flight evolution, $t > t_{nl}$, of various physical quantities like the current density and the number of created pairs is explained in section V using the Landau-Zener [20] or instanton approximation[23]. In the absence of any relaxation channels (for times below $t_L = L/v_g$), a neutral electron-hole plasma develops at ballistic times of order t_{nl}. This process cannot continue forever and is terminated by a relaxational recombination, which is the subject of section VI. We draw attention to a direct and unintrusive signature of the dynamical phase of Schwinger's pair creation: the flux of photons radiated from the surface of the sample is characterized by the creation rate since the photons are emitted via electron-hole pair annihilation and therefore proportional to $E^{3/2}$, a hallmark of Schwinger's process. In addition, the frequency, direction and polarization characteristics of the radiation generated by the electric field all bear footprints of the pair creation dynamics. A summary is given in section VII together with a brief discussion of the transport in finite samples, a comparison with the Landauer–Buttiker approach and several generalizations.

II. The Tight Binding Model and the Dynamical Approach to its Physical Properties

II.1. The tight binding model on the honeycomb lattice in an electric field

Electrons in graphene are described sufficiently accurately for our purposes by the 2D tight binding model of nearest neighbor interactions [4]. The Hamiltonian in **k** space is

$$\widehat{H} = \sum_{\mathbf{k}} \begin{pmatrix} c_{\mathbf{k}}^{A\dagger} & c_{\mathbf{k}}^{B\dagger} \end{pmatrix} H_{\mathbf{k}} \begin{pmatrix} c_{\mathbf{k}}^{A} \\ c_{\mathbf{k}}^{B} \end{pmatrix}, \tag{5}$$

where

$$H_{\mathbf{k}} = \begin{pmatrix} 0 & h_{\mathbf{k}} \\ h_{\mathbf{k}}^{*} & 0 \end{pmatrix}; \quad h_{\mathbf{k}} = -\gamma \sum_{\alpha} e^{i\mathbf{k}\cdot\delta_{\alpha}} \tag{6}$$

with $\gamma \simeq 2.7eV$ being the hopping energy; $\delta_1 = \frac{a}{3}\left(0, \sqrt{3}\right)$and $\delta_{2,3} = \frac{a}{3}\left(\pm\frac{3}{2}, -\frac{\sqrt{3}}{2}\right)$ are the locations of nearest neighbors separated by distance $a \simeq 3\mathring{A}$; and the pseudospin index A, B denotes two triangular sublattices. In most parts of the paper we keep the function $h_{\mathbf{k}}$ arbitrary. Let us first consider the system in a constant and homogeneous electric field along the y direction switched on at $t = 0$. It is described by the minimal substitution,

$$\mathbf{p} = \hbar\mathbf{k} + \frac{e}{c}\mathbf{A}, \tag{7}$$

with vector potential $\mathbf{A} = (0, -cEt)$ for $t > 0$ $(e > 0)$. Later it is generalized to more general fields including the ac field.

II.2. The dynamical "first quantized" approach

Since the crucial physical effect of the field is mainly a coherent creation of electron-hole pairs (though not always near the Dirac points, see below), a convenient formalism to describe the pair creation is the "first quantized" formulation described in detail in [30]. The tight binding model, Eq. (5), is an idealization of the experimental situation in several respects. The sample is considered "infinite" and absolutely homogeneous. This allows a convenient use of the axial gauge that is invariant under translations. Finite width (perpendicular to the electric field chosen to be oriented along the y axis) generally creates no complications and the whole discussion can be repeated for finite width W and variety of boundary conditions. On the other hand, finite length breaks the translation invariance and simplicity is lost. The latter case will be considered in other sections. The second idealization pertains the description of the leads. It assumed that the leads are absolutely unintrusive, namely one can imagine the Corbino disc geometry or a lead with no contact potential difference.

Before the electric field is switched on the spectrum is divided into positive and negative energy parts describing the valence and conduction band:

$$H_{\mathbf{k}}(E=0)\,u_{\mathbf{k}} = -\varepsilon_{\mathbf{k}} u_{\mathbf{k}}; \quad H_{\mathbf{k}}(E=0)\,v_{\mathbf{k}} = \varepsilon_{\mathbf{k}} v_{\mathbf{k}}; \tag{8}$$

$$u_{\mathbf{k}} = \frac{1}{\sqrt{2}}\begin{pmatrix} 1 \\ -z_{\mathbf{k}} \end{pmatrix}; \quad v_{\mathbf{k}} = \frac{1}{\sqrt{2}}\begin{pmatrix} 1 \\ z_{\mathbf{k}} \end{pmatrix}, \tag{9}$$

where $z_{\mathbf{k}} = h^*_{\mathbf{k}}/\varepsilon_{\mathbf{k}}$ is the phase and $\varepsilon_{\mathbf{k}} = |\mathbf{k}|$. Since all the momenta are independent in the Hamiltonian Eq. (6) due to the gauge used, in which electric field is represented via a homogeneous vector potential, a second quantized state is uniquely characterized by the "first quantized" amplitude,

$$\psi_{\mathbf{k}}(t) = \begin{pmatrix} \psi^1_{\mathbf{k}}(t) \\ \psi^2_{\mathbf{k}}(t) \end{pmatrix}, \tag{10}$$

which is a "spinor" in the sublattice space. It obeys the matrix Schrödinger equation in the sublattice space

$$i\hbar\partial_t \psi_{\mathbf{k}} = H_{\mathbf{p}} \psi_{\mathbf{k}}, \tag{11}$$

where $\mathbf{p}$ is defined in Eq. (7). The initial condition corresponding to a second quantized state at $T = 0$ in which all the negative energy states are occupied and all the positive energy states are empty is

$$\psi_{\mathbf{k}}(t=0) = u_{\mathbf{k}}. \tag{12}$$

A physical quantity is usually conveniently written in terms of ψ. We will be interested mostly in the current density (multiplied by factor 2 due to spin)

$$J_y(t) = -2e \sum_{\mathbf{k}\in BZ} \psi^\dagger_{\mathbf{k}}(t) \frac{\partial H_{\mathbf{p}}}{\partial p_y} \psi_{\mathbf{k}}(t), \tag{13}$$

as well as the energy and the density of the electron-hole pairs. The summation is over the honeycomb lattice Brillouin zone, see Figure 1, in which two Brillouin zones are outlined.

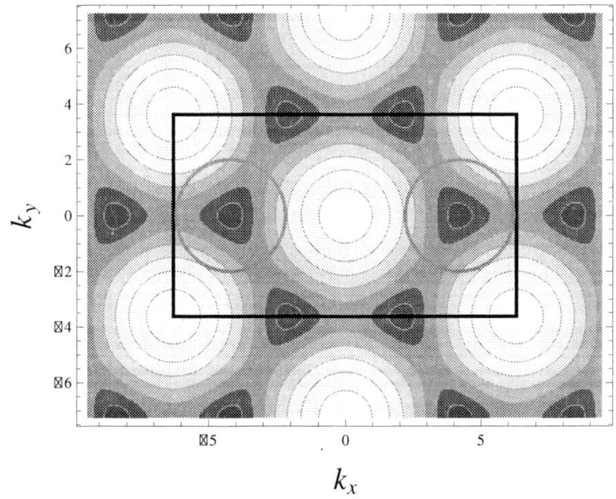

Figure 1. Constant energy map of the honeycomb-lattice Brillouin zone. The rectangle outlines two Brillouin zones, the area of integration in Eqs.(13 and 28). The circles show the vicinity of the two Dirac points with radius Λ for the integration in Eq.(31).

II.3. Units and conventions

The units of energy, time and distance are defined by the microscopic values γ, $t_\gamma \equiv \hbar/\gamma$ $\left(\simeq 2.4 \cdot 10^{-16} s\right)$ and a, correspondingly. The unit of the electric field will be $E_0 \equiv \frac{\gamma}{ea} \simeq 10^{10} V/m$, so that the dimensionless electric field is

$$\mathcal{E} = E/E_0 \tag{14}$$

and the unit of conductivity will be $e^2/\hbar$. Effectively one can set $\gamma = \hbar = a = e = 1$. In these units the first quantized equation reads

$$\begin{aligned} i\partial_t \psi_1 &= h_{\mathbf{p}} \psi_2; \\ i\partial_t \psi_2 &= h^*_{\mathbf{p}} \psi_1, \end{aligned} \tag{15}$$

where $p_y = k_y + A_y$, $A_y = -\mathcal{E}t$ for the dc field.

The next two sections deal with the perturbative treatment of the electric field and later (after having shown that it fails at a certain time scale) we will switch to a nonperturbative method.

III. Linear Response, the Pseudo-Diffusive Behavior and the Parity Anomaly

III.1. Expansion in the electric field

Expanding in the dimensionless electric field strength (assumed for simplicity homogeneous and constant after switching on the field at $t = 0$), $\psi_{\mathbf{k}} = e^{i\varepsilon_{\mathbf{k}} t} \left(u_{\mathbf{k}} + \mathcal{E}\xi_{\mathbf{k}} + ...\right)$, one obtains for the first correction the following equation:

$$i\partial_t \xi_{\mathbf{k}} = \begin{pmatrix} \varepsilon_{\mathbf{k}} & h_{\mathbf{k}} \\ h^*_{\mathbf{k}} & \varepsilon_{\mathbf{k}} \end{pmatrix} \xi_{\mathbf{k}} - t \begin{pmatrix} 0 & h'_{\mathbf{k}} \\ h^{*\prime}_{\mathbf{k}} & 0 \end{pmatrix} u_{\mathbf{k}}. \tag{16}$$

Writing the correction as $\xi_{\mathbf{k}} = \alpha_{\mathbf{k}} u_{\mathbf{k}} + \beta_{\mathbf{k}} v_{\mathbf{k}}$, it takes the form

$$\begin{aligned} i\partial_t \alpha_{\mathbf{k}} &= t\varepsilon'_{\mathbf{k}}; \\ i\partial_t \beta_{\mathbf{k}} &= 2\varepsilon_{\mathbf{k}} \beta_{\mathbf{k}} - t\varepsilon_{\mathbf{k}} I^-_{\mathbf{k}}, \end{aligned} \tag{17}$$

with initial conditions $\alpha_{\mathbf{k}}(t=0) = 1$, $\beta_{\mathbf{k}}(t=0) = 0$. We use the abbreviations $h'_{\mathbf{k}} \equiv \frac{\partial h_{\mathbf{k}}}{\partial k_y}$, $\varepsilon'_{\mathbf{k}} \equiv \frac{\partial \varepsilon_{\mathbf{k}}}{\partial k_y}$ and

$$I^-_{\mathbf{k}} \equiv \frac{h^*_{\mathbf{k}} h'_{\mathbf{k}} - h_{\mathbf{k}} h^{*\prime}_{\mathbf{k}}}{2\varepsilon^2_{\mathbf{k}}}. \tag{18}$$

The coefficient $\beta_{\mathbf{k}}$ denotes the amplitude of accumulation of electrons in the conduction band, whereas $\alpha_{\mathbf{k}}$ denotes the amplitude of remaining in the valence band. Solving the equations one obtains

$$\alpha_{\mathbf{k}} = 1 - \frac{i}{2} t^2 \varepsilon'_{\mathbf{k}}; \quad \beta_{\mathbf{k}} = \frac{i I^-_{\mathbf{k}}}{4\varepsilon_{\mathbf{k}}} \left(1 - e^{-2i\varepsilon_{\mathbf{k}} t} - 2i\varepsilon_{\mathbf{k}} t\right). \tag{19}$$

The expansion for the current, Eq. (13), in the electric field up to the first order contains the following momentum **k** contributions, $J_y = \sum_{\mathbf{k}} j_{\mathbf{k}}$:

$$j_{\mathbf{k}} = j^0_{\mathbf{k}} + j^p_{\mathbf{k}} + j^d_{\mathbf{k}}. \tag{20}$$

The zero order contribution is

$$j^0_{\mathbf{k}} = -u^\dagger_{\mathbf{k}} \begin{pmatrix} 0 & h'_{\mathbf{k}} \\ h^{*\prime}_{\mathbf{k}} & 0 \end{pmatrix} u_{\mathbf{k}} = 2\varepsilon_{\mathbf{k}} I^+_{\mathbf{k}}. \tag{21}$$

where

$$I^+_{\mathbf{k}} = \frac{h_{\mathbf{k}} h^{*\prime}_{\mathbf{k}} + h^*_{\mathbf{k}} h'_{\mathbf{k}}}{2\varepsilon^2_{\mathbf{k}}}. \tag{22}$$

The correction gives the 'paramagnetic' and 'diamagnetic' contributions to the current densities:

$$\begin{aligned} j^p_{\mathbf{k}} &= -2\mathcal{E} u^\dagger_{\mathbf{k}} \begin{pmatrix} 0 & h'_{\mathbf{k}} \\ h^{*\prime}_{\mathbf{k}} & 0 \end{pmatrix} \xi_{\mathbf{k}} + cc & (23) \\ &= 2\mathcal{E}\left(\alpha_{\mathbf{k}} - \alpha^*_{\mathbf{k}}\right) \varepsilon_{\mathbf{k}} I^-_{\mathbf{k}} = \mathcal{E}\left(I^-_{\mathbf{k}}\right)^2 \left[2\varepsilon_{\mathbf{k}} t - \sin\left(2\varepsilon_{\mathbf{k}} t\right)\right]; \\ j^d_{\mathbf{k}} &= \mathcal{E} t u^\dagger_{\mathbf{k}} \begin{pmatrix} 0 & h''_{\mathbf{k}} \\ h^{*\prime\prime}_{\mathbf{k}} & 0 \end{pmatrix} u_{\mathbf{k}} = -2\mathcal{E} t \varepsilon_{\mathbf{k}} I^+_{\mathbf{k}}. & (24) \end{aligned}$$

Both corrections for a specific momentum **k** diverge at large ballistic times as t, however one still has to integrate over the valence band momenta.

III.2. Integration over momenta and physical interpretation of the "quasi-Ohmic" behavior

To first order in electric field the current density is

$$J_y = J_0 + \sigma \mathcal{E}, \tag{25}$$

where the zero order contribution can be written as a derivative with respect to k_y of $\varepsilon_{\mathbf{k}}$,which is a periodic function

$$J_0 = -2 \sum_{\mathbf{k}} j^0_{\mathbf{k}} = 4 \sum_{\mathbf{k}} \varepsilon'_{\mathbf{k}}. \tag{26}$$

It vanishes upon integration, in accordance with the Bloch theorem, since the Brillouin zone can be chosen in such a way that it exhibits periodicity in the field (y) direction. For example, we can integrate over the rectangular area shown in Fig.1 containing two Brillouin zones:

$$\sum_{BZ} = \frac{1}{2} \frac{1}{(2\pi)^2} \int_{-2\pi}^{2\pi} dk_x \int_{-2\pi/3^{1/2}}^{2\pi/3^{1/2}} dk_y. \tag{27}$$

It should be noted that $\int_{-2\pi/3^{1/2}}^{2\pi/3^{1/2}} \varepsilon'_{\mathbf{k}} dk_y = \varepsilon_{k_x,2\pi/3^{1/2}} - \varepsilon_{k_x,-2\pi/3^{1/2}} = 0$ at every k_x, and due to the continuity of $\varepsilon_{\mathbf{k}}$, even at the Dirac points. Therefore one is left with the linear response.

The conductivity can be divided into an apparently linearly divergent part and a finite part, $\sigma(t) = \sigma_{BZ}(t) + \sigma_{DP}(t)$. As will be explained shortly, the contributions to the first term at large t stem mostly from the immediate neighborhoods of the two Dirac points, while contributions to the second come from whole Brillouin zone. The part diverging linearly with time can be written as a full derivative with respect to k_y:

$$\sigma_{BZ}(t) = t \sum_{\mathbf{k}} \varepsilon_{\mathbf{k}}^{-3} \left[\left(h_{\mathbf{k}} h^{*\prime}_{\mathbf{k}} - h^*_{\mathbf{k}} h'_{\mathbf{k}} \right)^2 - 2\varepsilon^2_{\mathbf{k}} \left(h_{\mathbf{k}} h^{*\prime\prime}_{\mathbf{k}} + h^*_{\mathbf{k}} h''_{\mathbf{k}} \right) \right] = -4t \sum_{\mathbf{k}} \varepsilon''_{\mathbf{k}}. \tag{28}$$

This too vanishes upon integration over the Brillouin zone, since it is again an integral of a derivative of a periodic function, albeit this cancellation is nontrivial. In Figure 2(a) the integrand of Eq. (28) is plotted within the integration area specified above, Eq. (27).The integrand is negative along the line connecting two Dirac points along the field direction, for example

$$\mathbf{K}_L = 2\pi \left(\frac{1}{3}, \frac{1}{\sqrt{3}} \right), \quad \mathbf{K}_R = 2\pi \left(-\frac{1}{3}, \frac{1}{\sqrt{3}} \right) \tag{29}$$

(recall that $a = 1$ in our units), and positive elsewhere. In Figure 2(b) several cross sections for various k_x values are shown. One observes, that as k_x approaches $2\pi/3$ the integrand goes to $-\infty$ at $k_y = 2\pi/\sqrt{3}$, corresponding to Dirac point $\mathbf{K}_L$. Consequently for all $k_x \neq 2\pi/3$,the integrand is a regular function, and thus the integral over k_y vanishes. At $k_x = 2\pi/3$ the integral over k_y is finite, yet does not influence the two-dimensional integral.

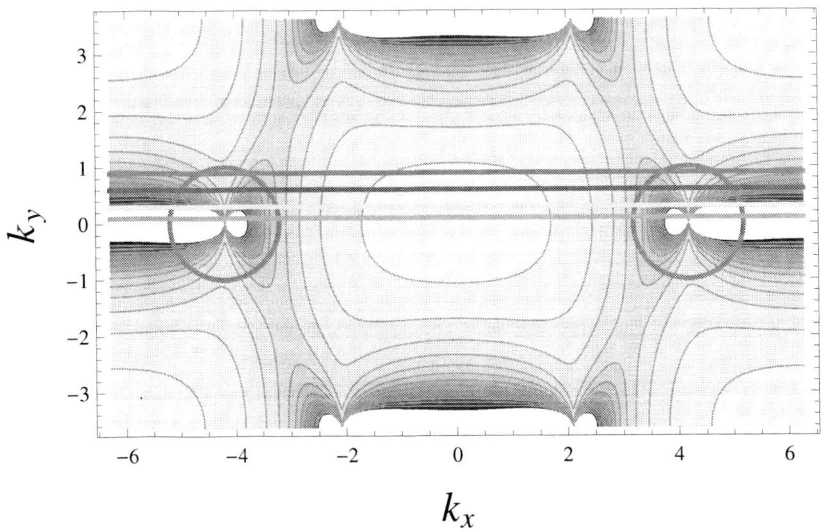

(a)

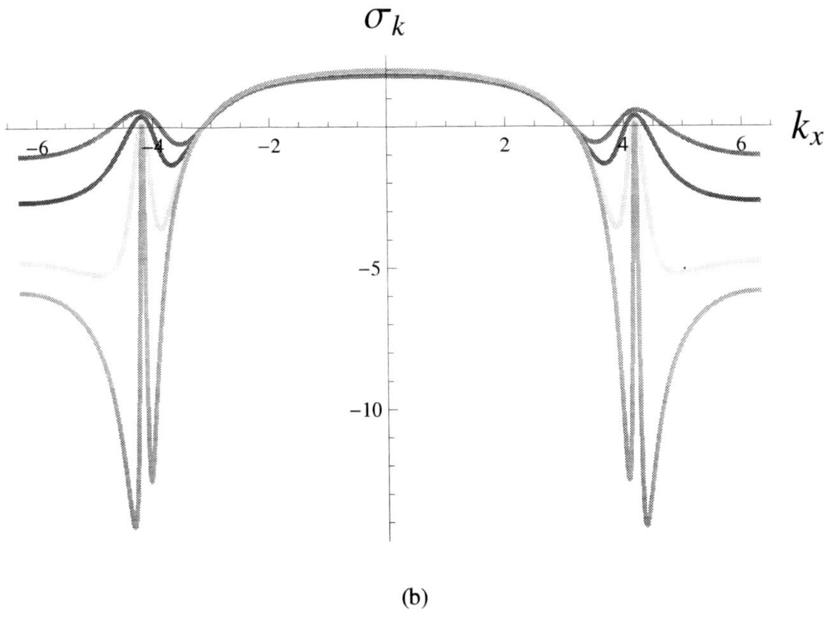

(b)

Figure 2. (a) The integrand of σ_{BZ} Eq. (28) is plotted within the integration area shown in Figure 1. The integrand is negative in the vicinity of the two Dirac points but positive elsewhere. (b) Several cross sections of the integrand of σ_{BZ} Eq. (28) are shown for the various k_y values indicated in Figure 2(a).

The remaining part

$$\sigma_{DP}(t) = -2\sum_{\mathbf{k}} \left(I_{\mathbf{k}}^{-}\right)^2 \sin\left(2\varepsilon_{\mathbf{k}} t\right), \tag{30}$$

gives a finite result. Unlike the divergent part of the conductivity, the integral for $t \gg 1$ (in units of t_γ) is dominated by contributions from the vicinity of the two Dirac points (circles in Fig. 2(a) of radius $\Lambda \simeq 1$ (in units of $\hbar/a$)). Indeed the prefactor $\left(I_{\mathbf{k}}^{-}\right)^2$ is bounded from above by Λ^{-2} and integral over the area outside the circles vanishes at $t \gg 1/\Lambda$. The contribution of a single Dirac point is obtained by integrating to infinity (in polar coordinates centered at the Dirac point),

$$\begin{aligned} \frac{\sigma_{DP}}{2} &= \frac{1}{(2\pi)^2} \int_{\varphi=0}^{2\pi} \int_{q=0}^{\Lambda} \sin(\varphi)^2 \frac{\sin(2v_g q t)}{q} \\ &= \frac{1}{4\pi} \int_{q'=0}^{2v_g\Lambda t} \frac{\sin(q')}{q'}. \end{aligned} \tag{31}$$

At long times the upper limit can be replaced by infinity which results

$$\sigma_{DP} = \frac{1}{2\pi}\mathrm{Si}\left(\infty\right) = \frac{1}{4}. \tag{32}$$

Here one can safely multiply the result by the valley degeneracy 2. Returning to physical units, one obtains $\sigma = \sigma_2$, Eq. (2).

The dependence on time was calculated numerically. After an initial increase on the natural time scale t_γ, the conductivity approaches σ_2 via oscillations, see Fig. 3. The amplitude of oscillations decays as a power $\frac{\sigma}{\sigma_2} \approx 1 + \frac{\sin(2t)}{2t}$ (for $t \gg t_\gamma$). A more detailed physical description of this resistivity without dissipation is as follows. The electric field creates electron-hole excitations in the vicinity of the Dirac points in which electrons behave as massless relativistic fermions with the graphene velocity v_g playing the role of the light velocity. For such particles the absolute value of the velocity is v_g and cannot be altered by the electric field and is not related to the wave vector $\mathbf{k}$. On the other hand, the orientation of the velocity is influenced by the applied field. The electric current is $e\mathbf{v}$, thus depending on orientation, so that its projection on the field direction y is increased by the field. An important observation, made for example in ref. [16], is that the electric field, while creating charge transport, does not change the overall momentum. As a consequence the effects of impurities do not affect the pair creation in the same way they affect "free" carriers. These two sources of current, namely creation and reorientation are roughly of the same size [20] in the linear response. As we will see below, the situation changes at ballistic times when the linear response breaks down.

III.3. ac field

A similar calculation within the dynamic first quantization formalism for the evolution of the current density can be performed for an arbitrary time dependence of the homogeneous

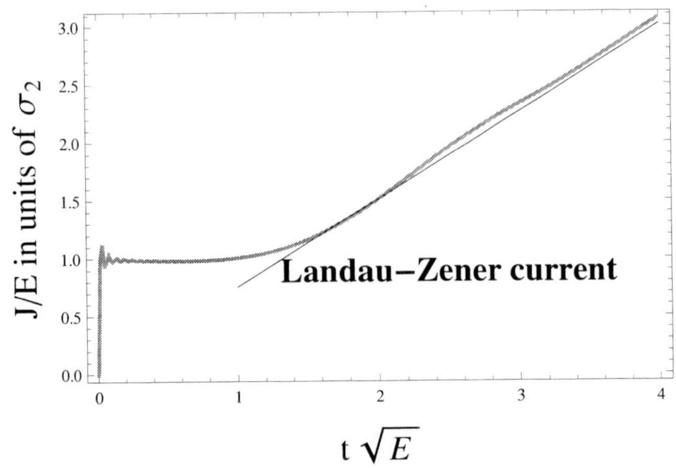

Figure 3. The time evolution of the dc conductivity in units of σ_2. The time is scaled with $\mathcal{E}^{-1/2}$.The simulations were performed in the field range $\mathcal{E} = 2^{-12} - 2^{-9}$ with UV cutoff equal to $1/a$. The straight line is the asymptotic linear behavior given by Eq. (51).

electric field. Let us consider an arbitrary time dependence of the y component of the vector potential $A_y = -\mathcal{E}a(t)$, subject to a "switching on" condition, $a(t) = 0$, for $t < 0$. Fourier transforms are defined by

$$a(t) = \int_\omega e^{-i\omega t} a(\omega); \quad \xi_{\mathbf{k}}(t) = \int_\omega e^{-i\omega t} \xi_{\mathbf{k}}(t), \tag{33}$$

and similarly for the current density and other quantities. The next to leading order in $\mathcal{E}$ first quantized tight binding equations are:

$$-a(\omega)\begin{pmatrix} 0 & h'_{\mathbf{k}} \\ h^{*\prime}_{\mathbf{k}} & 0 \end{pmatrix} u_{\mathbf{k}} + \begin{pmatrix} \varepsilon_{\mathbf{k}} - \omega & h_{\mathbf{k}} \\ h^*_{\mathbf{k}} & \varepsilon_{\mathbf{k}} - \omega \end{pmatrix} \xi_{\mathbf{k}}(\omega) = 0. \tag{34}$$

The switching on condition, $\xi_{\mathbf{k}}(t < 0) = 0$, can be taken into account by the $\omega + i0^+$ substitution regularizing the way the Fourier transform is defined for zero frequency. From Eq. (34) one obtains

$$\xi_{\mathbf{k}}(\omega) = -\frac{a(\omega)}{\sqrt{2}\omega(2\varepsilon - \omega)} \begin{pmatrix} \omega h^* h'/\varepsilon - h^* h' - h^{*\prime} h \\ -\omega h^{*\prime} + \varepsilon h^{*\prime} + h^{*2} h'/\varepsilon \end{pmatrix}. \tag{35}$$

In the same order the ac conductivity is an integral over the sum of the "paramagnetic" and the "diamagnetic" contributions given by

$$\frac{j_{\mathbf{k}}(\omega)}{a(\omega)} = 4\frac{(h^* h' - h^{*\prime} h)^2}{\varepsilon(2\varepsilon + \omega)(2\varepsilon - \omega)} - 2\frac{h^* h'' + h h^{*\prime\prime}}{\varepsilon}. \tag{36}$$

Subtracting a full derivative (independent of frequency), one can rewrite this as

$$\frac{j_{\mathbf{k}}(\omega)}{a(\omega)} = \frac{(h^* h' - h^{*\prime} h)^2}{\varepsilon^3} \frac{\omega^2}{4\varepsilon^2 - \omega^2}. \tag{37}$$

Real and imaginary parts, taking the $\omega + i0^+$ definition into account, are

$$\mathrm{Re}\,\sigma = \frac{4}{i\omega}\sum_{\mathbf{k}}\frac{(h^*h' - h^{*\prime}h)^2}{\varepsilon}\lim_{s\to 0}\frac{4\varepsilon^2-\omega^2+s^2}{\left[(2\varepsilon+\omega)^2+s^2\right]\left[(2\varepsilon-\omega)^2+s^2\right]} = \frac{1}{4}; \quad (38)$$

$$\mathrm{Im}\,\sigma = \frac{4}{i\omega}\sum_{\mathbf{k}}\frac{(h^*h' - h^{*\prime}h)^2}{\varepsilon}\lim_{s\to 0}\frac{2\omega s}{\left[(2\varepsilon+\omega)^2+s^2\right]\left[(2\varepsilon-\omega)^2+s^2\right]} = 0.$$

Therefore ac and dc conductivity are equal. This is consistent with both the Kubo formula derivations [14] and optical experiments [15].

IV. Relativistic Invariance: Weyl Fermions, Parity Anomaly and the Nielsen–Ninomiya Theorem

As will be discussed below, an electric field should not necessarily be very small in graphene experiments, so that corrections to linear response are of interest. In addition it is important to determine at which ballistic time scale perturbation theory breaks down. In this section we present a perturbative calculation of the leading nonlinear effect in both dc and ac, and discuss the ways to regularize the Weyl model "correctly" in order to calculate these corrections.

IV.1. Ultrarelativistic fermions near Dirac points

The tight binding model employed here has two Dirac points around which the spectrum becomes ultrarelativistic $\varepsilon = v_g\left|\Delta_{L,R}\mathbf{k}\right|$, $\Delta_{L,R}\mathbf{k} = \mathbf{k} - \mathbf{K}_{L,R}$, see Fig.4. In our units the graphene velocity is $v_g = \frac{\sqrt{3}}{2}$. The matrix element of the tight binding Hamiltonian can be expanded around K_L as

$$h^L_{\mathbf{k}} = v_g \exp\left(-i\frac{\pi}{3}\right)(\Delta k_x + i\Delta k_y). \quad (39)$$

The Weyl field describing the "left handed"[32, 31] fermions ψ^L, namely a field satisfying the relativistic Dirac equation with zero fermion mass

$$\begin{aligned} i\partial_t\psi_1^L &= v_g(\partial_x + i\partial_y)\psi_2^L; \\ i\partial_t\psi_2^L &= v_g(\partial_x - i\partial_y)\psi_1^L, \end{aligned} \quad (40)$$

can be constructed by the unitary transformation

$$\psi_1 = \psi_1^L, \quad \psi_2 = e^{-i\frac{\pi}{3}}\psi_2^L. \quad (41)$$

The matrix element around the second Dirac point $\mathbf{K}_R$ is different,

$$h^R_{\mathbf{k}} = v_g(\Delta k_x - i\Delta k_y), \quad (42)$$

and the Weyl fermions are "right handed" particles that obey

$$\begin{aligned} i\partial_t\psi_1^R &= v_g(\partial_x - i\partial_y)\psi_2^R; \\ i\partial_t\psi_2^R &= v_g(\partial_x + i\partial_y)\psi_1^R, \end{aligned} \quad (43)$$

without any unitary transformation required. They belong to a different representation of the 2+1 dimensional Lorentz group.

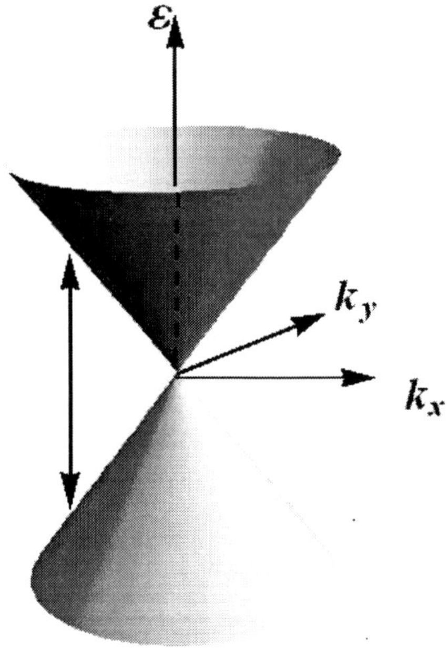

Figure 4. Pair creation and annihilation in effective relativistic Weyl model.

This is not just a peculiarity of the model, but a rather general feature of massless fermions on a lattice [31, 33]. It is well known that in order to get a massless spectrum of fermionic excitations with any ultraviolet cutoff (hexagonal lattice is an example), they come in multiple locations on the Brillouin zone (species "doubling"). In the Hamiltonian formulation the multiplicity is 2^D, where D is the dimensionality of space, $D = 2$ in graphene. In addition, the graphene fermions are "staggered", meaning the spinor components "live" on different sublattices of the honeycomb lattice. This reduces the doubling to $2^{D-1} = 2$. The doubling is intimately linked to the parity (discrete chiral symmetry) [31]. The two Dirac points have an opposite chirality so that there is no "parity anomaly" [34].

It is sometimes claimed in condensed matter literature that, at least while doing linear response, one can concentrate on the two neighborhoods of the Dirac points and neglect the rest of the Brillouin zone. Even more, often the calculation's result is just multiplied by the factor 2 (the valley degeneracy). Below we show that this is generally not an accurate description of what happens. This is not an academic question, since the low energy Weyl theory is simpler than the tight binding model and will be used in the next section to extend the calculations into higher orders in the electric field. One therefore needs a proper ultraviolet regularization of the Weyl model. Similar regularization issues are well known in field theory whenever chiral anomaly is encountered. Roughly, a "correct" regularization should respect the chiral symmetry leading to important constraints on the number and charges of massless fermions. Otherwise the model is called "anomalous" and results of perturbative calculations become arbitrary.

We therefore discuss in some detail the cancellation of infrared divergencies and the correct application of the "ultrarelativistic" approximation to the tight binding model.

IV.2. Cancellation of ultraviolet divergencies and the chiral symmetry

The ultrarelativistic approximation, Eq. (39), fails when applied to the first term in the expression for conductivity σ_{BZ} given in Eq. (28). At first glance the integral in Eq. (28) is dominated by the two Dirac points since the integrand diverges there, see Figure 2. For the both (widely separated) regions of the Brillouin zone, K_L and K_R, it has the same asymptotic form

$$\frac{\left(hh^{*\prime} - h^{*}h^{\prime}\right)^{2}}{2\varepsilon^{3}} - \frac{hh^{*\prime\prime} + h^{*}h^{\prime\prime}}{\varepsilon} \approx -\sqrt{3}\frac{(\triangle k_{x})^{2}}{|\triangle \mathbf{k}|^{3}}. \tag{44}$$

The integral over the neighborhood of each Dirac point converges in the "infrared" (here meaning $\mathbf{k} \rightarrow \mathbf{K}_{L,R}$) due to the Jacobian $|\triangle \mathbf{k}|$, but is linearly divergent in the "ultraviolet" and both have the same sign, see Figure 2. Hence the integration cannot be extended to infinity and the size of the Brillouin zone serves as a natural ultraviolet cutoff.

It is important to note that there is no cancellation of the divergence between the Dirac points since both have the same sign! The divergencies are however canceled by the contributions from regions of the Brillouin zone between the Dirac points in which the ultrarelativistic approximation is not valid. Therefore in the "ohmic" regime during ballistic times one is not allowed to neglect states far from the Dirac points and replacement by the Weyl equation is incorrect. Due to cancellation of the whole "divergent" term, Eq. (**??**), one can device an appropriate regularization in the UV in which these contributions are cancelled and even generalize the procedure to higher orders in the electric field. We propose and use such a scheme below in section V. Simple recipes like the momentum cutoff regularization (circles in Fig.1) or giving the fermions an infinitesimal mass, commonly used, may lead to unphysical results. As a consequence more sophisticated regularizations like the ζ-function regularization [9], "dimensional regularization and "Pauli–Villars" were developed for continuous field theories like the Weyl model. The tight binding model is very similar in this respect to the "lattice Hamiltonian" regularizations of the field theory (Hamiltonian meant here with time kept continuous) and also satisfies the chiral invariance criterion [31].

IV.3. Higher orders in electric field. Where does the linear response break down?

Since the current density is an odd function of the electric field, the leading nonlinear correction to it appears in the third order in the field. The calculation along the lines of subsection IIIA is quite straightforward albeit tedious. One can use the Weyl model instead of the tight binding model due to the following reason. It is well known in field theory that generally chiral anomaly effects, including the ultraviolet divergencies discussed in section IIID, appear only in one loop calculations [31]. We checked and found that indeed up to the third order the expression is finite in the ultraviolet. Within the dynamical approach it is natural to perform the calculations without Fourier transforming into the ω space.

Up to the third order the current density is

$$j\left(t\right)=\frac{\mathcal{E}}{4}\left[1+\frac{3}{64}t^{4}\mathcal{E}^{2}+O\left(\mathcal{E}^{4}\right)\right]. \tag{45}$$

The correction therefore is growing as a large power of the ballistic time. It becomes as large as the leading order for $t=2^{3/2}3^{-1/4}\mathcal{E}^{-1/2}t_{\gamma}$. Hence the perturbation theory breaks down on the time scale of

$$t_{nl}=\mathcal{E}^{-1/2}t_{\gamma}. \tag{46}$$

As will be seen in next section, this agrees well with the crossover time obtained from a nonperturbative calculation. Of course this time should be larger than any other possible relaxation time (due to impurities, phonons etc.) present in the system. A similar expansion was obtained for the number of electron-hole pairs [19].

IV.4. ac response and the third harmonic generation

An analogous calculation for the ac field $E=\mathcal{E}\cos\left(\omega t\right)$ switched on at $t=0$ results in

$$j\left(t\right)/\mathcal{E}=-\frac{1}{8\omega^{4}}\mathcal{E}^{2}+\operatorname{Re}\sigma\cos\left(\omega t\right)+\operatorname{Im}\sigma\sin\left(\omega t\right)+\sigma_{3\omega}\cos\left(3\omega t\right)+O\left(\mathcal{E}^{4}\right). \tag{47}$$

The first term is just the reflection of the initial conditions and is non-universal. In the limit $\omega\to0$ one recovers the dc result. The corrected value of the real (dissipative) part of the ac conductivity is

$$\operatorname{Re}\sigma=\frac{1}{4}\left[1+\left(\frac{63}{128\omega^{4}}-\frac{9}{64\omega^{2}}t^{2}\right)\mathcal{E}^{2}\right]. \tag{48}$$

One observes that the perturbation theory is inapplicable for

$$\omega^{2}<\mathcal{E}\ \text{ and }\ t>\omega/\mathcal{E}. \tag{49}$$

This is less restrictive compared to the dc condition, Eq. (46) for $\omega>\mathcal{E}^{1/2}/t_{\gamma}$.

The present formulas can be compared with the results of the dynamical calculation beyond linear response by Mishchenko [35] in which a phenomenological model of relaxation was employed. For the inverse relaxation time Γ, his nonperturbative result is

$$\operatorname{Re}\sigma=\frac{\omega^{2}\Gamma^{2}}{2v_{g}^{2}\mathcal{E}^{2}}\left[\left(1+\frac{v_{g}^{2}\mathcal{E}^{2}}{\omega^{2}\Gamma^{2}}\right)^{1/2}-1\right]\approx\frac{1}{4}\left[1-\frac{3\mathcal{E}^{2}}{16\omega^{2}\Gamma^{2}}+O\left(\mathcal{E}^{4}\right)\right].$$

This is consistent with the second correction terms in Eq. (48).

The inductive part was absent in linear response and now appears for the first time:

$$\operatorname{Im}\sigma=\frac{3^{3}}{2^{8}\omega^{3}}t\mathcal{E}^{2}. \tag{50}$$

Due to the two conditions, Eq. (49), it is much smaller than σ_{2}. The third harmonic is generated with the real part

$$\sigma_{3\omega}=\frac{1}{2^{9}\omega^{4}}\mathcal{E}^{2},$$

while the inductive part is absent at the present order.

V. Beyond Linear Response: Schwinger's Electron-Hole Creation Rate

Despite the fact that the tight binding model allows an exact solution (obtained, using the Floquet theory, in ref.[19]), here we start from the numerical solution of the Weyl approximation.

V.1. Numerical solution of the Weyl equations

The current density divided by the electric field (in physical units), $\sigma(t) \equiv J_y(t)/E$ practically coincides with that of the tight binding model, shown in Fig. 3, for various values of the dimensionless electric field $\mathcal{E} = E/E_0$. The (microscopic) unit of the electric field is very large $E_0 = \frac{\gamma}{ea} = 10^{10} V/m$, so that at realistic fields $\mathcal{E} \ll 1$. After an initial fast increase on the microscopic time scale t_γ, $\sigma(t)$ approaches the universal value $\sigma_2 = \frac{\pi}{2}\frac{e^2}{h}$ and settles there, consistent with the linear response, calculated below. Beyond the crossover time t_{nl} the conductivity rises linearly above the constant "universal" value σ_2:

$$\frac{J(t)}{E} = \sigma_2 \lambda \mathcal{E}^{1/2} \frac{t}{t_\gamma}, \tag{51}$$

where $\lambda = 2^{5/2} 3^{1/4} \pi^{-2} \simeq 0.75$ (the straight line in Figure 3). This is consistent with the Landau–Zener calculation (instantons) within the Weyl model [24, 20], see below. The crossover time, defined by $J(t_{nl})/E = \sigma_2$, is therefore $t_{nl} = \lambda^{-1} \mathcal{E}^{-1/2} t_\gamma$ and is consistent with the perturbation theory breakdown ballistic time, Eq. (46). This linear increase regime can be considered as a precursor of the Bloch oscillations, but is still universal in the sense that it depends solely on neighborhood of the Dirac points.

It is expected that the transition to the nonlinear regime is dominated by the neighborhoods of the Dirac points. Therefore it can be obtained also within the Weyl model provided it is properly regularized in the ultraviolet region consistent with chiral symmetry, as was discussed in section IV. For the calculation of the current density it suffices to renormalize the electric current by subtracting the UV divergent term Eq. (44):

$$J_y = -\frac{4v_g}{(2\pi)^2} \int_{|\mathbf{k}|<\Lambda} \left[\frac{1}{2} \left(\psi_2^* \psi_1 + \psi_1^* \psi_2 \right) + \frac{k_x^2}{|\mathbf{k}|^3} t \mathcal{E} \right]. \tag{52}$$

The results are practically indistinguishable compared to the tight binding model for times larger than a microscopic time scale t_γ and much smaller than the Bloch time. The asymptotic form is attained soon after t_{nl}, so it looks like quite a sharp crossover.

V.2. Semiclassical picture of the crossover to the nonlinear response

The calculation of the number of pairs can be approximately performed using the instanton approach initiated by Nussinov in the context of particle physics [23] and extended in [24, 36]. In condensed matter physics the method is known as the Landau–Zener probability [20, 25]. It provides an intuitive picture of Schwinger's pair creation rate. Unfortunately this picture cannot be extended to ballistic times smaller than t_{nl} in which one cannot use the asymptotic large time expressions, so the numerical solution of the previous subsection complements the current one.

The number of pairs is defined by

$$N(t) = 2\sum_{\mathbf{k}\in BZ} |\langle\psi(t)|v_{\mathbf{p}}\rangle|^2 = 2\sum_{BZ} |\psi_1^* + z_{\mathbf{p}}\psi_2^*|^2 . \tag{53}$$

The momentum is "shifted" a consequence of the fact that the Fermi level should be continuously "renormalized" as function of time. In the first quantized formalism this corresponds to a continuous modification of the "initial" state $v_{\mathbf{k}} \to v_{\mathbf{p}} = v_{\mathbf{k}-\mathcal{E}t}$ defining holes.

The density calculated using the simple Landau–Zener creation rate expression for one of the flavours is [19, 20, 23]:

$$N_{\mathbf{k}}(t) = \Theta(k_y)\,\Theta(\mathcal{E}t - k_y)\exp\left(-\pi v_g k_x^2/\mathcal{E}\right), \tag{54}$$

where Θ are the Heaviside functions. One considers the "tunneling" from the conduction band to the valence band, see Figure 4, at fixed k_x. The gap is given by $2|k_x|$. In Figure 5 the time evolution of the momenta p_y in our gauge are depicted, so that Eq. (54) corresponds to the initial to final states transition rate. Consequently the number pairs is (for four flavors)

$$N(t) = \frac{4}{(2\pi)^2}\int N_{\mathbf{k}}(t) = \frac{1}{\pi^2 v_g^{1/2}}\mathcal{E}^{3/2}t, \tag{55}$$

corresponding to creation rate

$$\frac{d}{dt}N_p = \frac{8}{\pi^2 v_g^{1/2}}\mathcal{E}^{3/2}. \tag{56}$$

The power dependence of the rate on electric field, $E^{3/2}$ is the same as the rate of the vacuum breakdown due to the pair production calculated beyond perturbation theory by Schwinger in the context of particle physics (when generalized to the $2+1$ dimensions and zero fermion mass [37, 36]). It is not surprising since the power $3/2$ is dictated by the dimensionality, assuming the ultra-relativistic approximation is valid. However the physical meaning is somewhat different.

There is a simple relation within the Weyl model between the rate and the current density, as was shown recently [20] (see also [30]):

$$\begin{aligned} J &= J_c + J_p \\ J_c &= 2v_g\frac{4}{(2\pi)^2}\int \frac{\mathcal{E}t - k_y}{\sqrt{k_x^2 + (\mathcal{E}t - k_y)^2}} N_{\mathbf{k}}(t) \\ J_p(t) &= 2v_g\frac{4}{(2\pi)^2}\int_k \frac{\sqrt{k_x^2 + (\mathcal{E}t - k_y)^2}}{\mathcal{E}}\frac{d}{dt}N_{\mathbf{k}}(t) \end{aligned} \tag{57}$$

Thus electric current consists of two parts: the conduction and the polarization. Substituting the instanton calculation results, Eqs. (55) and Eq. (56), the conduction current is

$$J_c(t) = \frac{2v_g}{\pi^2}\int_{k_x}\left(\sqrt{k_x^2 + (\mathcal{E}t)^2} - |k_x|\right)\exp\left(-\pi v_g k_x^2/\mathcal{E}\right), \tag{58}$$

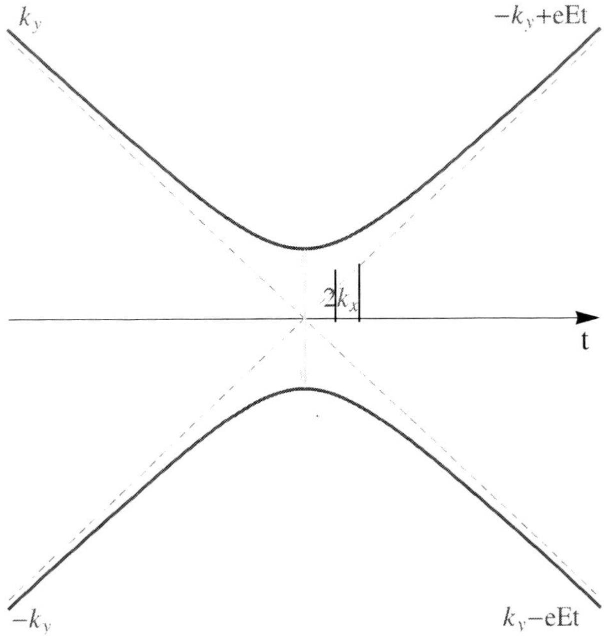

Figure 5. Landau–Zener tunneling for fixed value of k_x.

while the polarization is

$$J_p(t) = \frac{2v_g}{\pi^2} \int\limits_{k_x=0} 2k_x dk_x \exp\left(-\pi v_g k_x^2/\mathcal{E}\right). \tag{59}$$

Expanding in large $\mathcal{E}t$, one gets the conduction part as:

$$J_c \approx \frac{2v_g}{\pi^2}\mathcal{E}t \int\limits_{k_x} \exp\left(-\pi v_g k_x^2/\mathcal{E}\right) = 2v_g N(t) \tag{60}$$

Comparing this with linear response results one can write it as $J = \frac{1}{2\pi^2}\sigma_2 v_g^{1/2}\mathcal{E}^{3/2}t$ for $\sigma_2 = \frac{1}{4}$ or in physical units as in Eq. (4). The polarization current at large times approaches a constant

$$J_p(t) = \frac{2\mathcal{E}}{\pi^3} = \frac{8}{\pi^3}\sigma_2\mathcal{E}. \tag{61}$$

Therefore at large times the conduction dominates, while before t_{nl} both contributions are of the same order, see Figure 1 from ref. [20].

Let us note the relation between the dynamics on the time scale t_{nl} and quantum adiabatic transport near the quantum critical point [38, 25]. The ballistic evolution in graphene therefore can be considered as an example of the adiabatic quantum evolution which attracted much attention recently in connection to the Kibble–Zurek mechanism of phase transition dynamics. Graphene dynamics at large ballistic times offers an accessible system in which thee processes can be observed.

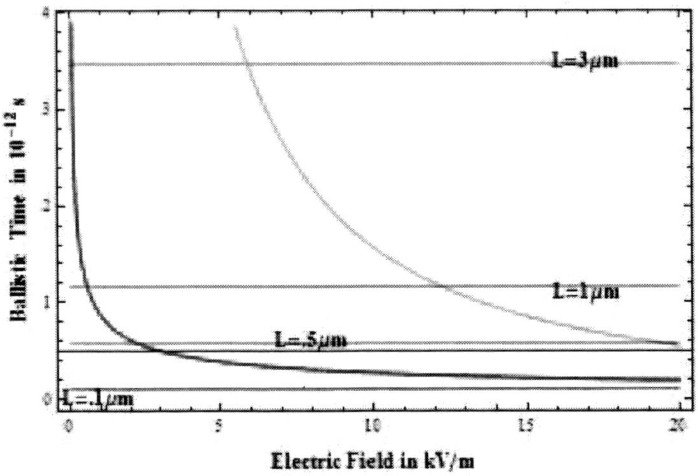

Figure 6. The nonlinear and the estimated electron-hole plasma time.

Is there a clear and unambiguous signature of the Schwinger's pairs creation phase? Next the optical signatures of the nonlinear behavior and possible new physics of the electron-hole plasma which might emerge are discussed.

VI. Emission of Photons and Transition Into Electron-Hole Plasma

VI.1. Coulomb interaction and the formation of the neutral electron-hole plasma

Even if the ballistic time and the relaxation time are very large, Schwinger's dynamical pair creation phase cannot persist for a long time at large field since density of charges (or both signs) becomes large. When the density of quasiparticles reaches the order of $\rho_p = 10^{11} cm^{-2}$ a neutral electron-hole plasma is created [39] (like in some semiconductor systems under irradiation). In this state electrostatic interactions (despite being screened at large distances) become dominant, as was discussed extensively in connection with electron-positron pairs creation in Quantum Electrodynamics[40]. When electrons and holes are close enough they strongly attract each other effectively facilitating the recombination process. The rate therefore far exceeds the one calculated within perturbation theory in Section III. Let us first estimate when this state is achieved at experimentally accessible situations.

The electrons and their electromagnetic interaction with photons are approximately described near a Dirac point by the Weyl Hamiltonian (we work in physical units since velocity of light enter in addition to v_g):

$$H = \int d^3r\, \psi^\dagger \left[\begin{array}{c} v_g \sigma \cdot \left(-i\hbar\nabla + \frac{e}{c}\mathbf{A}\right) \\ -\frac{\hbar^2}{2m}\left(\partial_z + i\frac{e}{\hbar c}A_z\right)^2 + V_{conf}(z) \end{array} \right] \psi. \tag{62}$$

Here ψ is the two component spinor second quantized field and $(\mathbf{A},A_z)$ is the vector potential (bold letters describe vectors in the graphene plane, while z is the perpendicular direction). Electrons (charge $-e$) and holes (charge e) in the graphene sheet are confined in this model to the $z = 0$ plane by a potential V_{conf} (small shape changes can be neglected for our purposes). The only requirement from this potential is that it is strong enough to "freeze" the motion along the z direction. In the single graphene sheet one has two left handed chirality Weyl fermions described by the above Hamiltonian in which σ denotes the in – plane Pauli matrices and two right handed Weyl fermions represented by $\sigma^{\dagger}$. To include the topological insulators case [26], we first concern ourselves with only one spinor. We consider the emission of a photon with wave vector $(\mathbf{k},k_z)$ and frequency $\omega = c\sqrt{\mathbf{k}^2 + k_z^2}$, described by a linearly polarized plane wave, $\mathbf{A}_{ph}=\frac{2E_0}{\omega}\mathbf{e}^{(\lambda)}\sin(\mathbf{k}\cdot\mathbf{r}+k_z z - \omega t)$. For regularization we make use of a finite box $W \times L \times L_z$, so that momenta are discrete and the single photon's electric field is $E_0^2 = \hbar\omega/(LWL_z)$. The unit vectors

$$\begin{aligned} \mathbf{e}^{(1)} &= (-\sin\varphi, \cos\varphi)\,;\; e_z^{(1)} = 0; \\ \mathbf{e}^{(2)} &= -\cos\theta\,(\cos\varphi, \sin\varphi)\,;\; e_z^{(2)} = \sin\theta, \end{aligned} \tag{63}$$

describe polarizations that are conveniently chosen similarly to a recent calculation of electromagnetic emission due to thermal fluctuations [41]. The vectors $\mathbf{e}^{(1)}$ and $\mathbf{e}^{(2)}$ represent the "in plane" and the "out of plane" polarizations, respectively. The electron and the hole wave functions are $\frac{1}{\sqrt{2LW}}e^{i\mathbf{p}\cdot\mathbf{r}}u_{\mathbf{p}}\psi_n(z)$ and $\frac{1}{\sqrt{2LW}}e^{i\mathbf{p}'\cdot\mathbf{r}}v_{\mathbf{p}'}\psi_n(z')$, with $\mathbf{p}+\frac{e}{\hbar c}\mathbf{A}_{ext}=p(\cos\phi,\sin\phi)$. $\psi_n(z)$ are wave functions of the confinement. The interaction with a photon at time t happens when the momentum is minimally shifted due to the DC field. The Golden rule photon emission rate (for an "initial" electron with momentum $\mathbf{p}$ and a "final" hole $\mathbf{p}'$ and a photon of polarization λ and momentum $(\mathbf{k}, k_z)$, Figure 4) is

$$W_{nn'}^{(\lambda)}(\mathbf{p},\mathbf{p}',\mathbf{k},k_z,t) = \frac{2\pi}{\hbar}\left|F_{nn'}^{(\lambda)}\right|^2 N_{\mathbf{p}}(t)\, N_{-\mathbf{p}'}(t)\,\delta\left(\hbar v_g\left(p+p'\right) - \hbar\omega\right). \tag{64}$$

Here $N_{\mathbf{p}}(t)$ is the density of electrons in a certain momentum range produced by the electric field $\mathbf{E}$, and $N_{-\mathbf{p}'}(t)$ the density of holes (equal to that of the electrons at the opposite momentum due to particle-hole symmetry). The two photon processes are suppressed by the factor of $\alpha_{QED} = \frac{e^2}{c\hbar} \approx \frac{1}{137}$ compared to the one photon due to an additional vertex, while the phase space of two diagrams is of the same order.

VI.2. Detailed characterization of the emitted radiation

Using the Landau–Zener creation rate expression, Eq. (54), the transition amplitude is given by

$$F_{nn'}^{(\lambda)} = i\frac{E_0}{\omega}\frac{ev_g}{2L^2}e^{i(v_g(p+p')-\omega)}\mathcal{F}_{\mathbf{p},\mathbf{p}'}^{(\lambda)}\int dz\; e^{ik_z z}\psi_n^*(z)\,\psi_{n'}(z)\,\delta\left(\mathbf{p}+\mathbf{p}'-\mathbf{k}\right), \tag{65}$$

where matrix elements $\mathcal{F}_{\mathbf{p},\mathbf{p}'}^{(\lambda)} \equiv v^{\dagger}(-\mathbf{p}')\,\sigma\cdot\mathbf{e}^{(\lambda)}u(\mathbf{p})$ are

$$\begin{aligned} \left|\mathcal{F}_{\mathbf{p},\mathbf{p}'}^{(1)}\right|^2 &= 2\left[1-\cos\left(2\varphi-\phi-\phi'\right)\right]; \\ \left|\mathcal{F}_{\mathbf{p},\mathbf{p}'}^{(2)}\right|^2 &= 2\cos^2\theta\left[1+\cos\left(2\varphi-\phi-\phi'\right)\right]. \end{aligned} \tag{66}$$

For tight confinement to the $z = 0$ plane one should consider only the ground state $n = n' = 0$. Note that the perpendicular component of the wave vector k_z, is "free" from conservation that prohibits the process in fully relativistic QED.

Let us define the spectral emittance per volume of the k-space (and area of the graphene flake) as

$$\mathcal{M}^{(\lambda)}(\mathbf{k}, k_z, t) = \frac{4\hbar\omega}{L^2}\sum_{\mathbf{p},\mathbf{p}'}\frac{d}{dk_z d\mathbf{k}}W^{(\lambda)}(\mathbf{p}, \mathbf{p}', \mathbf{k}, k_z, t) \tag{67}$$
$$= \frac{e^2 v_g^2}{(2\pi)^4}\int d\mathbf{p}\left|\mathcal{F}^{(\lambda)}_{\mathbf{p},\mathbf{p}-\mathbf{k}}\right|^2 N_{\mathbf{p}}N_{\mathbf{p}-\mathbf{k}}\delta\left(v_g\left(p + |\mathbf{k}-\mathbf{p}|\right) - \omega\right),$$

where the integration over $\mathbf{p}'$ was performed using the delta function expressing the conservation of momentum. We first study the frequency dependence of the radiation in the direction perpendicular to the graphene flake, $\mathbf{k} = \mathbf{0}$. Summing over the polarizations $\left(\sum_\lambda \left|\mathcal{F}^{(\lambda)}_{\mathbf{p},\mathbf{p}}\right|^2 = 4\right)$, the spin and valley degeneracy of graphene (4) and integrating over $\mathbf{p}$ one obtains, using $\omega = ck_z$, the spectral emittance

$$\mathcal{M}(\mathbf{k} = \mathbf{0}, \omega, t) = \frac{e^2 v_g^2 t_{nl}^2}{\pi^4}\int_{-t/t_{nl}}^{0} d\overline{p}\,\Theta\left(t_{nl}\omega/2 + \overline{p}\right)\frac{\exp\left[-2\pi\left(t_{nl}^2\omega^2/4 - \overline{p}^2\right)\right]}{\omega\left(t_{nl}^2\omega^2/4 - \overline{p}^2\right)^{1/2}}.$$

The spectral emittance, presented in Fig.6a for various frequencies as function of time, increases linearly for $t \ll \omega^{-1}$, $\mathcal{M}(\mathbf{k} = \mathbf{0}, \omega, t) = \frac{2e^2 t}{\pi^4 t_{nl}^2}e^{-\pi t_{nl}^2\omega^2/2}$, then rises sharply approaching a maximum at $t = \omega t_{nl}^2/2$ and stabilizes at

$$\mathcal{M}(\mathbf{k} = \mathbf{0}, \omega, t \gg t_{nl}) = \frac{e^2}{\pi^3}\omega e^{-\pi t_{nl}^2\omega^2/4} I_0\left(\frac{\pi t_{nl}^2\omega^2}{4}\right), \tag{68}$$

where I_0 is the modified Bessel function. The asymptotic value rises linearly with frequency, $\pi^{-3}\omega$, in the infrared, reaches its maximum at $\omega = t_{nl}^{-1}$ and falls slightly to $\frac{\sqrt{2}e^2}{\pi^4 t_{nl}}$ in the ultraviolet. In Fig.6b the emittance at various ballistic times is given as function of frequency. For each ballistic time the curve has two parts. The first follows the universal dependence given by Eq. (68). Therefore the frequency for observation of the Schwinger effect, not surprisingly, should exceed $\omega_{\min} = \sqrt{eEv_g/\hbar}$, that amounts to $3.6THz$ for $E = 10^4 V/m$. At a higher frequency $\omega_{\max} = 2t/\,t_{nl}^2$ the emittance sharply drops. Therefore the frequency does not exceed $2t_L/t_{nl}^2$.

Next we consider the angular and polarization dependence of the radiated power per unit area defined as the spectral intensity, Eq. (67), integrated over frequencies:

$$\mathcal{L}^{(\lambda)}(\theta, \varphi, t) \equiv \int_0^\infty d\omega\frac{\omega^2}{c^3}\mathcal{M}^{(\lambda)}\left(\mathbf{k}, \frac{\omega}{c}, t\right). \tag{69}$$

Performing integrations and simplifying utilizing the small parameter $v \equiv v_g/c \simeq$

$1/300 \ll 1$, one obtains

$$\mathcal{L}^{(1)}(\varphi,t) = \frac{e^4v^4E^2}{2^{5/2}\pi^4c\hbar^2}\left(\frac{t}{4\pi t_{nl}}\cos^2\varphi+\frac{t^3}{3t_{nl}^3}\sin^2\varphi\right) \tag{70}$$
$$\mathcal{L}^{(2)}(\theta,\varphi,t) = \frac{e^4v^4E^2}{2^{5/2}\pi^4c\hbar^2}\cos^2\theta\left(\frac{t}{4\pi t_{nl}}\sin^2\varphi+\frac{t^3}{3t_{nl}^3}\cos^2\varphi\right).$$

The radiant flux from a flake of a $\mu m \times \mu m$ size is $4.7 \cdot 10^{-21}W$, for $E = 10^4 V/m$ corresponding to the emission rate of just 10 photons per second, yet for the high current samples[22] with $E = 2 \cdot 10^6 V/m$ of the same area one gets a more significant output: the radiant flux is $1.3 \cdot 10^{-17}W$, corresponding to the emission rate of $3 \cdot 10^4$ photons per second.

The two quantities $\mathcal{L}^{(1,2)}$ and their sum are presented for $t = t_{nl}$ in the spherical plots Fig. 8(a)–(c), respectively. The radiated power is maximal in direction perpendicular to the graphene plane. For directions close to the azimuth angles $\varphi = 0°$ and $180°$ (perpendicular to the electric field or current) at small polar angles θ (perpendicular to the graphene plane) the intensities of the two polarizations are of the same order, while for $\theta \sim 90°$ (close to the in – plane direction) the "out of plane" polarization, $\lambda = 2$, dominates. On the other hand, for $0° \ll \varphi \ll 180°$ the picture is opposite. As expected, the unpolarized intensity, Fig. 8c, is less anisotropic; yet the radiation is somewhat depressed in the direction perpendicular to the current and close to the plane.

VI.3. Experimental evidence of the pair creation

In this subsection we use the above "radiation friction" scenario to discuss whether there is a clear and unambiguous signature of the Schwinger's pair creation phase in transport experiments. In a series of remarkable experiments the nonlinear I–V were measured at high electric fields of the order $2 \cdot 10^6 V/m$ at room temperature[22]. The samples were treated in such a way that, despite being non-suspended, the typical charge asymmetry did not appear and the Dirac point was accessed convincingly at zero gate voltage (this demonstrates high quality and is in variance with most samples on substrate). The I–V curves were studied in various high and low mobility samples (up to $\mu = 7000cm^2/(Vs)$) and effects of disorder were partially controlled by irradiating the samples. The samples were $L = 1 - 2\mu m$ long and rather narrow ($W = 0.5\mu m$) and the four-probe technique was applied. Although a nonlinear I–V dependence $I \propto V^\alpha$ with exponent $\alpha = 1.3 - 1.5$ was observed at Dirac point, surprisingly the nonlinearity disappeared in the highest mobility samples. In these experiments the ballistic time $t_{bal} = 2ps$ is much larger than $t_{nl} = 20fs$ at the highest applied voltage of $4V$. Unfortunately the crossover voltages, namely when $t_{nl}(V) = t_{bal}$ or

$$V_{nl} = \frac{\hbar v_g}{eL} = 0.32mV \tag{71}$$

were not probed since at room temperature $k_B T = 25meV$. As argued in the previous subsection the radiation friction causes a second Ohmic regime and the I–V curve in the clean samples is expected to be linear. It is disorder that might have caused the observed nonlinearity in irradiated samples. This requires an additional theoretical study that includes

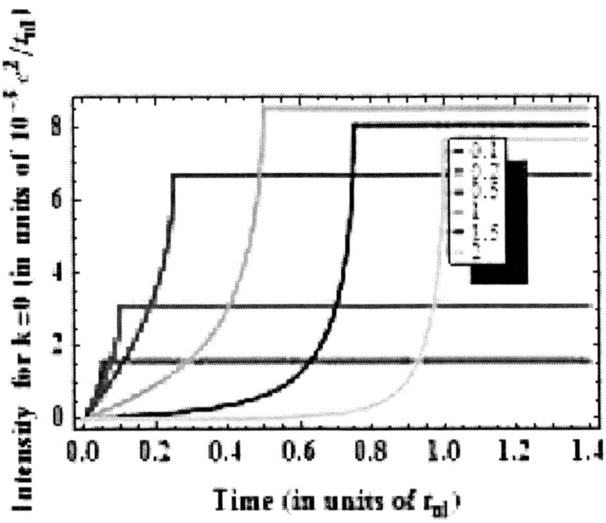

(a)

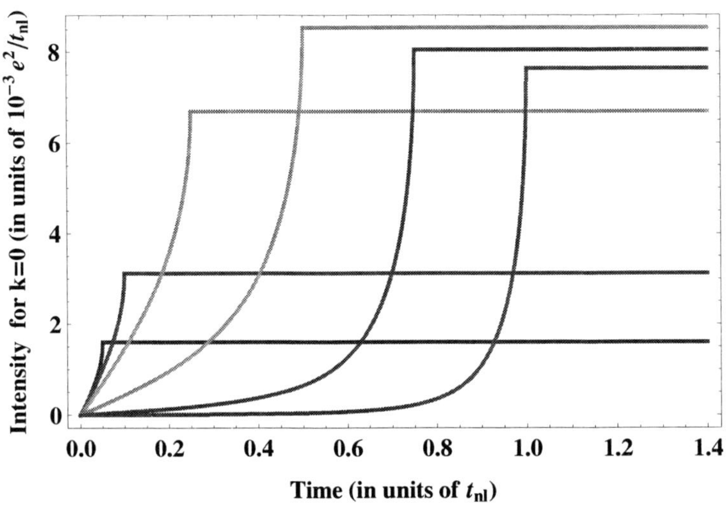

(b)

Figure 7. (a) The spectral emittance in direction perpendicular to the graphene plane, $k = 0$. Polarizations are summed over. The emittance (in units of e/t_{nl}) for various frequencies (in units of t_{nl}^{-1}) as function of ballistic time from $0.1t_{nl}$ to $1.4t_{nl}\}$. (b) The emittance at various times (in units of t_{nl}) as function of frequency.

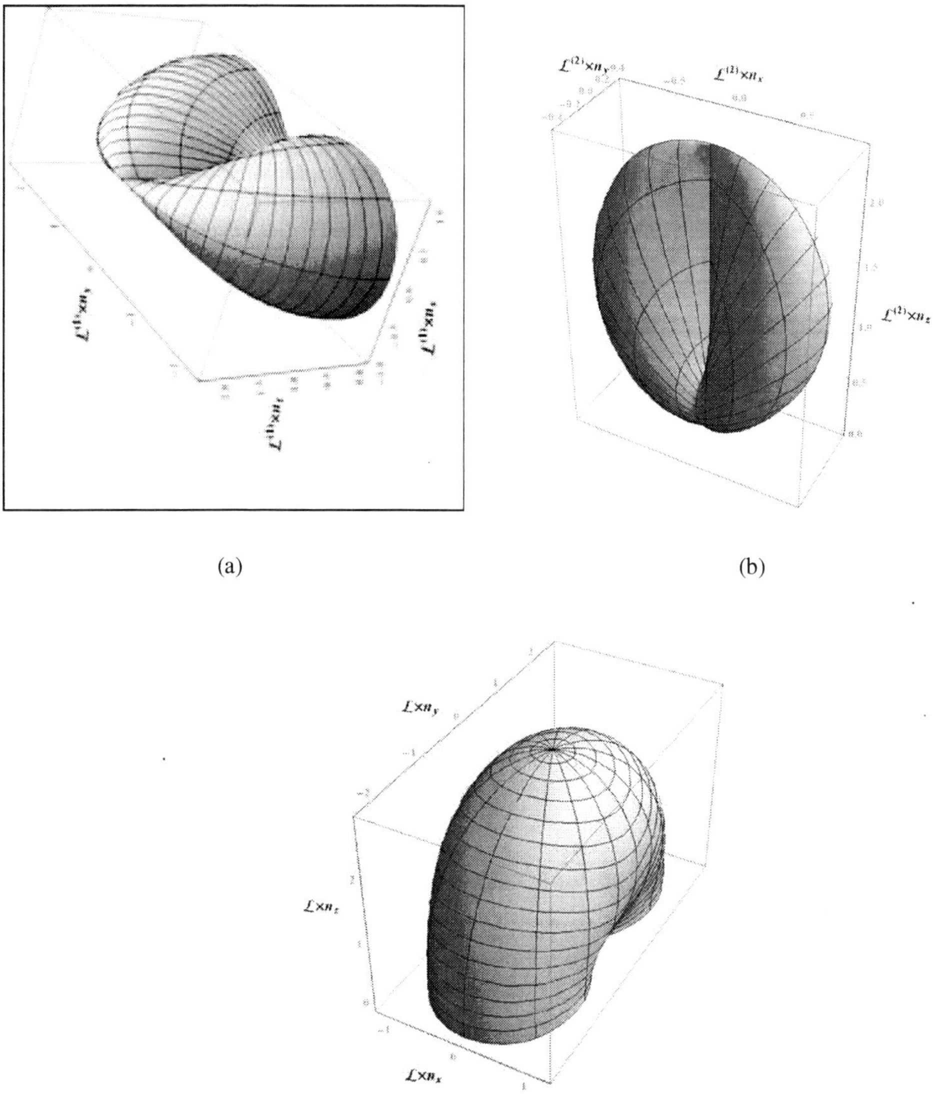

(a) (b)

(c)

Figure 8. (a) The angle dependence of the intensity as function of the photon spherical angles φ, ■ (half of the whole solid angle). Time is fixed at $t = t_{nl}$. The in – plane polarization. (b) The out of plane polarization. (c) Unpolarized light.

the effect of pair recombination. As argued above it becomes as important as the Landau–Zener–Schwinger pair creation process at such currents.

Detectors of light (photon counters) in the microwave-infrared which are sensitive enough have recently been developed [42]. Schwinger's pair creation rate formula can hence be directly tested using novel condensed matter materials endowed with relativistic fermion spectrum.

VII. Discussion, Conclusions and Generalizations

To summarize, the extraordinary ballistic transport in graphene near the Dirac points was reviewed focusing on the dynamics of the electron-hole pair creation by an electric field and their subsequent annihilation. The dynamics is "ultrarelativistic" with the graphene velocity v_g playing the role of the velocity of light. It was assumed that the transport is purely ballistic, thus effects of impurities, interaction with phonons, ripplons, as well as the screened interaction between electrons were neglected. The dynamical approach (which originated in elementary particle physics [30]) was adapted to the tight binding model of graphene and allowed us to calculate the evolution of the current density, energy and number of pairs beyond linear response. Several delicate issues within linear response were clarified, like the correct dc conductivity value and the proper use of the ultrarelativistic (zero mass) Weyl approximation to the tight binding model. The question of proper regularization within the Weyl approximation is remarkably linked to the chiral (parity) symmetry and the anomaly cancellation in the tight binding model. Using a proper regularization, the leading correction to the linear response in both dc and ac fields was calculated. The ac response which is purely pseudo-dissipative (no inductive part) and frequency independent in the linear response shows strong frequency dependence, third harmonic generation and inductive behavior.

New phenomena revealed include the ballistic time t_{nl}, Eq. (3), at which the behavior changes drastically. The linear response breaks down and the current grows linearly with time, Eq. (4). This is explained by Schwinger's creation rate, which in 2+1 dimensions is proportional to $E^{3/2}$. The copious production of electron-hole pairs by a sufficiently intense electric field eventually leads to certain remarkable new effects like recombination into photons and should terminate either when one of the dissipation channels opens up or finish as an electron-hole plasma.

We summarize the behavior of the conductivity and the number of electron-hole pairs in the various ballistic time regimes at finite electric field E in turn.

(i) $t < t_{nl}$. *Pseudo-dissipative behavior.* After a brief transient period (of order of several $t_\gamma = \hbar/\gamma$), the current density of the tight binding model approaches a finite value and stabilizes there. Therefore the minimal dc electric conductivity at zero temperature is $\sigma_2 = \frac{\pi}{2}\frac{e^2}{h}$. In this regime the pair creation rate is small enough, so that pairs can be considered independent. At this time the density of pairs is $\frac{eE}{a}$. It becomes comparable to the density of the electron-hole plasma, $n_{eh} = 10^{11}cm^{-2}$, only for $E = 10^5 V/m$. Let us assume that this does not happen and proceed to longer ballistic times.

(ii) $t_{nl} < t < t_{eh}, t_B$. *Schwinger's pair creation.* The pair creation is more intensive and the pair density asymptotically follows an analog of the Schwinger's formula for massless fermions in 2+1 dimensions, Eq. (15). The current above t_{nl} increases linearly

with time. In this regime most of the electrons are oriented along the field direction. Some electron-hole pairs annihilate with rate given in Eqs. (70). The annihilation rate however is not large enough to affect significantly the density of pair for achievable fields. Due to intensive pair creation the pairs accumulate and eventually can reach density of order n_{eh} at which the electron-hole plasma is formed. If not, Bloch oscillations may ensue.

(iii) $t > t_{eh}, t_B$. *Formation of the electron-hole plasma or Bloch oscillations.* One can only speculate what kind of dissipation process truncates the process of intensive pair creation and stabilizes the electron-hole plasma created by the electric field. Of course, the standard candidates are collisions with impurities, phonons, ripplons and the electron-electron interactions. We point out that the system is "open" and one should consider the "radiation friction" scenario: pairs annihilate emitting photons which take energy out of the graphene sheet. If the system is still ballistic at yet longer times, Bloch oscillations set in with a period of $t_B = \frac{\hbar}{eaE} = t_\gamma/\mathcal{E}$. This time scale is $2 \cdot 10^3 t_\gamma$ for relatively fields $E = 10^7 V/m$, see Fig. 2 in ref. [19]. Therefore Bloch oscillations are difficult to observe (see however a recent proposal [43]).

Now we discuss several generalizations of the present method and connection to other approaches. We start with the application to finite samples.

As mentioned above an independent approach giving the value σ_1 originates in mesoscopic physics. Leads are described by a large potential barrier U between them and the sample. Within the Landauer approach [27] one counts the "evanescent" modes in a ribbon of finite width W and length L leading to a result (for the armchair boundary conditions)

$$\sigma = 4\frac{e^2}{h}\frac{L}{W}\sum_{n=0}\cosh^{-2}\left(\pi n W/L\right) \underset{W/L\to\infty}{\longrightarrow} \sigma_1. \tag{72}$$

The dynamical approach recovers this values and it turns out that there are two contributions to conductance in the presence of a contact potential. One is the ultrarelativistic or the bulk contribution due to electron-hole pairs. At times smaller than both $t_L = L/v_g$ and $t_U = \hbar/U$ it dominates. However at larger times it diminishes. The second contribution is purely one particle, thus "nonrelativistic", approaches that of Eq. (72) at times larger than t_L, t_U.

The dynamic approach was generalized to bilayer graphene [44] in which similar questions exist for a long time. The correct dc conductivity for the N layered graphene is equal to the dynamical one $\sigma = N\sigma_2$, consistent with the vanishing frequency limit of the ac conductivity. The creation of the electron-hole plasma is even more likely in these systems. The calculated is trivially extended to any system with a Dirac point-like spectrum as double layer graphene and the recently synthesized family of materials called "topological insulators" [26] in which surface excitations are similar to those in graphene with the notable exception of chirality. Schwinger's mechanism is also expected in these materials since the mechanism does not involve chirality (left and right movers contribute equally to the emission rate of graphene). These materials have an advantage of not being strictly two dimensional, although ballistic times might be shorter at present. Very recently the nonlinear effects in magnetic field (including the effect of splitting due to spin very studied [45].

In this chapter only ballistic transport was considered. In principle, disorder and electron-electron interactions could be incorporated within the dynamical approach in the way the Boltzmann equation approach [46] can be extended beyond the linear response. In fact, using a phenomenological methodology, disorder has recently been incorporated for the ac field in ref. [35]. Similarly Coulomb interactions and the pair annihilations into photons, phonons etc. can be taken into account. Generally though, when the system has a large number of electron-hole pairs, the screening by the neutral plasma is more effective and the influence of impurities and interactions diminishes. Understanding these effects is crucial for investigating the (nonlinear) plasma waves and their damping [47].

Finally let us remark on the application of the dynamical approach to the calculation of the response to very short strong field pulses like the femtosecond (and an order of magnitude longer) laser pulses. A possibility of measuring the response to such fields was advocated by Rusin and Zawadzki [48]. For this purpose the dynamical linear response formulas, as presented in the Section III, can be directly applied for arbitrary time dependence of the pulse, while the nonlinear calculation of Section V describes the step-like pulse of finite duration only. Other shapes of the pulse can be calculated by breaking the pulse into several segments of different constant field.

Acknowledgments

We are indebted to D. Nguyen, B. Dora, J. Santoro, E. Andrei, M. Deshmukh, A. F. Morpurgo, Ziegler, V. Singh, E. Farber and W. B. Jian for valuable discussions. Work of B.R. and H.K. was supported by NSC of R.O.C. Grants No. 98-2112-M-009-014-MY3 and 98-2112-M-003-002-MY3, respectively, and MOE ATU program. H.K. acknowledges the hospitality of AUC during the visit. M.L. acknowledges the hospitality and support of the NCTS.

References

[1] S.V. Morozov et al, *Phys. Rev. Lett.* **100**, 016602 (2008).

[2] X. Du, I. Skachko, A. Barker and E. Y. Andrei, *Nature Nanotechnology* **3**, 491 (2008).

[3] K. I. Bolotin, K. J. Sikes, J. Hone, H. L. Stormer and P. Kim, *Phys. Rev. Lett.* **101**, 096802 (2008).

[4] A. H. Castro Neto, F. Guinea, N. M. R. Peres, K. S. Novoselov and A. K. Geim, *Rev. Mod. Phys.* **81**, 109 (2009); V. P. Gusynin, S. G. Sharapov and J. P. Carbotte, *Int. J. Mod. Phys.* B **21**, 4611 (2007); N.M.R. Peres, *J. Phys.: Condens. Matter* **21** 323201 (2009); *Rev. Mod. Phys.* **82**, 2673 (2010); E.R. Mucciolo and C. H. Lewenkopf, *J. Phys.: Condens. Matter* **22,** 273201 (2009).

[5] E.G. Mishchenko, *Phys. Rev. Lett.* **98**, 216801 (2007), *EPL* **83,** 17005 (2008), V. N. Kotov, B. Uchoa and A.H. Castro Neto, *Phys Rev B* **78**, 035119 (2008).

[6] V. P. Gusynin and S. G. Sharapov, *Phys. Rev. B* **73**, 245411 (2006).

[7] A.W.W. Ludwig et al, *Phys. Rev. B* **50**, 7526, (1994); K. Ziegler, *Phys. Rev. Lett.* **97**, 266802 (2006); S. Ryu, C. Mudry, A. Furusaki and A. W. W. Ludwig, *Phys. Rev. B* **75**, 205344 (2007); J. Cserti, *Phys. Rev. B* **75**, 033405 (2007).

[8] K. Ziegler, *Phys. Rev. B* **75**, 233407 (2007).

[9] C. G. Beneventano, P. Giacconi, E.M. Santangelo and R. Soldati, *J. Phys. A* **42**, 275401 (2009).

[10] K.S. Novoselov et al, *Nature* **438**, 197 (2005); Y. Zhang et al, *Nature* **438**, 201 (2005).

[11] E.A. Andrei, Talk at *Workshop on Nonequilibrium Phenomana*, Kanpur, January (2010).

[12] T. Ando, Y. Cheng and H. Suzuura, *J. Phys. Soc. Jap.* **71**, 1318 (2002); N.M.R. Peres et al, *Phys. Rev. B* **73**, 125411 (2006).

[13] E. Fradkin, *Phys. Rev. B* **33**, 3263 (1986).

[14] L.A. Falkovsky and A.A. Varlamov, *Eur. Phys. J. B* **56**, 281 (2007).

[15] R.R. Nair et al, *Science* **320**, 1308 (2008).

[16] L. Fritz, J. Schmalian, M. Müller, and S. Sachdev, *Phys. Rev. B* **78**, 085416 (2008).

[17] M. Lewkowicz and B. Rosenstein, *Phys. Rev. Lett.* **102**, 106802 (2009).

[18] H.C. Kao, M. Lewkowicz and B. Rosenstein, *Phys. Rev. B* **81**, (2010).

[19] B. Rosenstein et al, *Phys. Rev. B* **81**, 041416(R) (2010).

[20] B. Dora, R. Moessner, *Phys. Rev. B* **81**, 165431 (2010).

[21] V. Singh and M.M. Deshmukh, *Phys. Rev. B* **80**, 081404(R) (2009).

[22] A. Barreiro, M. Lazzeri, J. Moser, F. Mauri, and A. Bachtold, *Phys. Rev. Lett.* **103**, 076601 (2009); N. Vandecasteele, A. Barreiro, M. Lazzeri, A. Bachtold, and F. Mauri, *Phys. Rev. B* **82**, 045416 (2010).

[23] A. Casher, H. Neuberger and S. Nussinov, *Phys. Rev. D* **20**, 179 (1979); D. Allor, T. D. Cohen and D.A. McGady, *Phys. Rev. D* **78**, 096009, (2008).

[24] S.P. Gavrilov and D.M. Gitman, *Phys. Rev. D* **53**, 7162 (1996), S. P. Kim and D. N. Page, *Phys. Rev. D* **65**, 105002 (2002); T. D. Cohen and D.A. McGady, *Phys. Rev. D* **78**, 036008, (2008).

[25] F. Pellegrini, S.Montangero, G. E. Santoro, and R. Fazio, *Phys. Rev. B* **77**, 140404(R) (2008); D. Patan, L, Amico, A. Silva, R. Fazio, and G. E. Santoro, *Phys. Rev. B* **80**, 024302 (2009).

[26] B. A. Bernevig, T. L. Hughes and S-C. Zhang, *Science* **314,** 1757 (2006); X-L. Qi and S-C. Zhang, *Phys. Today* **63,** 33 (2010); C. Brune et al, *Nature Phys.* **6**, 448 (2010); L. B. Zhang et al, *N. J. Phys.* **12** 083058, (2010).

[27] J. Tworzydlo et al, *Phys. Rev. Lett.* **96**, 246802 (2006); E.B. Sonin, *Phys. Rev. B* **77**, 233408 (2008); *Phys. Rev. B* **79**, 195438 (2009); A. Rycerz, P. Recher and M. Wimmer, *Phys. Rev. B* **80**, 125417 (2009).

[28] M.I. Katsenelson, *Eur. Phys. J. B* **51**, 157 (2006).

[29] M. Lewkowicz, B. Rosenstein, and D. Nghiem, arXiv:1107.3039v1 [cond-mat.mes-hall]; E. Perfetto, G. Stefanucci, and M. Cini, *Phys. Rev. B* **82**, 035446 (2010).

[30] E.S. Fradkin, D.M. Gitman and S.M. Shvartsman, *Quantum Electrodynamics with Unstable Vacuum*, Springer-Verlag, Berlin (1991).

[31] L. Karsten and J. Smit, *Nucl. Phys. B* **183**, 103 (1981); J. Smit, *Introduction to Quantum Fields on a Lattice*, Cambridge Univesity Press, New York (2002).

[32] C. Itzykson and J.B. Zuber, *Quantum Field Theory*, Dover Publications, New York (2006).

[33] H.B. Nielsen and M. Ninomiya, *Nucl. Phys. B* **185**, 20 (1981); *B* **193**, 173 (1981).

[34] G.W. Semenoff, *Phys. Rev. Lett.* **53**, 2449 (1984).

[35] E.G. Mishchenko, *Phys. Rev. Lett.* **103**, 246802 (2009).

[36] T.D. Cohen and D.A. McGady, *Phys. Rev. D* **78**, 036008, (2008).

[37] J. Schwinger, *Phys. Rev.* **82**, 664 (1951); J. Dunne and T. Hall, *Phys. Rev. D* **58**, 105022 (1998); S.P. Kim, H.K. Lee and Y. Yoon, *Phys. Rev. D* **78**, 105013 (2008).

[38] S. Sachdev, *Quantum Phase Transitions*, Cambridge University Press (2000).

[39] O.G. Balev, F.T. Vasko and V. Ryzhii, *Phys. Rev. B* **79**, 165432 (2009).

[40] A.I. Ahiezer and V.B. Beresteckiy, "Quantum Electrodynamics", Nauka, Moscow (1981).

[41] M. Mecklenburg, J. Woo and B. C. Regan, *Phys. Rev. B* **81**, 245401 (2010).

[42] B. E. Kardynał et al., *Appl. Phys. Lett.* **90**, 181114 (2007); H. Shibata et al, *Appl. Phys. Lett.* **97**, 212504 (2010); T. Ueda and S. Komiyama, *Sensors* **10**. 8411 (2010); T. Yamashita et al, *Appl. Phys. Express* **3**, 102502 (2010).

[43] D. Dragoman and M. Dragoman, *Appl. Phys. Lett.* **93**, 103105 (2008).

[44] Y.-X. Wang, L.-P. Shi, S.-J. Xiong, *EPL* **87**, 57002 (2009).

[45] B. Dora, R. Moessner, *Phys. Rev. B* **83**, 073403 (2011).

[46] M. Auslender and M.I. Katsnelson, *Phys. Rev. B* **76**, 235425 (2007); L. Fritz et al, *Phys. Rev. B* **78**, 085416 (2008).

[47] A. Konar, T. Fang and D. Jena, *Phys. Rev. B* **82**, 115452, (2010).

[48] T.M. Rusin and W. Zawadzki, *Phys. Rev. B* **78**, 125419 (2008); *B* **80**, 045416 (2009).

In: Graphene: Properties, Synthesis and Applications
Editor: Zhiping Xu
ISBN: 978-1-61470-949-7

Chapter 3

PHYSICAL MECHANICS OF BILAYER GRAPHENE SYSTEM

Yufeng Guo*[*] *and Wanlin Guo

Institute of Nano Science, Nanjing University of Aeronautics and Astronautics, Nanjing, P. R. China

Abstract

Tuning properties of graphene system is a key issue for its actual application in nanoelectronic devices. Our first-principles calculations reveal surprisingly high sensitivity of the field-induced energy gap of bilayer graphene to changes in its interlayer spacing. Small adjustments in the interlayer spacing near its equilibrium value produce large modulations in the gap over a wide range of field strength. For epitaxial bilayer graphene grown on SiC substrate, the first C layer or the buffer layer is electronically active to the second C layers in the presence of electric and mechanical tuning. The second graphene on C-terminated SiC exhibits an *n*-to-*p*-type transition under electric field, while the Fermi level of the second graphene on Si-terminated SiC is significantly shifted by compressing interlayer distance with respect to its buffer layer. These results provide insights for fundamental understanding and innovative device applications of bilayer graphene materials.

1. TUNING FIELD-INDUCED ENERGY GAP OF BILAYER GRAPHENE VIA INTERLAYER SPACING

1.1. Introduction

Bilayer graphene, which is composed of two stacked graphene single planes, has recently attracted considerable attention. Previous studies have revealed many novel physical

[*] E-mail address: yfguo@nuaa.edu.cn

properties for bilayer graphene, such as the anomalous quantum Hall effect, peculiar band structure and ballistic electron transport behaviors [1-5]. These unconventional properties of bilayer graphene have potential technological importance in semiconductor and nanoelectronics applications. Different from single-layer graphene, one of the characteristic features of the bilayer graphene is its interlayer van der Waals bonding and unusual space contained inside [6]. The interlayer coupling indeed plays an important role in exceptional electronic properties of the bilayer graphene [5, 7]. However, interlayer structural deformation in bilayer graphene such as the change of the interlayer distance and stacking geometry can significantly affect the interaction between the two graphene layers and the corresponding electronic state [8, 9].

Another important property of bilayer graphene is that applying an external electric field on the two adjacent graphene sheets will open up an electronic gap in the spectrum. The bilayer graphene will behave as a semiconductor with tunable gap in the presence of electric field [10-12]. Recently, an experiment reported that the electronic gap of doped bilayer graphene films epitaxially grown on a silicon carbide substrate can be open and closed, if adjusting the electric field created by changing the carrier concentration in each layer [13]. Actually, chemical doping, substrate and external electric field effects not only influence the interlayer coupling and electronic properties of bilayer graphene but also may simultaneously lead to an adjustment in the interlayer distance due to the change of interlayer interaction [7, 13-16]. Modifying the electronic structure and conductance of bilayer graphene is a key issue for its potential applications. Therefore, an understanding of the coupled effects of interlayer deformation and electric field activities is necessary for controlling the peculiar properties of bilayer graphene.

Here we have investigated the coupled effects of electric field and interlayer space on the electronic structure and electron state of the bilayer graphene by *ab initio* density functional theory (DFT). It is found that both the external electric field and interlayer spacing can modulate the band structure and interlayer electron state of bilayer graphene. Under given electric field, the electronic gap and the interlayer coupling decrease with increasing interlayer distance. Change in interlayer distance can also modify the saturated energy gap of the bilayer graphene in the presence of electric field.

1.2. Results and Discussion

The bilayer graphene studied in this work is modeled by two 2-atom layers with Bernal (AB) stacking in a periodic supercell, and a vacuum region larger than 1.6 nm in the z direction is chosen to avoid any self-interaction of the slabs. The electronic properties are investigated using the DFT computational VASP code with the ultrasoft pseudopotential and local density approximation (LDA) for the exchange-correlation potential [17-19]. An energy cutoff of 400 eV and a sufficient number of special k points sampled on a $30\times30\times1$ Monkhorst-Pack mesh [20] are used to ensure the convergence less than 1meV/atom in our calculations. We have optimized the structure of each graphene layer before the electronic study. The external electric field E_{ext} is modeled by adding a sawtooth-like potential along the z direction, perpendicular to the graphene planes [21]. The numerical DFT technique in previous studies has been proved to produce correct equilibrium interlayer distance and to be accurate enough to describe the band structure of few-layer graphene with different interlayer distance, in spite of not explicitly accounting for the long-range van de Waals correlation [5, 7, 12, 15, 16]. In our DFT calculations, the equilibrium interlayer distance d of the bilayer graphene is 0.334 nm, which is consistent with other experimental and theoretical results [22, 23]. The obtained band structure of the d = 0.334 nm bilayer graphene also coincides with previous theoretical results [5, 7, 12]. Moreover, we find that the energy-minimized or equilibrium interlayer distance of the bilayer graphene in the presence of electric fields remains to be 0.334 nm.

Figure 1.1 shows the band structures of the valence band π and conduction band π^* near the K point of the bilayer graphene with different interlayer distance. In the absence of the electric field, the energy gap between the highest occupied valence band (HOVB) and the lowest unoccupied conduction band (LUCB) remains approximately zero with the change of interlayer distance, so that the bilayer graphene always behaves as a zero-gap semimetal. However, the split of the two additional lower energy bands with the π and π^* bands is obviously weakened with increasing interlayer distance, as shown in Figure 1.1(a), which indicates the decrease in interlayer coupling. At d = 0.5 nm, the band structure of the bilayer graphene is similar to that of single-layer graphene. When an external electric field of 2 V/nm is applied, each valley of the HOVB and LUCB is split and an energy gap is open up in the spectrum, as shown in Figure 1.1(b), and the bilayer graphene transforms into a semiconductor. Higher electric field of 4 V/nm improves not only the split of the two lower energy bands with the π and π^* bands but also the valley split of the HOVB and LUCB, as shown in Figure 1.1(c).

According to the band structures of the bilayer graphene, Figure 1.2(a) shows the variations of energy gap between the HOVB and the LUCB with the strength of the electric filed at different interlayer distance. The energy gap does not increase indefinitely but saturates when the electric field is high enough. The variation of the electric field induced energy gap and saturated gap of 0.27 eV for the d = 0.334 nm bilayer graphene are consistent with previous theoretical DFT study [12]. The saturated energy gaps for the d = 0.3 and 0.38 nm bilayer graphenes are 0.44 eV and 0.13 eV, respectively. It can be seen from this figure that both the saturated energy gap and corresponding critical electric field decrease with the interlayer distance increases.

To investigate the effects of varying interlayer distance, the variations of the energy gap with the interlayer distance d under different electric fields are plotted in Figure 1.2(b). The change of interlayer distance slightly influences the energy gap in the absence of electric field. Under an electric field of 0.5 V/nm, the energy gap decreases with increasing interlayer distance after d is larger than the equilibrated distance 0.334 nm. When the electric fields increase to 2 and 4 V/nm, the energy gaps become more sensitive to the change in interlayer distance and decrease sharply with increasing interlayer distance. At d = 0.5 nm, the energy gaps under 0.5, 2 and 4 V/nm decrease approximately to 0.01 eV, as shown in Figure 1.2(b) and Figure 1.1 (b) (c). Such small energy gap suggests that the bilayer graphene is gradually reversed from a semiconductor to a semimetal. The electronic structure of the bilayer graphene in the presence of a certain electric field then can be altered by adjusting the interlayer spacing.

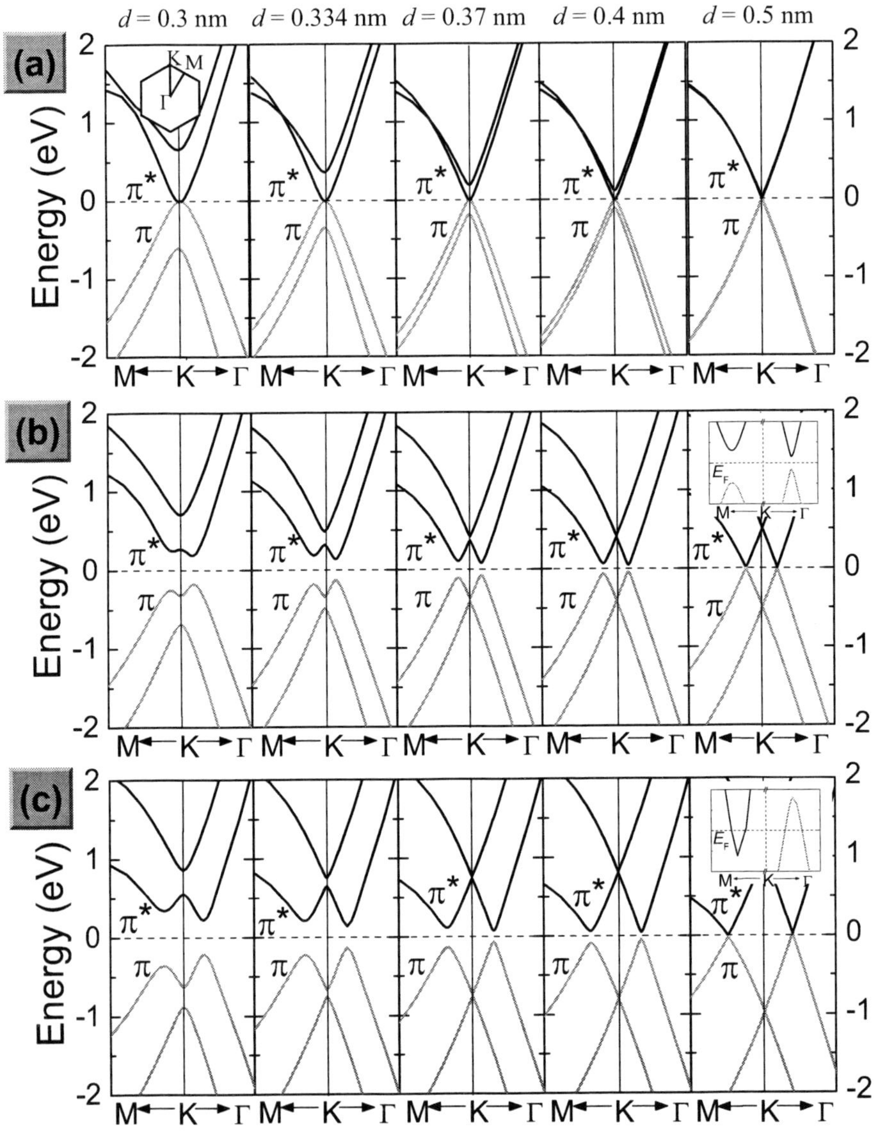

Figure 1.1. The valence bands π and conduction bands π^* near the K point of the bilayer graphene with interlayer distance d = 0.3, 0.334, 0.37, 0.4 and 0.5 nm in the presence of electric fields of (a) 0 V/nm, (b) 2 V/nm and (c) 4 V/nm. The Fermi energy is set to zero.

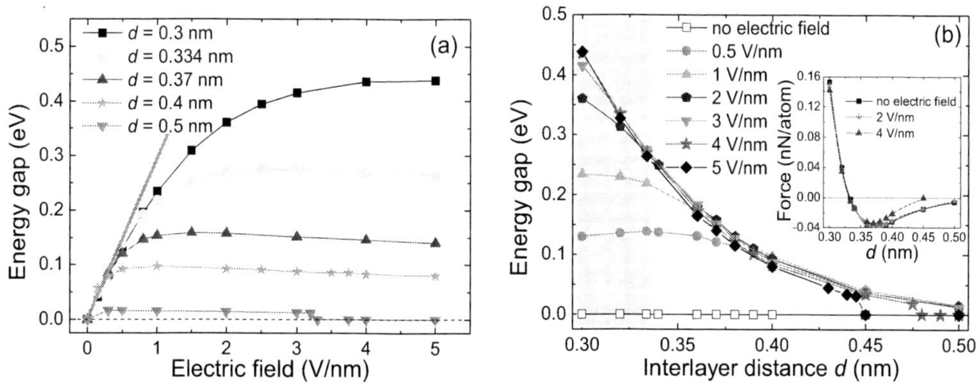

Figure 1.2. (a) Variations of energy gap between the highest occupied valence band (HOVB) and the lowest unoccupied conduction band (LUCB) with the strength of the electric filed at different interlayer distance *d*. (b) Variations of the energy gap between the HOVB and the LUCB with interlayer distance *d* under different electric fields. The inset shows the variations of interlayer force with interlayer distance under different electric fields.

The weak interlayer interaction of the bilayer graphene depends on its interlayer distance. The influence of the interlayer deformation is important to the electronic properties of the bilayer graphene. In Figure 1.3, the difference between the total charge density $\rho_{bilayer}$ of the bilayer graphene and the charge density ρ_s of two single graphene layers in the absence of electric field, $\Delta\rho = \rho_{bilayer} - 2\rho_s$, is shown. Here $\Delta\rho^-$ represents the area of conducting charge density increasing in the bilayer graphene compared with the free-standing single layers, and it directly relates with the transport properties of the bilayer graphene. In Figure 1.3 (b), $\Delta\rho^+$ represents the area of charge density decreasing. It is clear that the distribution of electron rich region $\Delta\rho^-$ is transferred into the space between the two graphene layers with increasing interlayer distance. From $\Delta\rho^+$ area shown in Figure 1.3(b), it is also seen that the electrons are transferred from the carbon atom sites to the interior space. Therefore, the special interlayer electron state can be considered as a main contribution for the novel electronic properties of bilayer graphene. Nevertheless increasing the interlayer distance reduces the interlayer coupling and leads to a nearly free electron (NFE) states [6] in the interior space of bilayer graphene.

To understand the effects of electric field on the bilayer graphene electronic structure, Figure 1.4 shows the difference between the total charge density $\rho_{bilayer}^{ext}$ of the bilayer graphene in the presence of electric field and that $\rho_{bilayer}$ of the bilayer graphene without electric field, $\Delta\rho_{bilayer} = \rho_{bilayer}^{ext} - \rho_{bilayer}$. Here $\Delta\rho_{bilayer}^-$ is the more negatively charged area, which denotes the area of conducting electrons increasing in contrast to the bilayer graphene in the absence of electric field. Under the electric field of 0.5 V/nm, the charge distribution in the bilayer graphene is slightly influenced by the electric field as shown by the small values of maximum $\Delta\rho^{-}_{bilayer}$ in Figure 1.4(a). The increased conducting electrons mainly distribute at the interior space and the carbon atom sites for the d = 0.3 nm bilayer graphene. At d = 0.334, 0.38 and 0.5 nm, the increased conducting electrons concentrate at the carbon atom sites. When the electric field increases to 2 V/nm, more conducting electrons are transferred from the carbon atom sites into the interior space of the bilayer graphene, as shown in Figure 1.4(b). However, the maximum $\Delta\rho^{-}_{bilayer}$ in the interior space decreases with increasing interlayer distance. From d = 0.38 nm to 0.5 nm, the distribution of conducting electrons in the carbon atom sites and exterior space is obviously improved. As the electric field further increases to 4 V/nm, the conducting electrons in the interior space are strengthened, as shown in Figure 1.4(c), but increasing the interlayer distance reduces this distribution as well. The external electric field strengthens the interlayer coupling, but interlayer deformation changes the distribution of conducting or valence electrons. Utilizing those coupled effects, the electronic properties of the bilayer graphene could be tuned more flexibly by mechanical and electric methods.

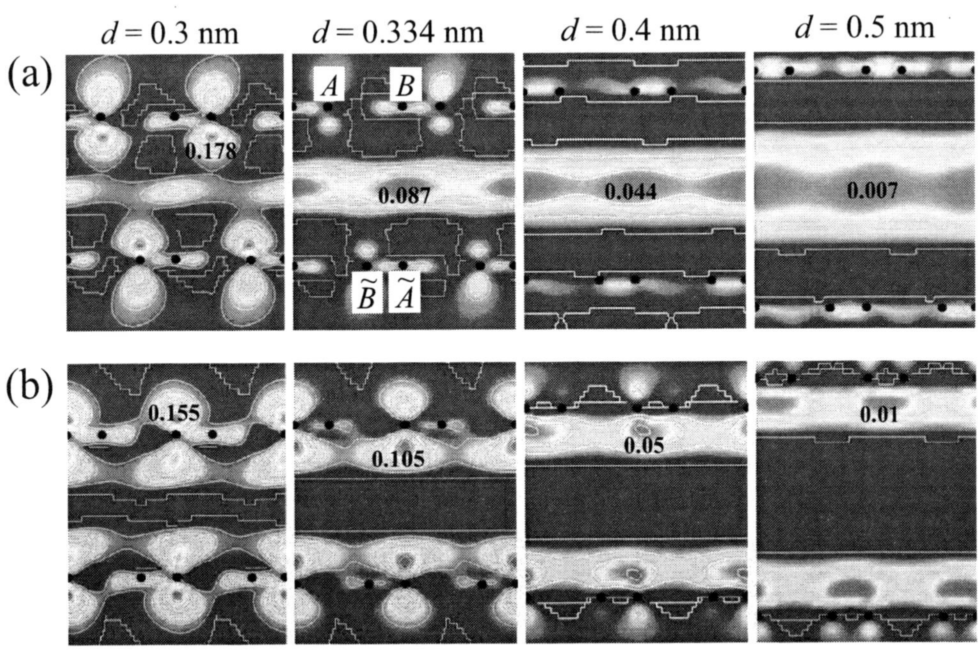

Figure 1.3. The difference between the total charge density of the bilayer graphene with different interlayer distance and that of two single graphene layers in the absence of electric field: (a) contour plot of the charge density of more negatively charged area, $\Delta\rho^{-}$, and (b) that of more positively charged area, $\Delta\rho^{+}$. Red, yellow, green and blue colors represent the magnitude of the charge density from high to low. The black dots are carbon atoms. The values shown in the figures are the largest charge density difference in units of $e/(\text{Å})^3$ and denote the locations.

The interlayer coupling depends on the interlayer spacing. Increasing the interlayer distance, the interlayer interaction is weakened but the binding of the carbon atomic nucleus on the conducting electrons becomes stronger. Although the electric field can improve the interlayer coupling, it is more difficult for the conducting electrons transferring into the interior space of the bilayer graphene, which has larger interlayer distance. Therefore, the lower saturated energy gap for the bilayer graphene of larger interlayer spacing, shown in Figure 1.2(a), can be attributed to the weaker interlayer coupling and less distribution of conducting electrons in the interior space.

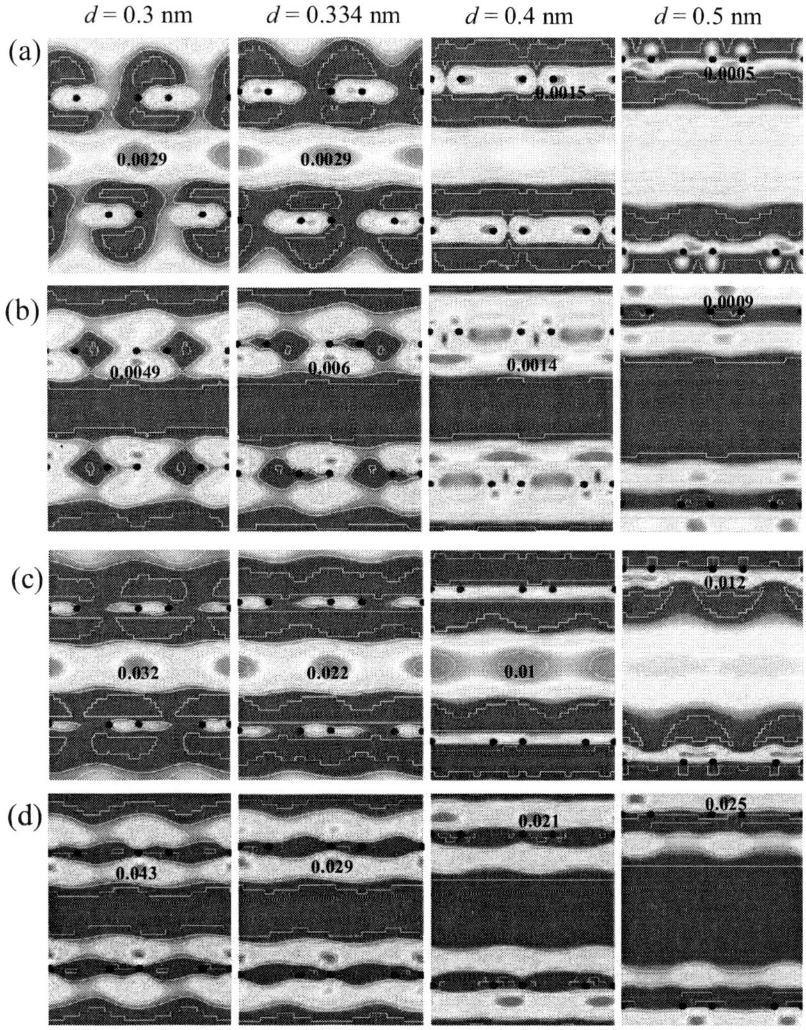

Figure 1.4. Contour plots of the charge density difference $\Delta\rho_{bilayer}^{-}$ (negatively charged area) between the total charge density of the bilayer graphene in the presence of electric fields (a) 0.5 V/nm, (b) 1 V/nm, (c) 2 V/nm, (d) 4V/nm and that of the bilayer graphene without electric field. Red, yellow, green and blue colors represent the magnitude of the charge density from high to low. The black dots are carbon atoms. The values shown in the figures are the largest charge density difference in units of $e/(\text{Å})^3$.

Recent experimental study has reported that graphitic layers attached on a tungsten tip can slide and rotate with other graphite surface in a frictional force microscope system [24]. Consequently the interlayer spacing between the graphitic layers becomes controllable in such device. Now it is not difficult to apply an electric field between a microscope tip and a substrate in experiments. Therefore, adjusting the interlayer distance and applying electric field at the same time are technically feasible for the few-layer graphene. On the other hand, we have also studied those coupled effects in ABA trilayer graphene. The energy band structures of ABA trilayer graphene can be modified by changing the interlayer distance, but the influence of electric field on the electronic properties is much weaker than that of bilayer graphene due to a larger screen effect.

In summary, we have shown that the interlayer deformation significantly affects the band structure and interlayer electron state of the bilayer graphene in the presence of electric field. The electric field induced energy gap as well as the maximum saturated gap decrease with increasing interlayer distance. The bilayer graphene can behave as semiconductor and semimetal at a suitable combination of external electric field and interlayer spacing. This modification of electronic structure in the bilayer graphene is mainly attributed to the change of interlayer coupling and the redistribution of the conducting electrons. The coupled effects of electric field and interlayer deformation on the electronic properties of the bilayer graphene may have potential applications in nanoscale mechanical-electric devices, sensors and functional electronic systems.

2. Electric and Mechanical Tuning on Epitaxial Bilayer Graphene

2.1. Introduction

When epitaxially grown on a substrate, as in most device applications, graphene systems exhibit new behaviors caused by the interaction with the substrate [15, 16, 25-28]. It has been shown that the first C layer of epitaxial bilayer graphene bonds strongly with substrates, and the π orbitals of carbon atoms in the first layer have hybridized with the dangling bonds of substrate surface atoms [15, 16]. Due to absence of π orbitals, the first C layer of epitaxial bilayer graphene is electronically inactive and decouples the second C layer from the substrate. The first C layer is then considered as a buffer layer or dead layer, and the intrinsic properties of graphene are recovered by the second C layer [15, 16, 25-28]. As a result, epitaxial bilayer graphene actually behaves as a single graphene.

How to control and modify graphene physcial properties is of importance in fundamental researches and applications. Applying an external electric field on a free-standing bilayer graphene can open up a gap in its energy spectrum, turning it from a semimetal to semiconductor with a tunable gap [10-12]. Meanwhile the field-induced gap of bilayer graphene can be effectively modified by mechanical method such as adjusting interlayer distance of bialyer graphene [29]. A recent experiment shows that epitaxial bilayer graphene on SiC exhibits *n* and *p* type transition under top-gated bias [30], but the underlying mechanism is not clear. Another experiment reports that the interlayer conductance of epitaxial bilayer graphene on ruthenium can be tuned by compressing interlayer distance between the top graphene and its buffer layer [25]. As electronic inactivity in buffer layer explored by previous studies, the effects of buffer layer to its top graphene layer are seldom investigated and considered for epitaxial bilayer graphene under electric and mechanical tuning.

We report first-principles density function theory (DFT) calculations that the buffer layer of epitaxial bilayer graphene grown on SiC plays an important role in electronic structure modifications of the second C layer under electric and mechanical tuning. The top or second graphene on C-terminated SiC (C-SiC) displays an *n*-to-*p*-type transition in the presence of different electric fields. In contrast, on Si-terminated SiC (Si-SiC) mechanical tuning by reducing interlayer distance between the top graphene and its buffer layer significantly changes the Fermi level of the top graphene. The energy gap of the top graphene both on C-SiC and Si-SiC increases with interlayer distance decreasing. We find that charge transfer and redistribution induced by electric and mechanical tuning occur between the top graphene and its buffer layer, which is the main contribution to electronic structure changes of the top grpahene. At this extent, the buffer layer is electronically active to its top graphene and can not be considered as a dead layer for epitaxial bilayer graphene.

2.2. Results and Discussion

Here we chose an ideal 6-layer $\sqrt{3}\times\sqrt{3}R30^\circ$ 6H-SiC (0001) or $(000\bar{1})$ cell as the substrate, which top is terminated by Si or C atoms and its bottom is terminated by hydrogen atoms. A (2×2) bilayer graphene cell with AB stacking on the top of the substrate is considered in our model. The graphene cell is expanded to fit the lattice constants of the substrate. The computations were performed using the VASP code with the ultrasoft pseudopotential and local density approximation (LDA) for the exchange-correlation potential [17-19]. An energy cutoff of 400 eV and special k points sampled on a 7×7×1 Monkhorst-Pack mesh [20] are used to ensure an energy convergence of less than 1 meV/atom. In the periodic supercell, there is a vacuum region larger than 1.6 nm in the direction perpendicular to the graphene planes to avoid any self-interaction of the slabs. External electric field E_{ext} is modeled by adding a sawtooth-like potential along the direction perpendicular to the graphene planes [21]. For each case, the whole system is relaxed by the conjugate-gradient algorithm until the force on each atom is less than 0.1 eV/nm.

After fully relaxation, we find that the first or bottom layer of the bilayer graphene bonds strongly with SiC substrate and behaves as a buffer layer, and their C-C bonds are expanded as well. This is consistent with previous studies [15, 16]. The equilibrium interlayer distances between the top graphene and the buffer layer on Si- and C-SiC are 0.33 and 0.34 nm, respectively. In our model, the mismatch between bilayer graphene and substrate lattice leads to an 8% stretch of graphene cell on SiC. The in-plane deformation of the second C layer is artificial. To clarify the effects of in-plane tension of the top graphene on its electronic structure, Figure 2.1 (a) plots the energy bands of the corresponding isolated (free-standing) graphene without and with in-plane tensile deformation. Due to the expansion of C-C bonds, the energy band of the deformed graphene is slightly broadened, and the pDOS slightly shifts away from that of the undeformed graphene (Figure 2.1 (b)). However, the electronic structure around the Fermi level of the deformed graphene is still consistent with that of the undeformed graphene. The structural expansion in our model imposes a slight influence on the graphene π bands.

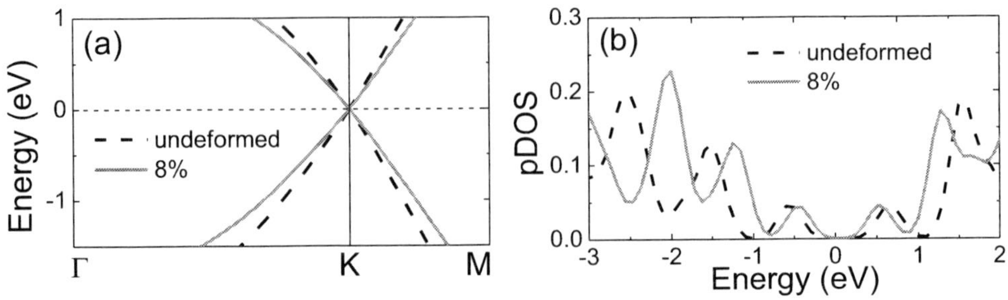

Figure 2.1. (a) The energy bands of the isolated graphene without and with in-plane deformation, and (b) corresponding projected density of states (pDOS) (in unit of states/atom).

Figure 2.2 shows the calculated energy band structures of bilayer graphene and projected density of states (pDOS) of the top graphene layer in the presence of electric fields. On Si-SiC, the Fermi level of the top graphene in the absence of electric field locals at 0.33 eV above the conical point (the Dirac level), resulting in an *n*-type doping in the graphene. The energy band structures shown in Figure 2.2(a) coincide with previous theoretical results of the same graphene cell on SiC [15, 16]. The electric field slightly influences the energy band and pDOS of the top graphene on Si-terminated surface, as shown by Figure 2.2 (a) and (c). In contrast, the Fermi level of the top graphene on C-terminated surface is significantly shifted by external electric field. The top graphene on C-SiC in the absence of an electric field is approximately neutral (Figure 2.2(b)). When an electric field of 2 V/nm is applied, the Fermi level of the top graphene locals at 0.26 eV above the Dirac level, indicating an *n*-type transition. However, at -2 V/nm the Dirac level moves above the Fermi level to 0.14 eV so that the top graphene becomes *p* type. From the pDOS of Figure 2.2(c), it can be also seen the shift of the Fermi level of the top graphene on C-SiC in the presence of different electric field. These results mean that the electron doping in the top graphene on SiC can be effectively tuned by applied electric field, but strongly depends on the type of substrate surface. On the other hand, external electric field just shifts the Fermi level but does not alter the energy band structures of the top graphene, as shown by Figure 2.2 (a)-(c). The in-plane tension will not change the field distribution on the top graphene either. Therefore, the expansion of the top graphene cell considered in our model has no qualitative effects.

The results in Figure 2.2(d) show the systematic trends in the variations of the Dirac level E_D of the top graphene at different electric field. On C-SiC, the change of electric field from -2 V/nm to 2 V/nm causes a remarkable influence on the Dirac level that decreases from 0.14 to -0.26 eV, indicating *p* to *n* type transition. This is in sharp contrast to the situation on Si-SiC where the Dirac level is slightly changed. It is also shown from Figure 2.2(d) that external electric field does not affect the energy gap of the top graphene on Si- and C-SiC. The overall change induced by electric field in the energy gap is less than 0.01 eV.

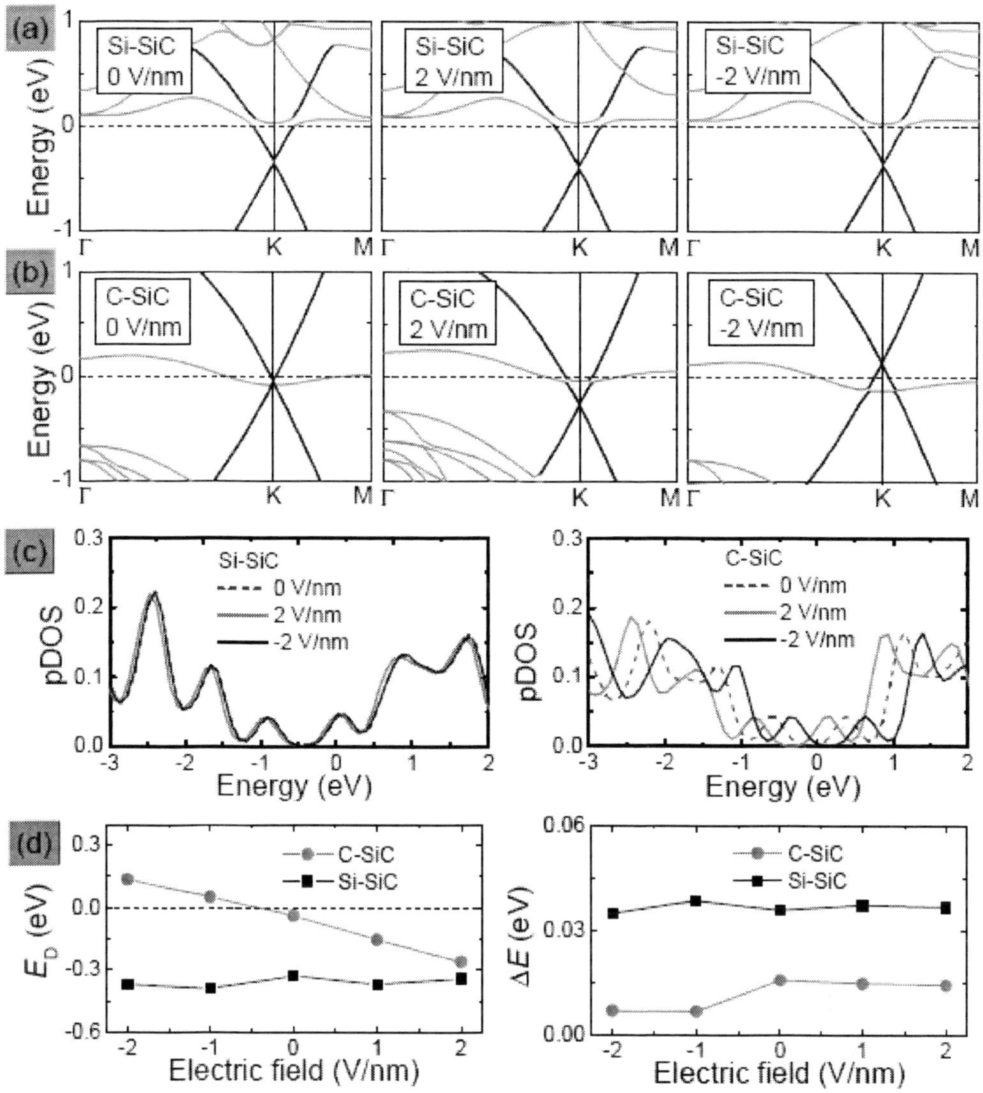

Figure 2.2. Band structures of bilayer graphene on (a) Si- and (b) C-terminated SiC substrates in the presence of electric fields. The top graphene energy bands are highlighted by dark lines. (c) Projected density of states (pDOS) of the top graphene layer on SiC (in unit of states/atom). (d) The variations of the Dirac level E_D and energy gap ΔE of the top graphene on SiC in the presence of electric fields. The Fermi level is set to zero.

There is a buffer layer formed between the substrate surface and the top graphene, thus the substrate is through the buffer layer to influence the top graphene. To understand the role of the buffer layer and the intriguing properties of the top graphene in response to applied electric field, in Figure 2.3 we examine the charge density difference $\Delta\rho^{ext} = \rho_{tot}^{ext} - \rho_{sub+buff}^{ext} - \rho_{grap}^{ext}$, where ρ_{tot}^{ext} is the total charge density in the presence of electric field, $\rho^{ext}_{sub+buff}$ is the charge density of isolated substrate and buffer layer, and ρ^{ext}_{grap} is the charge density of isolated top graphene. In the absence of electric field, there is a lot of accumulated charge localized at the top graphene on Si-SiC, while depleted charge localizes at the surface of buffer layer, as shown in Figure 2.3(a). This proves that the top graphene on Si-SiC is *n* doped and the accumulated charge is obtained from the buffer layer. On C-SiC, the charge depletion and accumulation all localize at the top graphene (Figure 2.3(b)) so that the graphene approximately remains neutral. When an electric field of -2 V/nm is applied, the top graphene on Si-SiC is still *n* doped, but the field changes the distribution of the accumulated charge at the top graphene. This leads to a slight modification on the Dirac level. For C-SiC under -2 V/nm, the charge depletion localizes at the top graphene while the charge accumulation localizes at the buffer layer. The charge transfers from the top graphene to the buffer layer, which causes the *p* type transition. Under a 2 V/nm electric field, the charge accumulation becomes more localized at the top graphene both on Si- and C-SiC, as shown by Figure 2.3. It can be also seen from Figure 2.3 that the field-induced charge exchange, which is vital to the electronic structure modification, mainly distributes between the top graphene and its buffer layer.

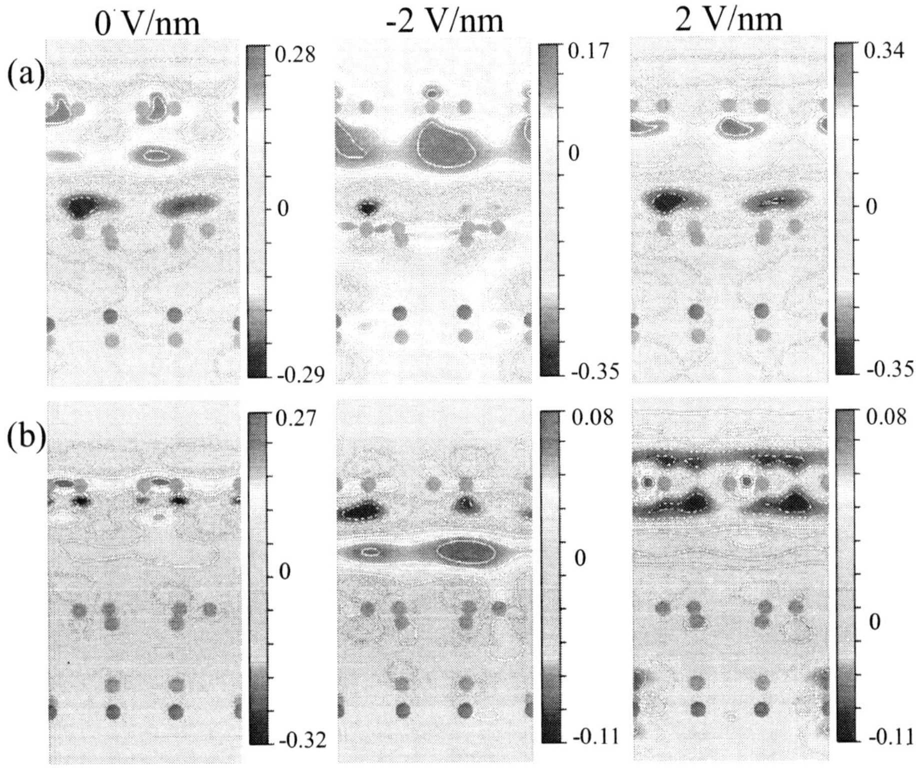

Figure 2.3. Contour plots of 2D projection of charge density change along (100) direction in the presence of electric fields. (a) On the Si-terminated SiC substrate, (b) C-terminated SiC substrate. The purple atoms in (a) and (b) denote the Si atoms, and the gray atoms denote the C atoms. Contour lines are represented in solid lines for positive values and in broken lines for negative ones, and the largest positive and negative values of the contour lines [in units of $e/(\text{Å})^3$] are shown in each panel. The contour line step is set to 1/10 of the range in each case.

For mechanical tuning, Figure 2.4(a) shows the band structures of the bilayer graphene with a compressed interlayer distance of 0.24 nm between the top layer and its buffer layer. Both on Si- and C-SiC, the energy bands of the top graphene are obviously deformed, and an energy gap is opened up because of stronger interlayer interaction. The corresponding pDOS of the top graphene (Figure 2.4(b)) is significantly altered from its initial state, indicating the change of π bands. Figure 2.4(c) plots the variations of the Dirac level E_D of the top graphene with compression of interlayer distance to its buffer layer. The change of interlayer distance slightly influences the Dirac level of the top graphene on C-SiC. For Si-SiC, the Dirac level of the top graphene decreases until the interlayer distance is reduced to 0.28 nm, and then increases with interlayer distance compressing. The energy gaps of the top graphene on SiC increase with interlayer distance reducing, as shown in Figure 2.4(d). Although compressing interlayer distance significantly changes the band structure and energy gap of the top graphene, according to Figure 2.4(a)-(c), the effects of mechanical tuning on the electronic structure of the top graphene still depend on the type of substrate.

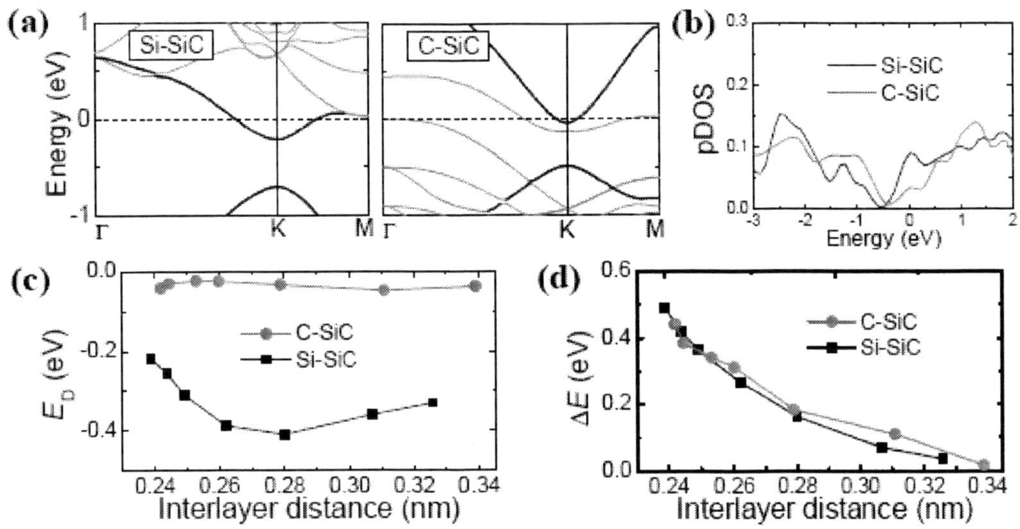

Figure 2.4. (a) The band structures of the bilayer graphene compressed to an interlayer distance of 0.24 nm on SiC, The top graphene energy bands on SiC are highlighted by dark lines. The Fermi level is set to zero. (b) Corresponding projected density of states (pDOS) (in unit of states/atom) of the top graphene. The variations of (c) the Dirac level E_D and (d) energy gap ΔE of the top graphene on SiC with interlayer distance.

In Figure 2.5 we examine the charge density difference $\Delta\rho$ in response to the compression of interlayer distance. Here $\Delta\rho = \rho_{tot} - \rho_{sub+buff} - \rho_{grap}$, ρ_{tot} is the total charge density of the system, $\rho_{sub+buff}$ is the charge density of isolated substrate and buffer layer, and ρ_{grap} is the charge density of isolated top graphene at the corresponding position. When interlayer distance decreases to 0.28 nm (Figure 2.5(a) and (c)), the charge accumulation becomes more localized and intense at the A site of the top graphene on SiC substrate, and no accumulated charge observed at the B site. The symmetry at the A and B sites is then broken by the charge redistribution and this leads to an energy gap opened up. The magnitude of symmetry breaking of the top graphene increases at an interlayer distance of 0.24 nm (Figure 2.5(b) and (d)), so that the energy gap increases with interlayer distance reducing. At the interlayer distance of 0.24 nm, the charge accumulation at the top graphene on Si-SiC is more concentrated, as shown by Figure 2.5(b). On the contrary, the distribution of accumulated charge on C-SiC is similar to that at 0.28 nm (Figure 2.5(d)). This explains the increase in the Dirac level on Si-SiC (Figure 2.4(c)) after the interlayer distance is less than 0.28 nm. As the same as electric field, mechanical tuning induced charge accumulation and depletion mainly occur between the top graphene and its buffer layer. Thus the buffer layers play a key role in tuning electronic properties of epitaxial graphene and actually not behave as a dead layer in the presence of external fields.

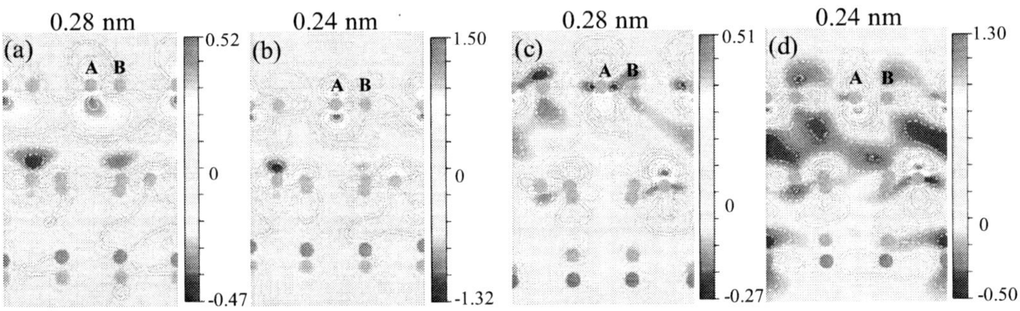

Figure 2.5. Contour plots of 2D projection of charge density difference along (100) direction with compression of interlayer distance. (a) (b) On the Si-SiC, (c) (d) C-SiC. The notations are the same as in Figure 2.3.

Previous studies show that the first C layer epitaxially grown on SiC has structural ripples in the actual geometry [31]. However, theoretical calculations and experiments prove that those C layers do not have the electronic structure of graphene but act as a buffer layer, since the strong bonding region is dominant [27, 28, 31]. Therefore, our small-cell model assuming that the buffer layers completely bond with the substrates will not influence the main results.

In summary, we demonstrate that the buffer layer of epitaxial bilayer graphene grown on SiC substrate is electronically active to it top graphene when it is subjected to electric and mechanical fields. The modifications in electronic structures of epitaxial bilayer graphene strongly depend on its buffer layer and the substrate type. We elucidate the underlying mechanisms by examining the charge redistribution and transfer between the graphene and its buffer layer. These results provide important insights for fundamental properties and application of epitaxial graphene under electric and mechanical tuning.

REFERENCES

K. S. Novoselov, A. K. Geim, S. V. Morozov, D. Jiang, Y. Zhang, S. V. Dubonos, I. V. Grigorieva and A. A. Firsov, *Science* 2004 306, 666.

A. K. Geim and K. S. Novoselov, *Nat. Mat.* 2007 6, 183.

E. McCann and V. I. Fal'ko, *Phys. Rev. Lett.* 2006 96, 086805.

J. Nilsson, A. H. Castro Neto, F. Guinea, and N. M. R. Peres, *Phys. Rev. Lett.* 2006 97, 266801.

S. Latil and L. Henrard, *Phys. Rev. Lett.* 2006 97, 036803.

M. Posternak, A. Baldereschi, A. J. Freeman, E. Wimmer and M. Weinert, *Phys. Rev. Lett.* 1983 50, 761.

T. Ohta, A. Bostwick, J. L. McChesney, T. Seyller, K. Horn and E. Rotenberg, *Phys. Rev. Lett.* 2007 98, 206802.

S. V. Morozov, K. S. Novoselov, M. I. Katsnelson, F. Schedin, L. A. Ponomarenko, D. Jiang and A. K. Geim, *Phys. Rev. Lett.* 2006 97, 016801.

J. C. Meyer, A. K. Geim, M. I. Katsnelson, K. S. Novoselov, T. J. Booth and S. Roth, *Nature* 2007 446, 60.

E. McCann, *Phys. Rev. B* 2006 74, 161403.

F. Guinea, A. H. Castro Neto and N. M. R. Peres, *Phys. Rev. B* 2006 73, 245426.

H. Min, B. Sahu, S. K. Banerjee and A. H. MacDonald, *Phys. Rev. B* 2007 75, 155115.

T. Ohta, A. Bostwick, T. Seyller, K. Horn and E. Rotenberg, *Science* 2006 313, 951.

J. Nilsson and A. H. C. Neto, *Phys. Rev. Lett.* 2007 98, 126801.

A. Mattausch and O. Pankratov, *Phys. Rev. Lett.* 2007 99, 076802.

F. Varchon, R. Feng, J. Hass, X. Li, B. Ngoc Nguyen, C. Naud, P. Mallet, J.Y. Veuillen, C. Berger, E. H. Conrad, and L. Magaud, *Phys. Rev. Lett.* 2007 99, 126805.

G. Kresse and J. Hafner, *Phys. Rev. B* 1993 47, 558.

G. Kresse and J. Hafner, *Phys. Rev. B* 1994 49, 14251.

D. Vanderbilt, *Phys. Rev. B* 1990 41, 7892.

H. J. Monkhorst and J. D. Pack, *Phys. Rev. B* 1976 13, 5188.

J. Neugebauer and M. Scheffler, *Phys. Rev. B* 1992 46, 16067.

Y. Baskin and L. Mayer, *Phys. Rev.* 1955 100, 544.

A. N. Kolmogorov and V. H. Crespi, *Phys. Rev. B* 2005 71, 235415.

M. Dienwiebel, G. S. Verhoeven, N. Pradeep, J. W. M. Frenken, J. A. Heimberg and H. W. Zandbergen, *Phys. Rev. Lett.* 2004 92, 126101.

P. W. Sutter, J. Flege, and E. A. Sutter, *Nat. Mat.* 2008 7, 406.

D. Martoccia *et al.*, *Phys. Rev. Lett.* 2008 101, 126102.

S. Y. Zhou *et al.*, *Nat. Mat.* 2007 6, 770.

S. Kim, J. Ihm, H. J. Choi, and Y. W. Son, *Phys. Rev. Lett.* 2008 100, 176802.

Y. F. Guo, W. L. Guo, and C. F. Chen, *Appl. Phys. Lett.* 2008 92, 243101.

Y. Q. Wu *et al.*, *Appl. Phys. Lett.* 2008 92, 092102.

F. Varchon, P. Mallet, J. Y. Veuillen, and L. Magaud, *Phys. Rev. B* 2008 77, 235412.

In: Graphene: Properties, Synthesis and Applications ISBN: 978-1-61470-949-7
Editor: Zhiping Xu

Chapter 4

STUDY OF ELECTRONIC PROPERTY OF GRAPHENE SHEET AND ITS EDGES USING CTAFM AND DFT

***S. Banerjee*[1]*, *D. Bhattacharya*[2], *M. Sardar*[3], *N. Gayathri*[3], *A. K. Tyagi*[3] *and Baldev Raj*[3]**
[1]Surface Physics Division
Saha Institute of Nuclear Physics
1/AF Bidhannagar, Kolkata, India
[2]Biophysics Division, Saha Institute of Nuclear Physics
1/AF Bidhannagar, Kolkata, India [3]Materials Science Group
Indira Gandhi Centre for Atomic Research
Kalpakkam, India

Abstract

We have presented here an extensive study on electrical spectroscopy using conducting tip atomic force microscopy (CTAFM) of graphene sheets and its edges. We have tried to address in the present study two main issues: (1) How does the electrical property of the graphene sheet changes when the graphene sheet is displaced by shear forces over highly oriented pyrolytic graphite (HOPG) substrate? and (2) How does the electrical property of the graphene sheet change across it edges? While addressing these two issues we observed that the conductivity of graphene sheets is higher whenever they are loosely bound to the underlying bulk HOPG substrate. We also observe that certain edges of the graphene layers show sharp rise in current when biased, indicating enhanced electronic density of states spatially localized near those edges. In certain edges, we do not observe this phenomenon. These two observations are substantiated by a theoretical study of these edges (passivated with hydrogen) using molecular orbital calculations based on density functional theory (DFT) with B3LYP functional. Structure of edges of 2D graphene sheets are either in zigzag (trans) or armchair (cis) forms. In our theoretical investigation we observed high electron density along the zigzag edge and low density at the armchair edge. We also found out that the electron density at the central region of the sheet increases as the interlayer separation among the graphene layers are increased which supports our experimental

*E-mail address: sangam.banerjee@saha.ac.in

data. We also find that the HOMO–LUMO gap of the graphene sheet varies as a function of its size . We have shown here that the electron density around a mono-vacancy in a graphene sheet is high, which further leads to bending of the graphene sheet.

1. Introduction

A sheet of carbon atoms arranged in a two-dimensional hexagonal lattice is known as graphene as shown in Figure 1(b). It is the thinnest elastic and hardest material known. These graphene sheets can be transformed into various dimensional carbon materials by self-assembly and is thus the (2D) building block for many carbon materials. For example:- 2D graphene sheets can be periodically stacked to form 3D graphite, it can be rolled to form 1D nanotubes, it can be wrapped into 0D fullerenes (buckyballs), nanocones etc., and all these formations depend on the growth conditions [1–4]. Recently, electronic properties of these graphene nanostructures have attracted much attention from the point of view of basic research and applications [5]. It has been noted that graphene can act as field effect transistors (FET) [6], its conducting property can be changed under certain conditions [7, 8]. Graphene sheet can also show magnetic moment [9]. All these properties are presumable due to electronic modification and confinement in 2D plane of graphene. It is now well established that properties of most nanomaterials are quite different from bulk due to confinement effect and this is also true for nano-graphene. Graphene is a semi-metal or zero-gap semiconductor. The electronic band structure of graphene exhibits linear E-k relation for low energies near the six corners of the two-dimensional hexagonal Brillouin zone where the valence band (VB) and the conduction band (CB) touches as shown in Figure 1(c) and (d), leading to zero effective mass for electrons and holes. Due to this linear (or conical) dispersion relation at low energies, electrons and holes near these six points behave like relativistic particles described by the Dirac equation for spin 1/2 particles. Hence, the electrons and holes in graphene are called Dirac particles, and the six corners of the Brillouin zone are called the Dirac points. The Fermi velocity is $v_f \sim$ 1/300 the speed of light c. We have slow ultrarelativistic electrons in graphene.

Highly oriented pyrolytic graphite (HOPG) is a periodical stack of two-dimensional (2D) graphene sheets (layers) along the c axis. The hexagonal lattice of carbons in graphene are bonded by strong σ bonding (sp^2) in the a-b plane (see Figure 1(a)). The perpendicular π-orbital electrons (along the c axis) are responsible for the conductivity along the a-b plane. The conduction occurs by the quantum mechanical hopping of these electrons. Each of these layers is weakly bonded to their neighboring layers by interlayer interaction forces in graphite. Because of the weak interlayer interaction forces, the graphene layers in graphite can easily slide against each other and result in easy peel off. The edges of the graphene sheets are of two types: (1) zigzag (trans) and (2) armchair (cis) edges as shown in Figure 1(b). The nature of the π electronic states are strongly affected at the edges of a graphene layer [10]. We shall show here that the graphene sheets have high electron density edge states localized [7, 8, 10–18] at the zigzag edges. In contrast, the armchair edges have no edge states, hence making the armchair edge less conducting than the zigzag edge.

The atomic force microscope (AFM) with a conducting tip (CTAFM) can be used to study the local electronic properties of graphene surface and its edges. Mapping conductivity across step edges using scanning tunneling microscope (STM) has some difficulties.

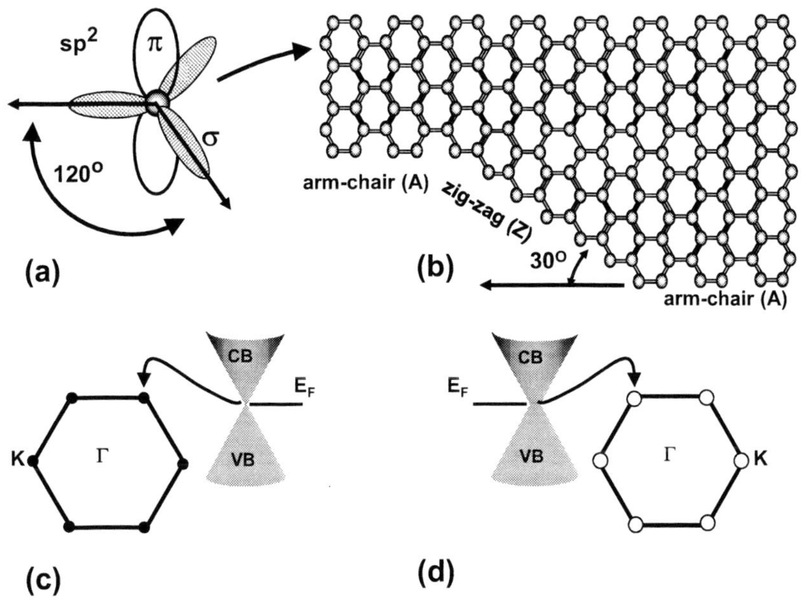

Figure 1. (a) Carbon atoms showing sp^2 hybridized σ orbitals in the a-b plane and nonhybridized π orbitals along the c-axis of graphite layer. (b) Graphene layer showing armchair (cis) edge and zigzag (trans) edge when cut at 30° with respect to each other. (d) Six points in the Brillouin zone showing zero-gap and having zero-overlap of the conduction band (CB) and the valence band (VB) at Fermi energy (E_F) i.e., zero density of state (DOS) at E_F and (e) finite overlap of CB and VB leading to finite DOS at E_F [7].

In STM eventhough one measures the local density of states, at the graphene edges, the difference in the electron density at the edges is very small compared to the bulk surface. The current varies exponentially as a function of tip-sample distance in STM and hence it becomes difficult to detect these small differences. One has to perform a I-V characteristics measurements at every point by stopping the feed-back signal to map the local conductivity with high resolution in STM. This can be overcome by using CTAFM where one can directly obtain the local conductivity over the graphene surface and at its edges in direct contact during scanning.

In the theoretical front most of the earlier calculations of graphene were carried out using tight binding approach utilizing translational periodicity of infinite 2D sheet. Due to the recent enhanced interest on graphene and other nano-materials, many quantum chemical analysis of graphene are being carrid out, mostly using DFT studies following plane wave pseudopotential method [19–21] or semi-empirical methods [22]. In order to use the intrinsic translational symmetry of both the method and the molecule, the edges were generally ignored in most of the studies. A few recent studies, however, have pointed out the need of terminal hydrogens in such calculations [11, 23]. However, those models also have some inherent translational symmetry in some of the directions and hence a true electronic picture of a nano-meter size sheet is still unknown. It is also beleived that the carbon atoms at the edges can be self-passivated by adsorption of atomic hydrogen from molecular atmosphere. We have therefore attempted to study theoretically the properties of real finite size graphene sheet with atomic hydrogen at its edges and also looked into the electronic property of graphene with point defect and evolution of electronic properties as a function of layer separation and compared the results with our experimental findings [24]. In order to calculate electronic properties of small graphene sheet, we have adopted the methodology mostly used to measure properties of small chemical compounds in vacuum, i.e. Hartree–Fock (HF) or density functional theory (DFT) with generalised gradient approximation (GGA). We shall try to understand the following phenomena using molecular orbital calculations based on density functional theory i.e., (1) Observation of high electron density along the zigzag edge and low electron density at the armchair edge. (2) Spatial variation of the electron density in nanographene as a function of interlayer separation. (3) Variation of HOMO−LUMO gap of graphene sheet as a function of its size. (4) Density of electron orbitals around a monovacant site. Phenomena 1 and 2 has been observed experimentally [7, 8] by us while phenomenon 3 and 4 are important for device fabrication using nano-graphene.

2. Conducting Tip Atomic Force Microscopy (CTAFM) Studies of Graphene Layers

For the present investigation we have used freshly cleaved highly oriented pyrolytic graphite HOPG. By simply peeling off the surface layers using scotch tape, we were able to obtain ribbons and terraces with step edges. The peeling off process dislocates the layers laterally and vertically. We have studied only monolayer steps to avoid complexity in the electrical analysis. Commercial AFM from NT-MDT, Russia was used for the present investigation. Cantilevers used were platinum coated (CSG 10/Pt, NT-MDT) with a radius of curvature of the tip $\sim$ 35 nm and the cantilever elastic constant of 0.1 N/m. For imaging in the contact mode the normal force between the surface and the tip was kept around $\sim$25 nN (constant force mode). The sample was biased with 10 to 15 mV for measuring the local conductance (spreading resistance imaging) of the sample surface with the conducting tip. By measuring the tip-sample contact conductance (resistance) during the scanning process, we measure the local conductance of the sample surface. For purely ohmic contacts (i.e., metal-metal contacts), the I-V is of the typical spreading resistance form

$$R_{spreading} = \frac{\rho}{2a}, \tag{1}$$

where ρ is the mean resistivity of the tip-sample contact and a is the contact area. All the measurements were conducted at room temperature and ambient conditions. For topographic images we have only subtracted the linear line fit along the scan direction from the scanned images and presented without any further filteration or averaging. For the conductance map we show only the raw results without any processing (processing the data was avoided as this can sometimes be misleading).

Figure 2 shows the schematic diagram of the CTAFM showing the principle of the technique used in this study. The CTAFM technique is also called as the spreading resistance imaging mode. It is employed for various measurements, such as to determine defects in conductive and low conductive films, materials characterization in terms of local resistance etc. Realization of this mode is based on the use of a conducting tip, which comes in contact with the sample surface. A bias voltage is applied between the sample and the conducting tip and the current through the tip is monitored while the scan is under progress.

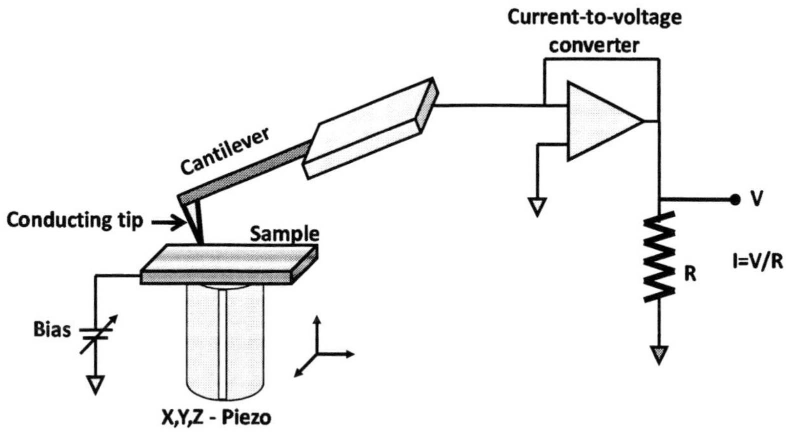

Figure 2. Schematic diagram of the CTAFM showing the principle of the technique

In Figs. 3(a) and 3(b) we show the topographic image and local conductance (spreading resistance) map of a freshly cleaved HOPG sample. We see that the contrast in the local conductance map is much better than the topographic image. Each seperate graphite ribbons can be easily distinguished in the local conductance map. The local conductance image Figure 3(b) of the HOPG sample clearly indicates that the conductivity of these ribbons are different since the conductance was mapped in a constant force mode (which ensures that the contact area remains constant during the whole scan over the sample surface). Hence the tip-sample contact conductivity measured is a direct measure of the local conductivity of the sample and is not due to change in the area of the contact (see Eq. 1). Since we have used a 35 nm radius of a curvature tip, we can safely say that the tip-sample contact area is limited by this and will be less than this value.

In Figure 4(a) we show a topographic two-dimensional (2D) image of a selected region containing two monolayer steps/edges (staircase like) on the surface of the HOPG sample. In Figure 4(b) we show the line profile of the line marked as A in Figure 4(a). The sawtooth-like line profile is an artifact of the subtraction of a linear fit to the data to remove the tilt of the sample. We can clearly see the step heights to be around ~0.6 nm for the intermediate step (layer II) and ~0.7 nm for the top step (layer I), both values higher than the reported c-axis value of 0.35 nm for the HOPG sample. During the peeling-off process, the top few layers become loosely bonded to the sample surface resulting in the increased step heights.

In the inset of Figure 5(a) we show conductance maps taken simultaneously along with topography (Figure 4(a)) showing higher conductance (bright) for the top layer labeled as layer I. The intermediate (layer II) shows intermediate conductance (gray) and the lowest layer (layer III) shows the least conductance (dark). We also observe a bright current streak appearing on the top layer edge indicated by an arrow. The intermediate edge does not show this bright current streak on its edge. This observation clearly indicates the presence

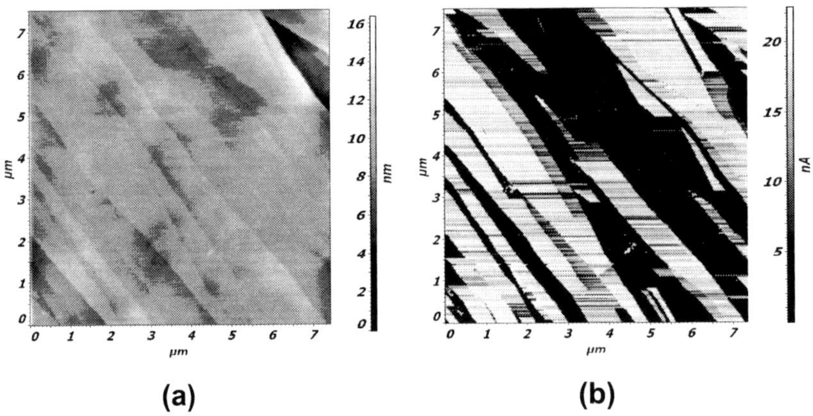

Figure 3. Atomic force microscopic image (a) topographic (b) conductivity landscape of highly oriented pyrolytic graphite (HOPG) [7].

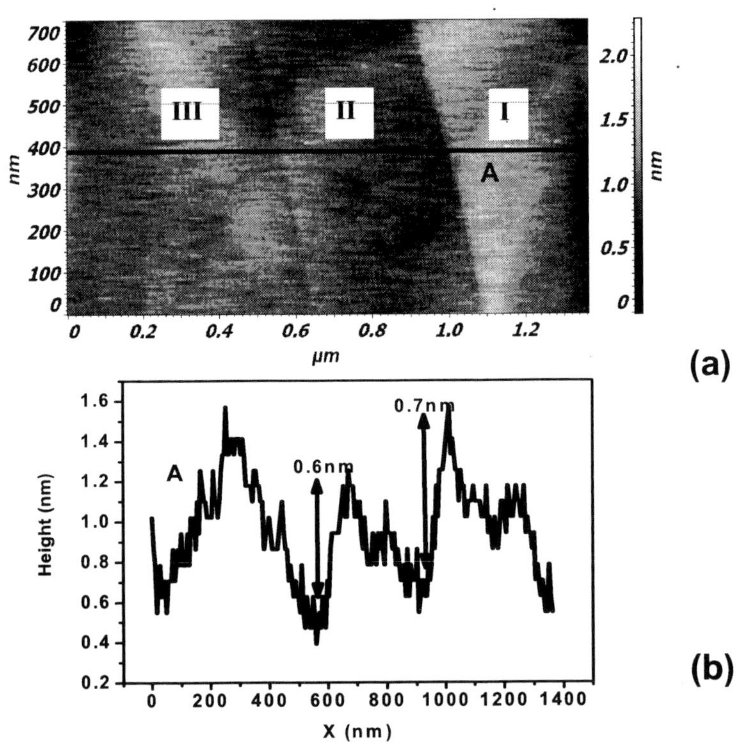

Figure 4. (a) Topographic two-dimensional (2D) image of a selected region containing two monolayer steps/edges (staircase like) on the surface of the HOPG sample. (b) Two height profiles along the lines marked in (a) as A. The arrows indicate the step heights. (Note: The tilt in the step (line profile) is an artifact of processing (linear line fit along the scan x-direction)) [8].

of two distinct types of edges on these two graphite layers. This is more illustrative in the line profile of the conductance current shown in Figure 5(a) for the lines marked A and B in the inset. The current at the edges showing the bright streaks are 2 to 4 times the value measured on the terrace. In Figure 5(b) we show a single ribbon edge containing two types of edges labeled as A and Z. The angle between A and Z is 30° as schematically shown in Figure 1(b). The edge labeled as Z shows bright current streak and the edge labeled as A does not show this phenomenon. We have also performed current-voltage (I-V) measurements as shown in Figure 6 at various regions of steps (terraces) marked in Figure 5(a). We clearly see that the layers which are brighter in the local conductance image show larger slope in the I-V measurement near zero bias. Combining topographic image, conductance map and I-V measurements, we find that the top layers which are dislodged more from the bulk have higher conductivity.

The first observation is the variation of conductance of graphite ribbons on the surface of HOPG. The top layer always exhibits more conductance than the lower layers (Figure 5). During the peeling off process the top few layers of graphite peels off inhomogeneously forming a ribbonlike structure as shown in Figure 3. The ribbon edges formed will have monolayered and multilayered steps at the edges. So, during the peeling off process it will exert vertical and lateral stresses on the ribbons. Due to applied vertical and lateral stresses during the peeling process, these ribbons will get dislocated vertically and laterally. It is known that the π electrons not only take part in conductivity, but also contribute substantially to the polarization cloud (without electron transfer) that gives the bonding between the layers. If the top layer is loosely held, then the π electrons of this layer do not participate much in the bonding with the layer underneath. This makes the π electrons more mobile leading to higher conductivity of the top layer. Note: The carrier density does not change and it is only the mobility which increases thus leading to a higher conductivity. Another important way in which the top layer may be more conducting could be due to an unavoidable crumpling/ripple formation on the top graphene sheet (small displacement of carbon atoms in and out of the plane) especially whenever such distortions are unrelaxed. This effectively introduces the next nearest neighbor hopping amplitude if one thinks of the π electrons in terms of the tight binding approximation. The six isolated points in the Brillouin zone (BZ) where the valence band (VB) and the conduction band (CB) touches for only the nearest neighbor hopping case is shown in Figure 1(c). With the introduction of the next nearest neighbor hopping (induced due to the displacement of the carbon atoms out of the plane due to crumpling/rippling) expands these isolated points to a finite area as shown in Figure 1(d). This means larger density of low energy current carrying states at the Fermi level, resulting in an increased conductivity. In other words, the likeliness of having local unrelaxed crumpling/rippling of sheets is larger for the top layers and any crumpling automatically dopes the zero-gap semi-metal with added carriers. The peeling of the layers was performed manually in an uncontrolled fashion. This can lead to different degrees of applied stress on different layers and cause different amounts of vertical displacement. The one which has a larger step height (c-axis distance) will be less bound (causing large orbital overlap along the plane) than the ribbon which has a lower step height (c-axis value). We motivate this schematically in Figure 7. The difference in conductance among the steps can be explained invoking the argument that the loosely bound ribbons are more conducting than the tightly bound layers. We shall show below using quantum chemical DFT calcula-

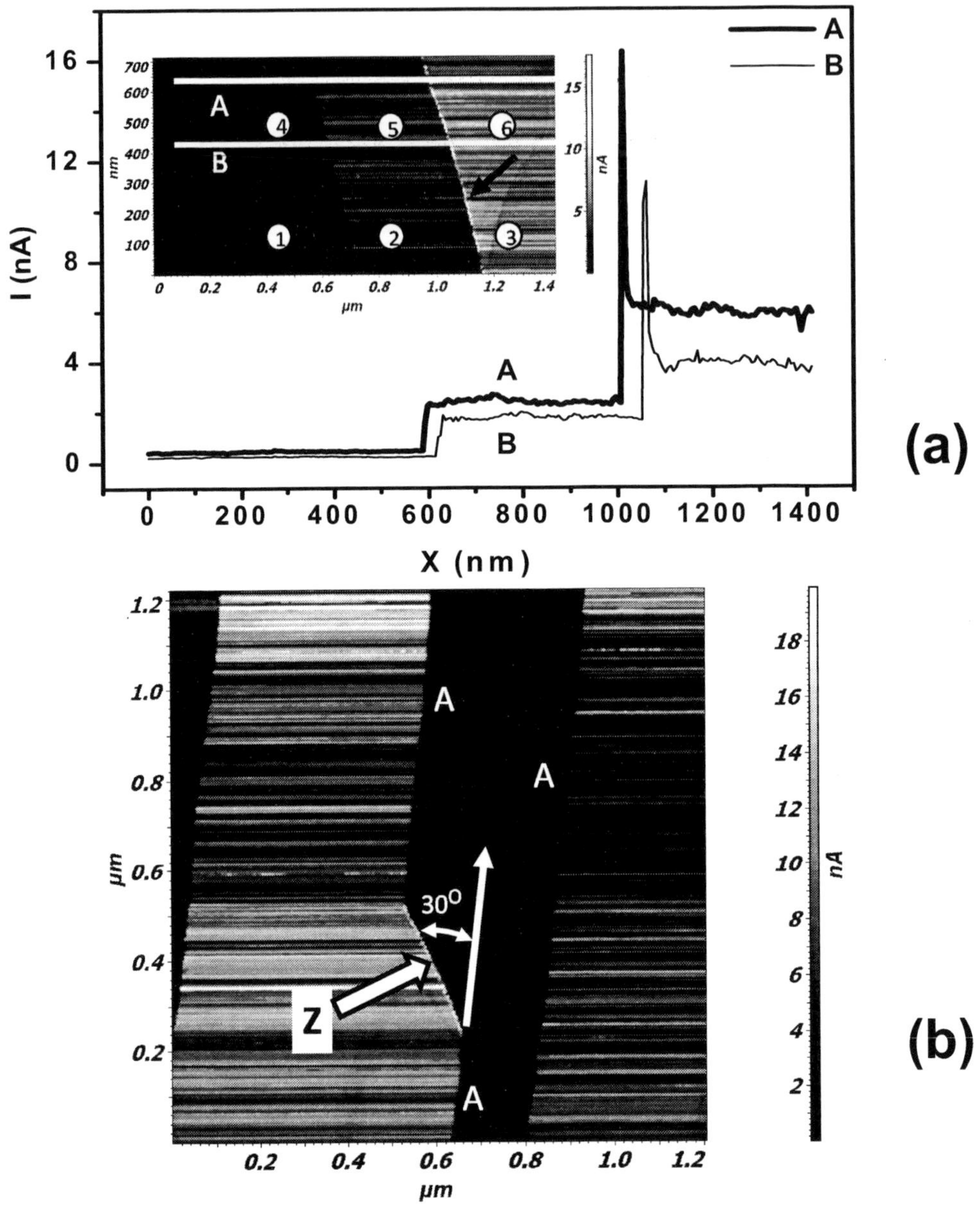

Figure 5. (a) Shows line profiles of the conductance current of the lines marked A and B respectively in the inset. Inset: conductance map showing top layer having high conductance (bright – layer I), the intermediate step showing intermediate conductance (grey – layer II) and the lowest layer showing lowest conductance (dark – layer II). Bright current streaks appearing on certain edges is shown by an arrow. (b) Shows a single ribbon edge containing two types of edges labeled as A and Z. The angle between A and Z is 30° as schematically shown in Figure 1(b). The edge labeled as Z (marked by arrow) shows bright current streak and the edges labeled as A does not show bright current streak at its edges [8].

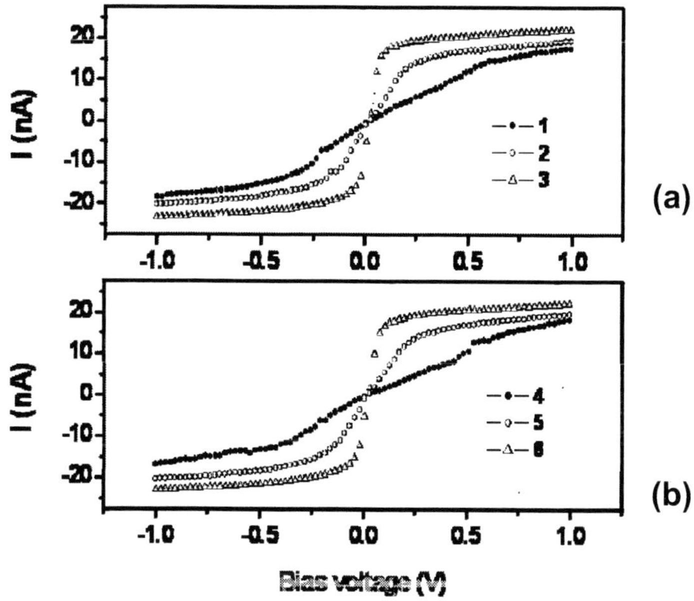

Figure 6. Current-voltage (I-V) measurement carried out at various regions of steps/terraces marked in the inset of Figure 5(a). Low slope value in the I-V curve at zero-bias indicates lower conducting region than the region having higher slope. (Note: all the I-V measurement was carried at same value of constant force) [8].

tion that indeed as the separation of the layers increases the electronic population of inner carbon atoms in creases.

The second observation is the presence of sharp current peaks at the edges (Figure 5(a) and the Z region marked by arrow in Figure 5(b)) of the graphene ribbon sheets. As mentioned earlier, there are mainly two types of edges formed on graphene sheets when it is cut (see Figure 1(b)). Recently, detailed theoretical calculations have shown that zigzag edges have edge states localized at the edges having a higher electron density [11–17]. Hence these edges will have higher electrical conductance. Thus, during the mapping of the local conductance these edges will show up as brighter streaks. The layers can be oriented in a different direction with respect to each other and during the tearing process it will tear along any suitable (either zigzag or armchair) direction as we observe in Figure 5(a). As mentioned earlier, the armchair edges have no electronic edge states and thus show no sharp peak in the current at its edges. Thus, we can say that the edges A are armchair edges and the edge Z is the zigzag edge in Figure 5(b) (see Figure 1(a) which shows the angle between the armchair and the zigzag is 30°).

A more careful understanding of the stability of the edge states when exposed to passivating atmosphere such as hydrogen is to be understood. In the next section, the electronic properties of nano-graphene with hydrogen terminated edges are presented determined by quantum chemical Density Functional Theory calculations [24]. This becomes necessary because in ambient condition the graphene edges will be passivated by hydrogen and to

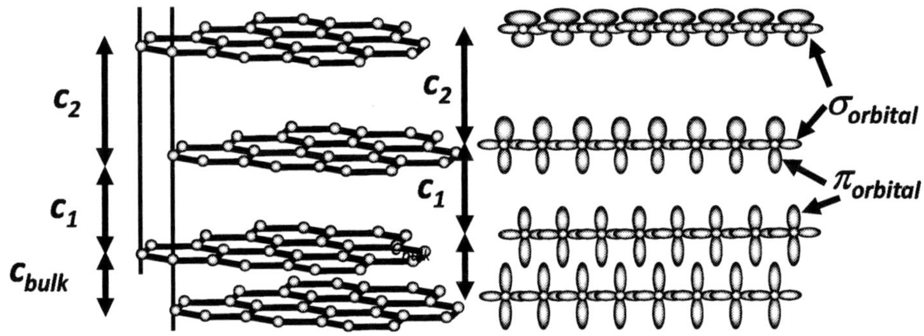

Figure 7. Schematically showing graphite layer seperations $C_2 > C_1 > C_{bulk}$. The top layer is loosely bound showing schematically the π electrons/orbitals are spread out more along the a-b plane causing increase in the overlap of the orbitals along a-b plane and thus leading to higher conductivity. In the bulk the orbitals have lesser degree of overlapping and are extended more towards the z-direction responsible for interlayer bonding [8].

know whether the electronic property still remains same on hydrogen passivation.

3. Quantum Chemical Density Functional Theory Calculations on Hydrogen Terminated Graphene

The edges of graphene are known to spontaneously get reduced by atmospheric hydrogen to neutralize the valencies of all the carbons at the edges. We have performed detailed calculation to theoretically understand the electronic properties of these passivated edges using molecular orbital calculations based on hybrid density functional theory with B3LYP functional. The following methodology was adopted to do the calculations. Atomic coordinates of all the carbon atoms arranged in an hexagonal lattice having zigzag and armchair edges with the edges terminated by hydrogen atoms, were generated by the molecular modeling software MOLDEN [25] using standard bond lengths (b(C−C) = 1.421 $\mathring{A}$ and b(C−H) = 1.009 $\mathring{A}$) and all angles equal to 120°. The edges were terminated by hydrogens to form C−H bonds for neutralizing valencies of all the carbon atoms. All the electronic calculations were performed by the free software GAMESS-US [26] using hybrid density functional theory (DFT) with B3LYP/6-31G(d,p) method [27]. Three graphene sheets were generated with zigzag and armchair edges for different sizes (i) smallest graphene sheet (a pyrene molecule) with 16 carbon atoms and 10 hydrogen atoms, (ii) intermediate model with 32 carbons and 14 hydrogens and (iii) largest model with 73 carbons and 21 hydrogens (Figure 8). Another sheet was generated with all bonds in armchair orientation in one side with 68 carbons and 22 hydrogens (see Figure 9). We also generated a sheet with a vacancy at the center (Figure 10), while the valencies of the carbons at the vacant site were fulfilled by adding three hydrogens. This molecule has 72 carbons and 24 hydrogens. We have energy minimized structures of all these molecules after initial model building by Hartree–Fock method with RHF/4-21G basis set by GAMESS-US. All the molecular orbitals were

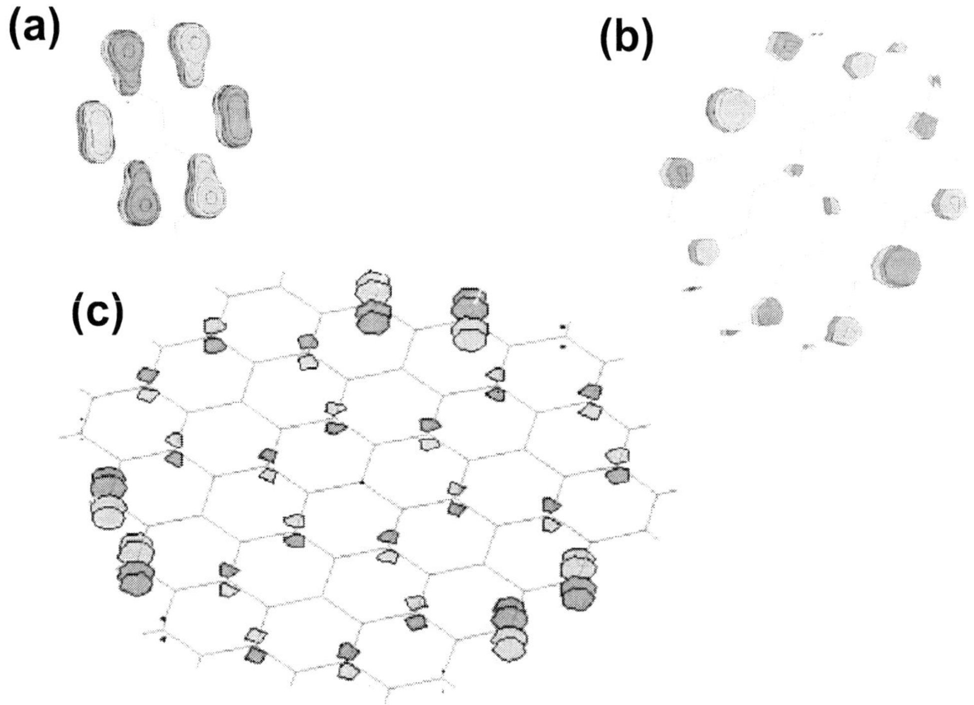

Figure 8. Structures of three nano-graphene sheets of different dimension along with molecular orbitals corresponding to HOMO are shown. The cis edges are located at the corners while the trans edges line up along the sides of the sheets: (a) smallest model (pyrene), (b) intrmediate model and (c) largest model [24].

visualized by MOLDEN software.

Graphene is an infinite two-dimensional sheet, hence quantum chemical calculations using plane wave pseudo-potential and making use of crystalline translational periodicity is quite suitable to understand its electronic properties. However we are interested in understanding structure and electronic properties of the edges or its defects due to vacancy creation in finite size graphene sheets of nanometer dimension. Hence we have adopted the hybrid-DFT formalism, which is suitable for small molecular systems without periodicity. We have performed geometry optimizations of three perfect graphene models with both armchair and zigzag edges, of different size as shown in Figure 8 using HF (Hartree–Fock) method and compared their orbitals calculated by DFT procedures. The three models of graphene sheets of different sizes have 106 electron in 53 doubly filled orbitals for pyrene (smallest model), 206 electrons in 103 doubly filled orbitals for the intermediate model and we used 460 paired electrons in 230 doubly filled orbitals for the largest model. We used 290, 508 and 1200 basis functions for DFT calculation to describe the electron distribution in the three models consisting of zigzag and armchair edges. This clearly suggests that thorough geometry optimization using DFT is almost impossible. Comparison of energy Eigen values for the highest occupied molecular orbitals (HOMO) or valenceband (VB) and lowest unoccupied molecular orbitals (LUMO) or conduction band (CB), which provides direct estimation of band−gap, indicates that band−gap of pyrene (the smallest model) to be 0.14880H (4.05 eV), the intermediate model has band gap of 0.0944 H (2.57 eV) which reduces to 0.0680 H (1.85 eV) for the largest model suggesting that these nanographene can be used as a wide band−gap semiconductors for device application. Our analysis indicates that finite size graphene will have a band gap which decreases on increasing the dimension of the sheet to near metallic situation (band gap vanishes) [28, 29].

Computer graphics representation of the highest occupied molecular orbitals (HOMO) for the smallest, intermediate and largest models with zigzag and armchair edges are shown in Figure 8. We can clearly see that the molecular orbitals located at pure zigzag edges are more prominent than the orbitals at the armchair edges. A very similar property was noted by AFM (as discussed in the earlier section) [7, 8] and STM [30] studies. In order to characterize total electronic population on all the atoms we have calculated the Mulliken charges [31] on each of the atoms and their values for carbons at the edges are shown in Figure 9(a) and (b). In Figure 9(b) we show a nanographene with armchair carbons at a stretch on one side unlike the three models considered earlier. In Figure 9(b) we also show the partial electronic (Mulliken) charges on the carbon atoms at the edges to compare it with the orbitals of the largest model shown in Figure 9(a). Negative values indicates electron density and positive values are the hole density indicating they are electron deficient. These absolute negative values are clearly much smaller on the carbon atoms at the armchair edges than at the zigzag edges.

The graphene sheet with a vacancy at its center, undergoes significant structural alteration from planarity to accommodate three extra hydrogens at the place of one carbon atom. Electronic population analysis of the HOMO (Figs. 10) clearly shows that significantly more electronic crowd is located near the vacant site. Many groups [9, 32–34] have recently studied magnetisation of graphene introduced by edge defect or cavity creation. It was noted that irradiation of graphene sheet by proton (hydrogen ion) gives rise to magnetic moment while those irradiated by alpha particle does not show magnetism [32]. Hence it gives an indication that reduction of carbon by hydrogen around the vacancy giving rise to extra electron occupation around the vacancy may be responsible for observed magnetism in graphite. We can argue with a conjecture that when two hydrogen atoms come close together to form a H_2 molecule the total spin goes to singlet state, whereas when three H atoms come close to form a complex, then the orbital state is antisymmetric and the spin state is symmetric to make the total wave function antisymmetric. Hence, this complex would presumably give rise to high moment state leading to magnetism. Thus, one can guess why proton (H) irradiated samples only shows magnetism and not the alpha particle irradiated samples.

We have further attempted to understand the role of interlayer separation between two graphene sheets in terms of electron accumulation. We considered two graphene layers, each having 54 carbons and 18 hydrogen atoms in two models: (1) with $3.4\mathring{A}$ separation representative of stacked 3D graphite and (2) with $10.0\mathring{A}$ separation emulating practically unstacked graphene layers. The lower layer is shifted in plane so that the carbons of the lower layer are juxtaposed to the centers of the top carbon hexagons. We required 1800 basis functions to represent wave function of all the electrons, which is close to the limiting number of basis functions in the software adopted. In order to understand the enhancement

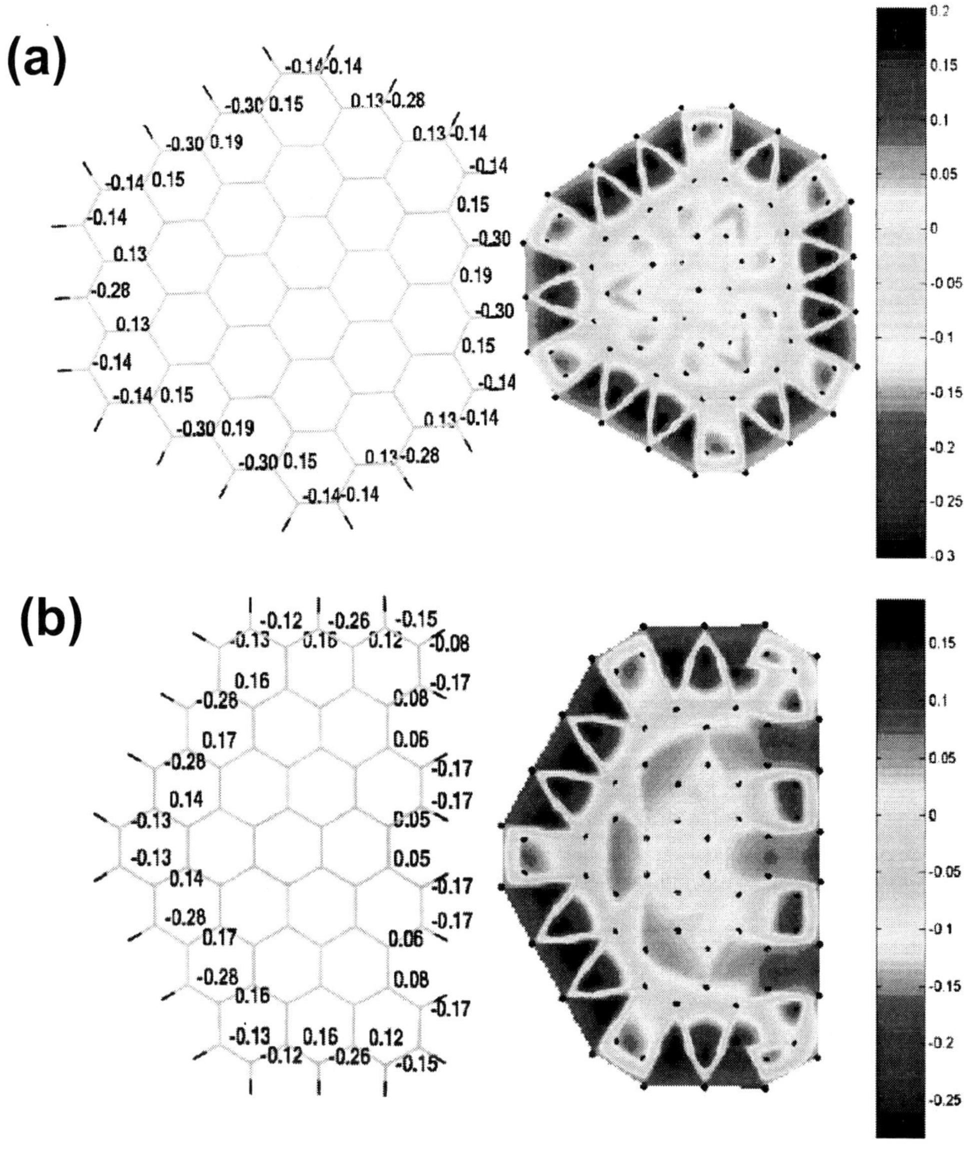

Figure 9. Total partial charges on the carbon atoms along the edges of (a) largest model and (b) the model containing only cis edge on one side. The number on the edges indicates electron population subtracted from the nuclear charge and the colour representation gives indication of the inner carbon atoms also where positive value/red color represents electron defficiency and the negative value/blue color are representative of excess electron accumulation [24].

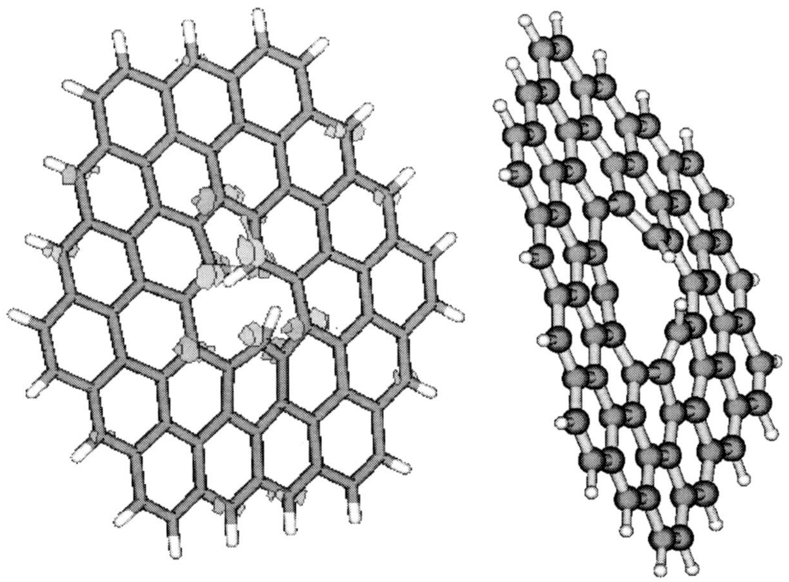

Figure 10. Structure of a nano-graphene with cis and trans edges and a reduced mono-vacancy at the center along with the molecular orbital of the HOMO is shown [24].

of electrical conductivity as a function of sheet separation, the above two systems were considered. We compared electronic populations of HOMOs of these two systems. It is found that average population of electrons considering only the inner carbons on the closely separated layers (3.4Å) is 0.0123 (std =0.011), while that on the well separated layers (10Å) is 0.0243 (std = 0.0198) and is depicted schematically in Figure 11. Considering the small values of partial charges, we have also calculated these by a better method namely electrostatic surface potential fit (ESP fit) by MOLDEN. Average value of the ESP fit charges are -0.00083 (std = 0.036) and -0.06993 (std = 0.006) for the 3.4Å and 10.0Å layers, respectively. Average value of charge, even smaller than its standard deviation, possibly indicates effectively zero charge density on the interior carbon atoms for the closely spaced layers. These results further indicates that there is a net increase in electron population in the inner region of the sheet as a function of sheet separation. This will lead to increase in the electrical conductivity and this has been observed experimentally (as detailed in the earlier section) [7,8]. This charge distribution on the sheet as a function of sheet separation arises as a result of localization/confinement of electron on transformation from 3D graphite to 2D graphene sheet.

4. Conclusion

Quantum chemical calculation using density functional theory of hydrogen atom passivated edges of graphene sheets reveals that zigzag edges have a higher electron density compared to the armchair edges. This calculation also revealed an increase in the electron density at

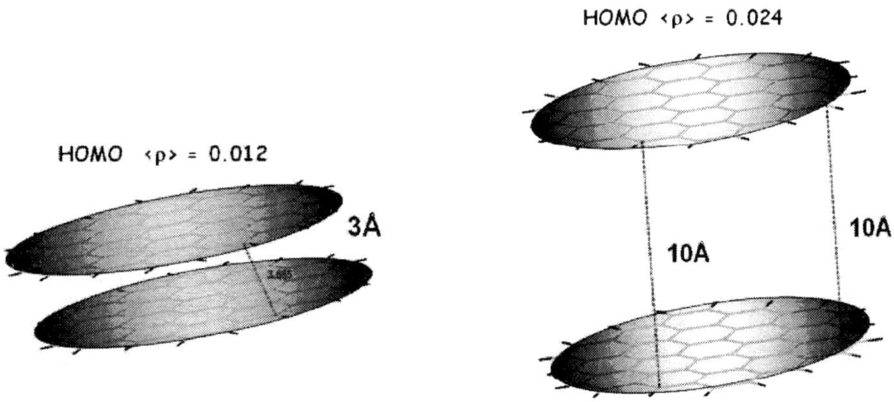

Figure 11. Schematic depiction of average population of electrons considering only the inner carbons on the closely separated layers (3.4$\mathring{A}$) is 0.012 (std = 0.01), while that on the well separated layers (10$\mathring{A}$) is 0.024 (std = 0.02).

the central region of the a sheet (compared to the edges) when two sheets are considered. As the interlayer separation is increased between the two graphene sheets, the electron density at the central region of the sheets also increases. We also observed decrease in the HOMO−LUMO gap of graphene sheet as the size increases and large enhancement of density of electron orbitals around a monovacant site. The experimental results of electrical conductivity of graphene sheets using CTAFM could be explained by the results obtained from the theoretical calculations i.e., sharp increase in the conductivity at certain edges could be attributed to the zigzag edges and when the graphene layers are strained and dislodged from the surface of the bulk graphite sample then the electrical conductivity of that layer enhances. The enhancement of the conductance depends on the degree of dislodgement of the layers. The larger the dislodgement of the layer the more enhanced is the electrical conductivity.

References

[1] For review see R. Setton, P. Bernier, S. Lefrant (Eds.), Carbon Molecules and Materials, Taylor and Francis, New York, 2002.

[2] V.N. Popov, Mater. Sci. Eng. R. 43 (2004) 61.

[3] S.P. Jordan, V.H. Crespi, Phys. Rev. Lett. 93 (2004) 255504.

[4] R. Saito, G. Dresselhaus, M.S. Dresselhaus, Physical Properties of Carbon Nanotubes, Imperial College Press., London, 1998.

[5] K.S. Novoselov et al., Nature Phys. 2 (2006) 177.

[6] K.S. Novoselov et al., Nature (London) 438 (2005) 197.

[7] S. Banerjee, M. Sardar, N. Gayathri, A.K. Tyagi, Baldev Raj, Phys. Rev. B 72 (2005) 075418.

[8] S. Banerjee, M. Sardar, N. Gayathri, A.K. Tyagi, B. Raj, Appl. Phys. Lett. 88 (2006) 062111.

[9] P. Esquinazi, A. Setzer, R. Hohne, C. Semmelhack, Y. Kopelevich, D. Spemann, T. Butz, B. Kohlstrunk, M. Losche, Phys. Rev. B 66 (2002) 024429;P. Esquinazi, D. Spemann, R. Hohne, A. Setzer, K.-H. Han, T. Butz, Phys. Rev. Lett. 91 (2003) 227201.

[10] M. Fujita, K. Wakabayashi, K. Nakada, and K. Kusakabe, J. Phys. Soc. Jpn. 65, 1920 (1996)

[11] K. Nakada, M. Fujita, G. Dresselhaus, and M. S. Dresselhaus, Phys. Rev. B 54, 17954 (1996).

[12] K. Wakabayashi, M. Fujita, H. Ajiki, and M. Sigrist, Phys. Rev. B 59, 8271 (1999).

[13] S. E. Stein and R. L. Brown, J. Am. Chem. Soc. 109, 3721 (1987).

[14] H. Hosoya, Y.-D. Gao, K. Nakada, and M. Ohuchi, in New Functionality Materials, edited by C. T. Tsuruta, M. Doyama, and M. Sena Elsevier, New York, (1993), p. 27.

[15] K. Tanaka, S. Yamashita, H. Yamabe, and T. Yamabe, Synth. Met. 17, 143 (1987).

[16] M. Fujita, M. Yoshida, and K. Nakada, Fullerene Sci. Technol. 4, 565 (1996).

[17] S. Ryu and Y. Hatsugai, Physica E, 22, 679 (2004).

[18] M. Elstner, D. Porezag, G. Jungnickel, J. Elsner, M. Haugk, T. Frauenheim, S. Suhai, and G. Seifert, Phys. Rev. B 58, 7260 (1998).

[19] H. Lee, N. Park, Y.-W. Son, S. Han, J. Yu, Chem. Phys. Lett. 398 (2004) 207211.

[20] H. Lee, Y.-W. Son, N. Park, S. Han, J. Yu, Phys. Rev. B 72 (2005) 174431.

[21] A. Montoya, T.N. Truong, A.F. Sarofim, J. Phys. Chem. A 104 (2000) 6108.

[22] H. Sabzyan, M. Babajani, J. Mol. Struct.: Theochem 76 (2005) 155.

[23] Frauenheim, S. Suhai, G. Seifert, Phys. Rev. B 58 (1998) 7260.

[24] S.Banerjee and D. Bhattacharyya, Comp. Mat. Sc., 44, 41 (2008).

[25] G. Schaftenaar, J.H. Noordik, J. Comput. Aid. Mol. Des. 14 (2000) 123–134.

[26] M.W. Schmidt, K.K. Baldridge, J.A. Boatz, S.T. Elbert, M.S. Gordon, J.J. Jenson, S. Koseki, N. Matsunaga, K.A. Nguyen, S.Su, T.L. Windus, M. Dupuis, J.A. Montgomery Jr., J. Comput. Chem. 14 (1993) 1347–1363.

[27] C. Lee, W. Yang, R.G. Parr, Phys. Rev. B 37 (1988) 785–789; A.D. Becke, J. Chem. Phys. 98 (1993) 5642–5648.

[28] B. Obradovic, R. Kotlyar, F. Heinz, P. Matagne, T. Rakshit, M.D. Giles, M.A. Stettler, D.E. Nikonov, Appl. Phys. Lett. 88 (2006) 142102.

[29] L. Yang, C.-H. Park, Y.-W. Son, M.L. Cohen, S.G. Louic, Cond-mat 0706.1589v1 (2007).

[30] Y. Niimi, T. Matsui, H. Kambara, K. Tagami, M. Tsukada, H.Fukuyama, Phys. Rev. B 73 (2006) 085421.

[31] R.J. Mulliken, J. Chem. Phys. 23 (1955) 1833.

[32] P.O. Lehtinen, A.S. Foster, Yuchen Ma, A.V. Krasheninnikov, R.M. Nieminen, Phys. Rev. Lett. 93 (2004) 187202.

[33] Yuchen Ma, P.O. Lehtinen, A.S. Foster, R.M. Nieminen, New J. Phys. 6 (2004) 68.

[34] K. Kusakabe, M. Maruyama, Phys. Rev. B 67 (2003) 092406.

In: Graphene: Properties, Synthesis and Applications
Editor: Zhiping Xu
ISBN: 978-1-61470-949-7

Chapter 5

CHEMICAL METHODS FOR GRAPHENE SYNTHESIS

Xi Yang, Hongzheng Chen, Junjie Li and Mingsheng Xu*

State Key Lab of Silicon Materials
MOE Key Laboratory of Macromolecule Synthesis and Functionalization
& Department of Polymer Science and Engineering
Zhejiang University, Hangzhou, P. R. China

Abstract

Graphene, a two-dimensional sp^2 hybridized lattice of carbon atoms, has attracted intense interest from both theoretical and experimental scientists due to its unique electronic, mechanical, chemical, and catalytic properties. Recent synthetic efforts have been paid in diverse directions. For instance, chemical vapor deposition on catalytic metal surfaces, carbon segregation from silicon carbide or carbon-doped metal substrates, micromechanical cleavage of graphite, exfoliation of graphite in solvents, and chemical conversion from graphite have been developed to produce single-layer or few-layer graphene sheets. Among them, chemical conversion from graphite and exfoliation of graphite in solvents appears to be efficient approaches to produce graphene or chemically modified graphene (CMG) on large scale and at low cost. Although chemical modification partly destroys the sp^2 crystalline structure of graphene, the functional groups or defects of CMGs may provide them with new properties or functions. Compared to graphene films produced from physical methods, graphene from this chemical way is readily dispersible in solvents and easily processable. Thus, the further manipulation, modification, or applications is possible. As a result, CMG have a variety of technological applications for including nanoelectronic and photoeletronic devices, energy storage materials, polymer composites, and electrocatalysis. However, many limitations exist in the chemical synthesis of graphene layers. It is still insurmountable challenges to synthesize graphene with controlled thickness, high crystalline structure, large size, or high concentration suspension. Therefore, it is of significance to understand the influence of synthesis parameters on the resulting graphene files in order to control the synthesis process to realize the desired properties.

* E-mail address: msxu@zju.edu.cn

The content of this chapter mainly focuses on recent progress in chemical methods for graphene synthesis (*i.e.*, chemical conversion from graphite and exfoliation of graphite in solvents), the unique properties and functions of graphene/CMG generated from different chemical approaches, and the potential applications. Moreover, future developments are also addressed.

INTRODUCTION

Graphene is an atomically thin layer of sp^2 hybridized carbon atoms arranged in a honeycomb lattice [1]. Observation of graphitic carbon monolayers by chemists and surface scientists should date back to the 1960s even earlier [2, 3]. In 1986, Boehm *et al.* referred to such a single layer of graphite as graphene [4]. However, exfoliation of graphite into individual graphene sheets remained a curiosity until 2004, when isolated graphene sheets using the simple Scotch tape method were reported [5, 6]. The subsequent discoveries of its unusual electronic properties have led to an extraordinary amount of interest amongst researchers across virtually all scientific disciplines [7-10]. Graphene has a large theoretical specific surface area (2630 m^2g^{-1}) [11], high intrinsic mobility (200 000 $cm^2v^{-1}s^{-1}$) [12], high Young's modulus (~1.0 TPa) [13], and thermal conductivity (~5000 $Wm^{-1}K^{-1}$) [14], and its optical transmittance (~97.7%) and good electrical conductivity [15]. Graphene is thus usually considered as an attractive candidate for next-generation electronic materials and holds great promise for widespread applications for such as field-effect transistors [16], photovoltaic cells [17, 18], sensors [19], nanocomposites [20], and supercarpacitors [21].

As the first step to investigate its properties and further to realize these practical applications, graphene must be synthesized in a controllable manner, thus the controllable synthesis of graphene is of great importance for both fundamental research and practical applications [22]. Recent synthetic efforts have been paid in different directions. For instance, micromechanical cleavage [6], liquid-phase exfoliation [23-27], reduction of graphene oxide [28-32], chemical vapor deposition (CVD) [33-37], carbon segregation [38-40], chemical synthesis [41-43], and deterministic placement [44-45] have been developed to produce single-layer (SLG) or few-layer graphene (FLG). The current efforts on synthesis is mainly focused on the control over its area, structural quality, and number of layers [46].

In the case of micromechanical cleavage, graphene is produced by peeling off a piece of highly oriented pyrolitic graphite (HOPG) with an adhesive tape [6]. SLG up to millimeters in size and with highly structural quality can be produced. Although this is still the method of choice for laboratory research to exploit its unique properties, it remains difficult to prepare large-scale samples needed for the widespread applications of this material.

Liquid-phase exfoliation (LPE) is to make dispersions of graphite in various solvents with assistance of ultrasonication. LPE can be probably viewed as an alternative to the solid-phase micromechanical cleavage while with more facile and larger yields but difficult to control the numbers of graphene layers [24, 47]. In this micromechanical approach, interaction between solvent such as N,N-dimethylformamide (DMF) or N-methylpyrrolidone (NMP) with graphene may lower the energy barrier of exfoliation by providing a better match of surface energies between the two components [24]. Moreover, appropriate organic molecules (both small molecules and polymers) can also be used to help exfoliating graphite into graphene by inserting into graphite layers [48, 49].

Reduction of graphene oxide is the most popular wet chemical approach to produce graphene colloidal suspensions, which is defined as reduced graphene oxide (rGO) [27]. Graphene oxide (GO) in aqueous solution, generated by the oxidization (resulting in oxidation of graphite, *i.e.*, graphite oxide) and subsequent exfoliation of graphite, can be reduced to graphene by reducing agents [50], thermal treatment [51] or electrochemical methods [52], transforming electrically nonconductive GO to conductive graphene. Because of the facile solution processing, large amount yields, and low cost, this wet chemistry approach is especially attractive in the fields of chemistry and biomaterials. However, the transport properties of the resulted rGO are moderate due to the remaining of defects and oxygen groups on the graphene plane [30, 32].

SLG and FLG can be grown on metal substrates *via* CVD method. When a catalytic metal surface is heated to a suitable temperature, gaseous hydrocarbons are decomposed into carbon atoms and hydrogen gas (or oxygen gas), and the carbon atoms then form a graphene monolayer or multilayer [34, 53-57]. Although large area SLG or FLG films have been generated by CVD method, and graphene-like carbon materials have been produced by substrate free CVD, radio-frequency plasma-enhanced CVD, aerosol pyrolysis and solvothermal synthesis, the uniform growth of SLG or graphene layers with predefined numbers of layers is still a major challenge possibly due to the large amount of carbon sources absorbed by catalytic metals [58-60].

Carbon segregation has also been exploited to prepare graphene layers from silicon carbide (SiC) or carbon-doped metal substrates [38-40]. SLG or FLG can be generated *via* the sublimation of Si of SiC or the catalytic decomposition of hydrocarbon following high temperature annealing. Similar to CVD method, this segregation method also faces the challenge of how to achieve high quality graphene layers with fine thickness control over the entire substrates.

The outstanding electronic and optical properties of graphene accompanied with recent breakthrough of its synthesis methods have attracted the attention on applications for ultrafast electronic circuits and detectors [61], clean and renewable energy [62], and rapid DNA sequencer [63]. Take organic photovoltaics (OPVs) as an example, it has been demonstrated that the efficiency prediction of graphene-based OPVs (graphene serves as photoactive materials) can show the potential for single-cell efficiencies exceeding 12% (and 24% in a stacked structure) [64]. Since graphene can be solution-processed and made to be either metallic (in the pristine state) or semiconducting, it is possible to fabricate an all-graphene OPV, where semiconducting graphene is used as the photoactive material and metallic graphene as the conductive electrodes.

We will mainly concentrate on the recent chemical progress in the production of graphene and chemically modified graphene (CMG) from colloidal suspensions, including reduction from GO, liquid-phase exfoliation, and other preparation methods. As compared to other methods, this chemical processing approach is more attractive with the respect to yield scale, cost, and further chemical functionalizations towards development of flexible electronic devices.

GRAPHENE FROM GRAPHENE OXIDE

Definition

Before the discussion of graphene oxide (GO) and reduced graphene oxide (rGO), we recall the definition of graphene based materials. The term "graphite" designates that modification of the chemical element carbon, in which planar sheets of carbon atoms, each atom bound to three neighbors in a honeycomb-like structure are stacked in a three-dimensional regular order [65]. It is not correct to use the term "graphite" for a single carbon layer. The term "graphene" should be used only when the reactions, structural relations or other properties of individual layers are discussed [66]. Chemically modified graphene (CMG) is the result of chemical exfoliation of graphite [27, 50, 67]. CMG is also a 2-dimensional (2D) form of carbon sheet but with chemical moieties that render new functionalities while preserving some of the unique properties of the pristine graphene. Graphite oxide is the oxidation form of graphite while still retains the layered structures. After exfoliation, graphite oxide can be exfoliated into individual sheets [67], which is defined as graphene oxide. So the graphite oxide can be viewed consisting of a layered structure of graphene oxide. In this chapter, we define GO as graphene oxide instead of graphite oxide.

Graphene Oxide (GO)

Graphite oxide was first prepared ~150 years ago by Brodie [29], who used fuming nitric acid and potassium chlorate under cooling to oxidize graphite. Later, Staudenmaier in 1859 [68] and Hummers in 1958 [69] oxidized graphite firstly with a mixture of concentrated H_2SO_4 and HNO_3 to form graphite hydrogen sulfate and then performed final oxidation with $KClO_3$ or $KMnO_4$. The graphite oxide consists of layered structure of graphene oxide (GO) sheets. The interlayer distance between the GO layers was determined by X-ray diffraction in 1932 to be 0.6 nm, [70] which is thicker than that of graphite (0.335 nm). In addition, the layer distance can be up to 1.2 nm with increasing relative humidity [71]. More importantly, the graphite oxide can be completely exfoliated to aqueous colloidal suspensions of GO sheets by simple sonication [67] and by stirring for appropriate time [72].

The chemical exfoliation of graphite *via* oxidation leads to covalent functionalization of the conjugated aromatic plane, which dramatically alters the bonding structure of graphene. Several structure models of graphene oxide have been proposed on the basis of extensive studies though the real chemical structure is still under debate. The investigations with ^{13}C NMR spectroscopy of graphite oxide and ^{13}C-labelled graphite oxide demonstrated a favorable model as shown in Figure 1 [73]. Individual sheets of GO can be viewed as graphene decorated with oxygen functional groups on both sides of the graphene basal plane (*e.g.*, epoxy and hydroxyls) and at the edges (*e.g.*, carboxyl and carbonyl). Due to ionization of carboxyl groups, which are primarily present at the sheet edges, GO can be electrostatically stabilized to form a colloidal suspension in water, alcohols, or some organic solvents without a surfactant [74]. The measurement of surface charge (zeta potential) of GO sheets shows that they have negative charges when dispersed in water, which suggests that it is the electrostatic repulsion between negatively charged GO sheets that generate a stable aqueous suspension of them [75]. Although exfoliation of graphite oxide into individual sheets can be facilitated by ultrasonic agitation or rapid heating, excessive ultrasonication and high temperature can result in decrease of lateral dimensions [76].

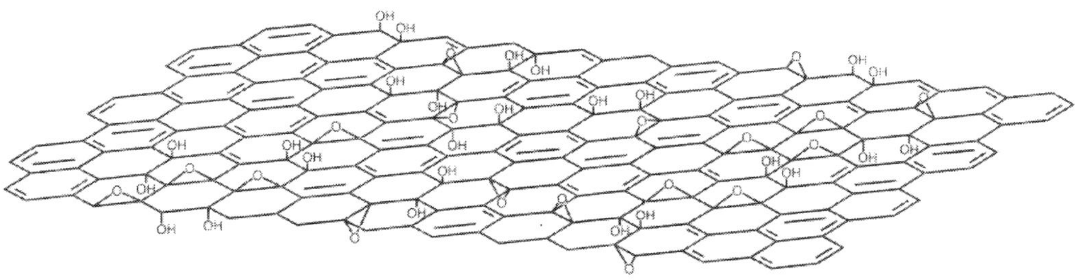

Figure 1. Proposed chemical structure of the graphene oxide. For clarity, minor functional groups, carboxylic groups and carbonyl groups have been omitted at the edges. Reproduced with permission from ref. [73] © 1998 Elsevier.

Oxidation of graphite results in a brown-colored viscous slurry, which contains graphite oxide and exfoliated sheets along with nonoxidized graphitic particles and residue of the oxidizing agents. Centrifugation, sedimentation, or long-term dialysis can remove salts and ions resulted from the oxidation process from GO suspensions. Homogeneous colloidal suspensions of monolayer GO in aqueous and various organic solvents can be achieved by simple sonication of graphite oxide, and followed centrifugation to remove nonoxidized graphitic particles and thick graphite oxide platelets. The resulting hydrophilic graphene oxide can be easily dispersed in water (at concentrations up to 3-4 mg ml^{-1}), affording brown/dark-brown suspensions [9]. Furthermore, monodispersed GO suspensions based on the lateral size of the sheets can be obtained by density-gradient centrifugation [77].

In general, GO can be synthesized by either Brodie, Staudenmaier, or Hummers methods with more or less variation of these methods. Despite the so-called varied methods, little change of the basic reaction has been actually made since Brodie. One development in the present methods is that the sonication step is now usually included to exfoliate the graphite oxide into single layer GO sheets in suspensions [9]. For example, the most common synthesis method of GO nowadays is a variation based on Hummers method, which usually be called as modified Hummers method in literatures [67]. All three methods involve oxidation of graphite to various levels. Brodie [29] and Staudenmaier [68] used a combination of potassium chlorate ($KClO_3$) with nitric acid (HNO_3) to oxidize graphite, and the Hummers method [69] involves treatment of graphite with potassium permanganate ($KMnO_4$) and sulfuric acid (H_2SO_4). Graphite salts made by intercalating graphite with strong acids such as H_2SO_4, HNO_3 or $HClO_4$ have also been used as precursors for the subsequent oxidation to GO [78]. Figure 2 shows the typical AFM image of GO sheets resulted from Hummers method. The height profile shows that the thickness of single layer GO sheets is approximately 1 nm [50].

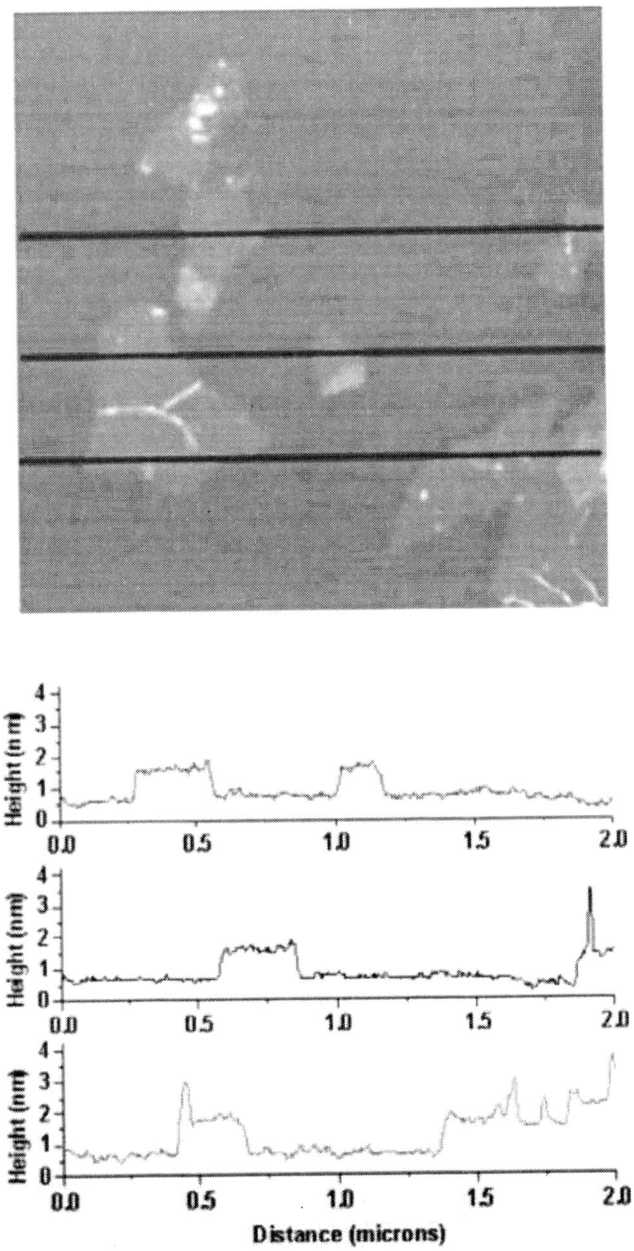

Figure 2. A non-contact mode atomic force microscopy (AFM) image of exfoliated GO sheets (left) with cross-sectional profiles (right). Reproduced with permission from ref. [50] © 2007 Elsevier.

In the past decades, researchers modified and developed the synthesis of GO towards better structures, more uniform morphologies, new functions as well as controlled size and large quantities for their broad range of applications.

It is well known that the properties of graphene films strongly depend on its structural quality and number of the graphene layers or morphology. For example, the size of individual graphene sheets of a graphene film or the defective centers of graphene films plays an important role in their conductivities [76]. Accordingly, it is highly demanded to develop simple and efficient methods for the production of graphene sheets with controlled size of individual graphene sheets and large quantities. Chen *et al.* [79] studied the effect of oxidation time and the amount of oxidants on the size control of GO sheets by chemical exfoliation. GO sheets with different sizes, namely GO-1, GO-2, GO-3 and GO-4, were prepared using the modified Hummers method from flake graphite, where GO 1-4 refer to the products with progressive oxidation steps and have statistical mean sizes of 59,000, 22,000, 3500 and 550 nm^2, respectively (shown in Figure 3a). Thus with increasing reaction time and oxidants, the mean sizes of GO sheets, which exhibit Gaussian distribution, show a regular decrease in the order from GO-1 to GO-4, and the maximum distribution of size is 20,000, 15,000, 2400 and 550 nm^2 correspondingly (shown in Figure 3b). However, further study should be performed to understand the size-dependent physical and chemical properties such as conduction of graphene films consisting of different sizes of individual graphene sheets.

a)

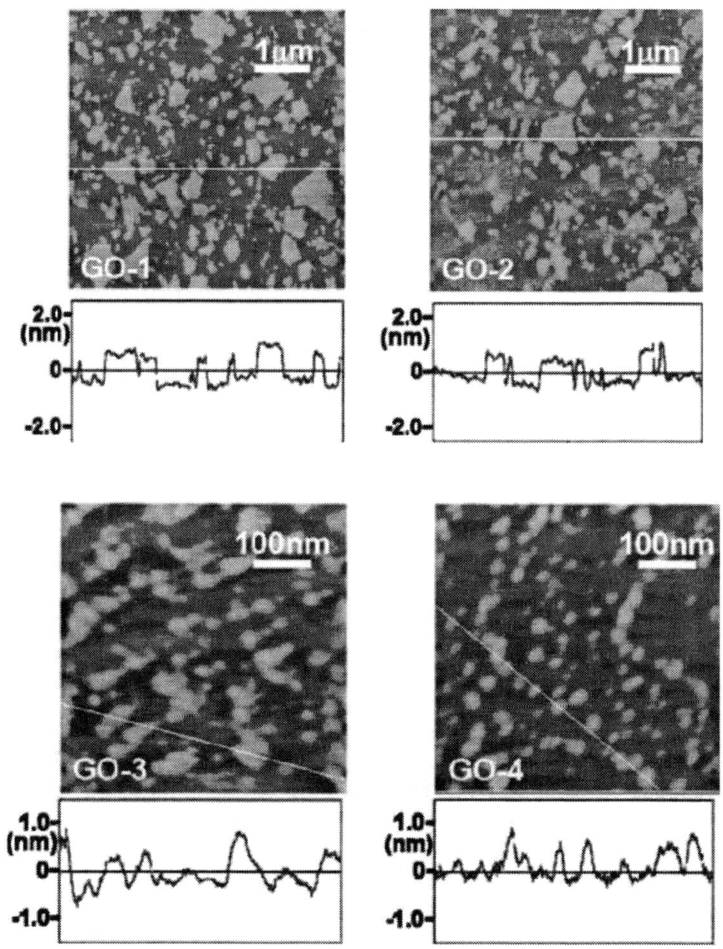

Figure 3. (Continued).

b)

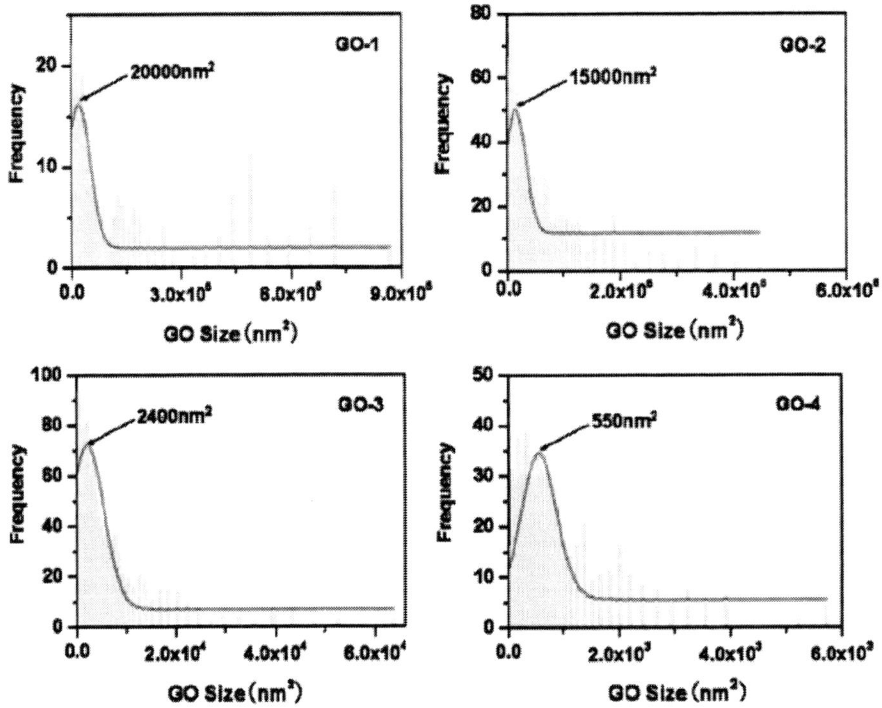

Figure 3. a) Typical tapping mode AFM images (top) and the corresponding cross-sectional profiles (bottom) of GO-1, GO-2, GO-3 and GO-4 on mica substrates. The heights of most GO sheets in four samples are in the range of 0.8–1.0 nm. The sheet size decreases from GO-1 to GO-4. b) Histograms of the size distribution of GO-1, GO-2, GO-3 and GO-4 membranes, obtained from more than 1300 sheets for each sample. The red curve is the Gaussian fit and the maximum distribution of size is 20,000, 15,000, 2400 and 550 nm^2, correspondingly. Reproduced with permission from ref. [79] © 2009 Elsevier.

For a typical synthesis process, the graphene sheets are not only derivatized with oxygen-containing groups but also torn up into smaller pieces. As a result, the lateral sizes of the as-synthesized GO sheets are usually very polydisperse, ranging from a few nanometers to tens of micrometers, which may even vary from one to another synthesis. Such random, top-down, size reduction approaches are unlikely to result in uniform size distribution. Direct synthetic method for making small GO sheets with a more uniform size distribution is still a major challenge. Huang *et al.* [80] reported a size-controlled synthesis of GO nanosheets with lateral dimension smaller than 100 nm using graphite nanofibers, in which the graphene planes are coin-stacked along the length of the nanofibers, as the precursor (see Figure 4). Since the upper size limit is predetermined by the diameter of the nanofiber precursor, the size distribution of the GO nanosheets is much more uniform than that of common GO synthesized from graphite powders. The size can be further tuned by the oxidation time. The nano-sized GO sheet have very similar spectroscopic characteristics and chemical properties with the micrometer-sized counterparts but different solution properties, such as surface activity (see Figure 5) and colloidal stability. Due to higher charge density originating from their higher edge-to-area ratios, aqueous GO nanocolloids are more stable than their micrometer-sized counterparts made from graphite powders.

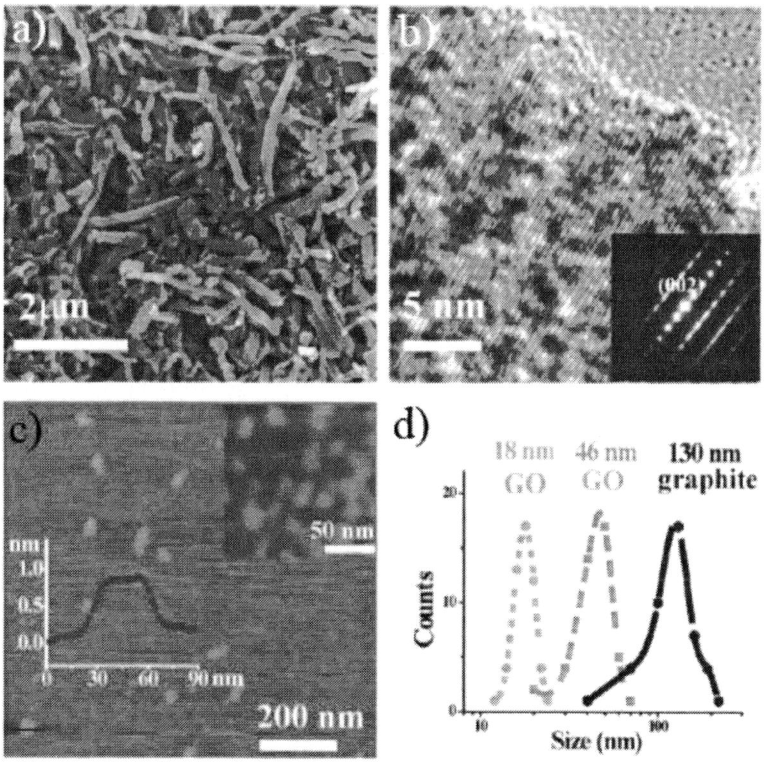

Figure 4. GO nanosheets synthesized from graphite nanofibers. (a) Scanning electron microscopy (SEM) image of the graphite nanofiber precursor. (b) High-resolution transmission electron microscopy (HRTEM) image and electron diffraction (ED) pattern (inset), showing that the graphene planes are coin-stacked along the ⟨001⟩ fiber growth direction. (c) AFM image of GO nanosheets obtained after 2 and 12 h (inset) of reaction, deposited on mica. Their apparent thickness was found to be around 1 nm, as shown in the height profile taken along the blue line. (d) Size distribution of the starting graphite nanofibers (solid black line) and GO nanosheets obtained after 2 (dashed green line) and 12 h of oxidation (dotted orange line). Reproduced with permission from ref. [80] © 2010 ACS.

2D graphene is widely considered to be an ideal material for the use as transparent conductive films (TCFs) because of its optical transparency to visible and near-infrared light (380-1300 nm), remarkable electrical conductivity, as well as excellent mechanical properties. [74] However, the size of individual chemically derivated GO sheets currently used for the fabrication of TCFs is very small, typically in the order of 100 μm^2, because of the difficulty in the production of large-area GO sheets, *i.e.,* unavoidable breaking of GO sheets during the vigorous oxidation and exfoliation processes. The small size of GO sheets results in high contact resistance of the whole rGO films due to a large amount of intersheet junctions, *i.e.*, defects. Hence, an ability of producing large-area graphene sheets is highly desirable for TCF applications. Cheng *et al.* [76] introduced a modified chemical exfoliation technique to prepare large-area graphene oxide (GO) sheets. Three modifications were made in their experiments compared to common methods:

(1) Using graphite with a large lateral size (nature flake graphite with an average size of 500-600 μm) as starting material;
(2) Applying mild oxidation and sonication to avoid the overcracking of graphite during oxidation and exfoliation;
(3) Using a two-step centrifugation to successively remove thick multilayer flakes and small flakes.

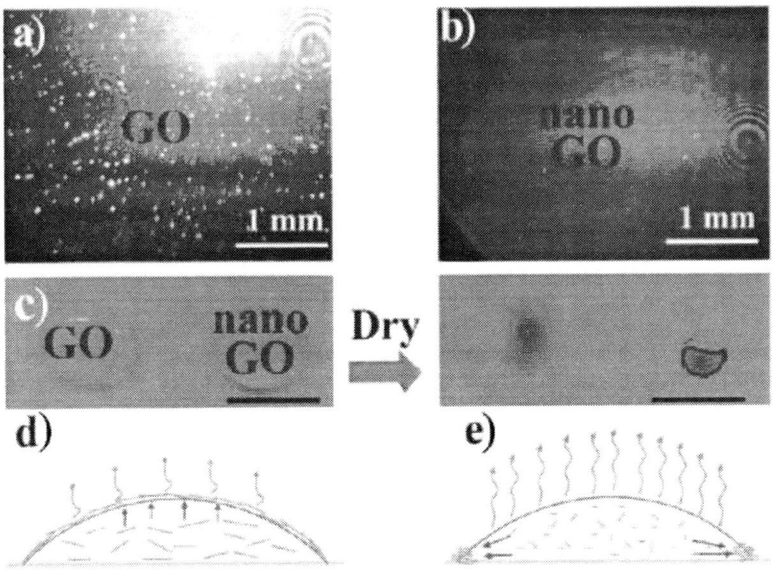

Figure 5. Surface activity of micrometer and nanometer GO. (a, b) *In situ* Brewster angle microscopy (BAM) observation of air-water interface during evaporating a GO and GO nanocolloidal solutions, respectively. The micrometer-sized GO sheets tend to enrich at the water surface over time, but nano-sized GO is much less surface active. Therefore, micrometer GO colloids tend to develop a coating on the water surface during evaporation and leave a continuous film after drying (c, d). In contrast, GO nanocolloids tend to leave the typical "coffee ring stain" type of drying marks commonly seen for aqueous colloidal dispersions (c, e). The scale bars in panel c represent 1 cm. Reproduced with permission from ref. [80] © 2010 ACS.

The maximum area of the GO sheets obtained in their experiment could reach ~ 40000 μm^2 as shown in Figure 6. It was found that the GO area is strongly correlated with the C-O content of the graphite oxide, which enables the area of the synthesized GO sheets controllable. The higher C-O content in graphite oxide makes layer cracking much easier due to the presence of more stretched C-C bonds or C-O-C bonds while sonication as shown in Figure 7a. By simply changing oxidation conditions, GO sheets with an average area of ca. 100-300, ca. 1000-3000, and ~ 7000 μm^2 were selectively synthesized. The rGO film made from GO sheets with an average area of ~ 7000 μm^2 shows a sheet resistance of 840 Ω/sq at 78% transmittance, which is much lower than that (19.1 KΩ/sq at 79% transmittance) of a rGO film made from small-area GO sheets of ca. 100-300 μm^2 (Figure 7b).

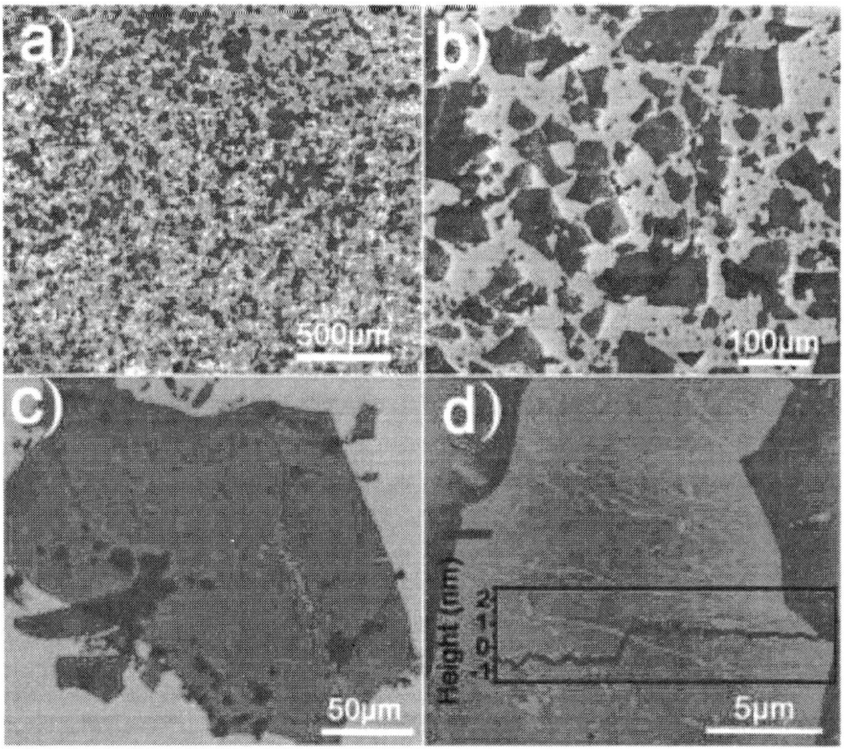

Figure 6. (a-c) SEM and (d) AFM images of large-area GO sheets. The inset of panel d shows that the thickness of the measured GO sheets is ~ 1.2 nm. Reproduced with permission from ref. [76] © 2010 ACS.

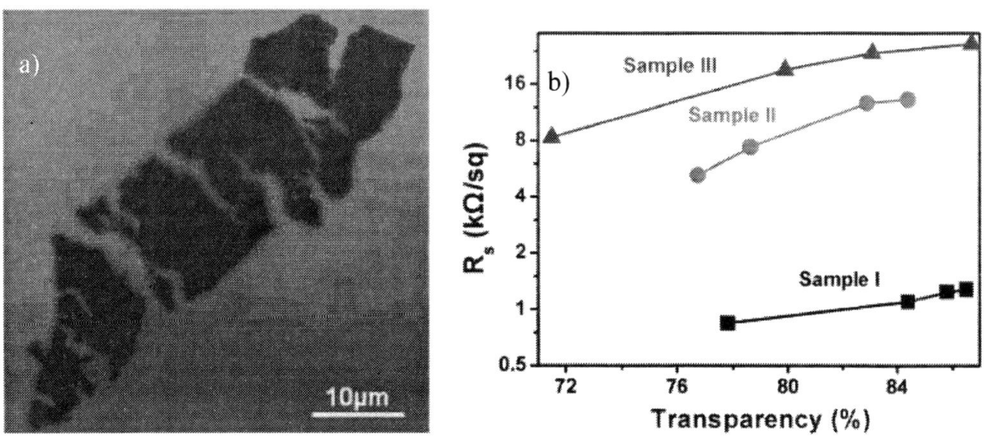

Figure 7. a) SEM image of several pieces of GO sheets with a high C-O content, showing that the small sheets come from the cracking of a big sheet. b) Sheet resistance and transmittance at 550 nm of TCFs assembled with sample I, sample II, and sample III GO sheets. Reproduced with permission from ref. [76] © 2010 ACS.

Reduced Graphene Oxide (rGO)

Although GO sheets have attracted vast interests by providing reactive sites for chemical modification or functionalization using sophisticated surface chemistry and ability to generate homogeneous colloidal suspensions, it is not suitable for active materials of electronic devices because GO is electrically insulating owing to disruption of the aromatic networks [30]. Interestingly, GO can be reduced by chemical methods [9], thermal methods [51], electrochemical methods [52], or ultraviolet-assisted methods [81] to produce electrically conducting CMG, *i.e.*, reduced graphene oxide (rGO).

The reduction of GO by chemical methods mainly involves exposure of GO to reductants such as hydrazine (hydrazine monohydrate [50], dimethyldrazine [27], or anhydrous hydrazine [82]), hydrides (sodium borohydride [83] or sodium hydride [84]), hydroquinone [85] and *p*-phynylene diamine [86]. However, none of the existed methods can completely remove the oxygen containing groups at the graphene plane, resulting in only partial restoration of sp^2 conjugated graphene network. Consequently, the obtained rGO is not the same as pristine graphene. This is also the reason why we cannot refer rGO to graphene.

Compared to other methods such as thermal treatment, the reduction of GO by exposure of GO to a reductant in solution is considered as the most efficient reduction method because both sides of the sheets can interact with the reducing agent. Hydrazine is one of the most used reducing agents for reduction of GO. When the aqueous GO suspension is reduced by hydrazine at the pH value of the as-made GO suspension, the rGO will precipitate from the suspension as black solid (shown in Figure 8) [50]. After the agglomerated rGO sheets are dried, a black powder that is electrically conductive (powder conductivity, $\sim 2\times10^2$ S m^{-1}) can be achieved. The reduction degree can be evaluated by elemental analysis (atomic C/O ratio, ~10), which reveals the existence of a significant amount of oxygen, indicating that reduced graphene oxide is not the same as pristine graphene. Theoretical calculation of the reduction of graphene oxide (the model used for graphene oxide has the graphene decorated with hydroxyl and epoxide groups) suggests that reduction below 6.25% of the area of the graphene oxide (C/O = 16 in atomic ratio) may be difficult in terms of removing the remaining hydroxyl groups [87]. Furthermore, studies revealed that hydrazine can only remove the in-plane functional groups such as epoxy and hydroxyls while leave the edge moieties such as carboxyl and carbonyl intact [75, 88]. Interestingly, these residual edge groups can be further removed by additional exposure to concentrated H_2SO_4 [88].

In order to obtain homogeneous colloidal suspensions of electrically conducting rGO, the reduction of GO by reductants was usually conducted in the presence of polymer or surfactant. Stankovich *et al.* [67] reduced GO suspension in the presence of poly(sodium 4-stryrenesulphonate) to get a black aqueous suspension of rGO sheets coated in the polymer. Tour *et al.* [89] first reduced GO sheets modified with sodium dodecylbenzenesulfonate (SDBS) and then chemically functioned it with aryl diazonium salt leading to SDBS-wrapped rGO that can be dispersible in DMF, N,N'-dimethylacetamide, and NMP. In addition, small organic molecules were used to modify rGO to afford colloidal suspensions. For instance, *Shi et al.* [90] prepared stable aqueous dispersions of rGO sheets using a water-soluble pyrene derivative, 1-pyrenebutyrate (PB^-), as a stabilizer. The pyrene moiety has strong affinity with the basal plane of graphite *via* π-π stacking. On the basis of this dispersion, large area graphene sheets with a conductivity of $\sim2\times10^2$ S/m were obtained by filtration. Park *et al.* [91] also used hydrazine to reduce KOH-modified GO under basic conditions. It was believed that the K^+ ions and the carboxylate anions at the edges of CMG sheets formed ion pairs. The electrical conductivity of the prepared rGO paper was ~ 690 Sm^{-1}.

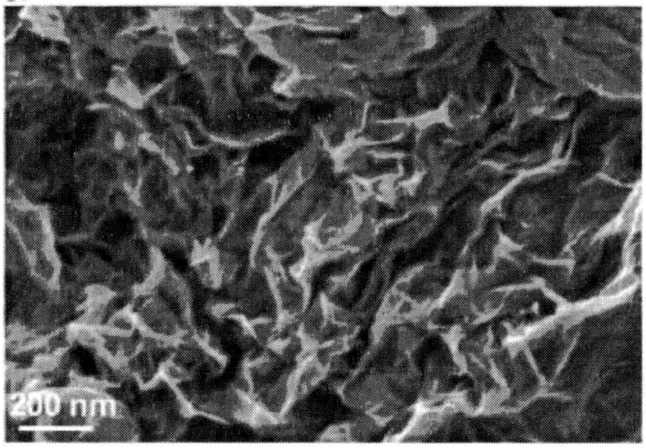

Figure 8. A SEM image showing aggregates of graphene oxide sheets chemically reduced with hydrazine monohydrate. Reproduced with permission from ref. [50] © 2007 Elsevier.

As addressed above, the resulting rGO sheets are all in the form of composites or modification forms. This can efficiently improve dispersibility of the material in various solvents. However, it is not desirable for applications where pure rGO is required. Li *et al.* [75] reported that chemically converted graphene sheets can readily form stable aqueous colloids without the help of stabilizers or surfactants. They reduced GO sheets under basic conditions (pH 10) using hydrazine and the excess hydrazine was removed by dialysis. Since carboxylic acid groups of GO are unlikely to be reduced by hydrazine under the given reaction conditions, these remained groups after reduction could be negatively charged, which led to the formation of well-dispersed rGO colloids (see Figure 9). The rGO paper made by filtration and then dried the rGO colloids in air showed good electrical conductivity (~ 7200 Sm^{-1}). This is a facile approach to large-scale production of aqueous rGO dispersions without the need for polymeric or surfactant stabilizers, making it possible to process graphene materials using low-cost solution processing techniques, opening opportunities to use this unique rGO for many technological applications.

Despite of this progress producing rGO suspensions without stabilizers, the solubility of rGO sheets is still an obstacle to practical applications. The method of Li *et al.* can only lead to rGO suspension with limited solubility in water (< 0.5 mg/mL). Samulski *et al.* [83] reported a chemical route to aqueous solutions of isolated graphene sheets by reducing graphene oxide using three-step reduction. This method removed residual oxygen functionality and introduced sulfonic acid groups in partially reduced GO in a controlled manner. The charged SO^{3-} units prevent the graphitic sheets from aggregating in solution after the final reduction stage of the graphene oxide. This resulted in isolated sheets of lightly sulfonated graphene with improved water solubility (~2 mg/mL in the pH range of 3-10). The electrical conductivity of the rGO thin film on a glass slide after drying at 120 °C was ~1,250 Sm^{-1}.

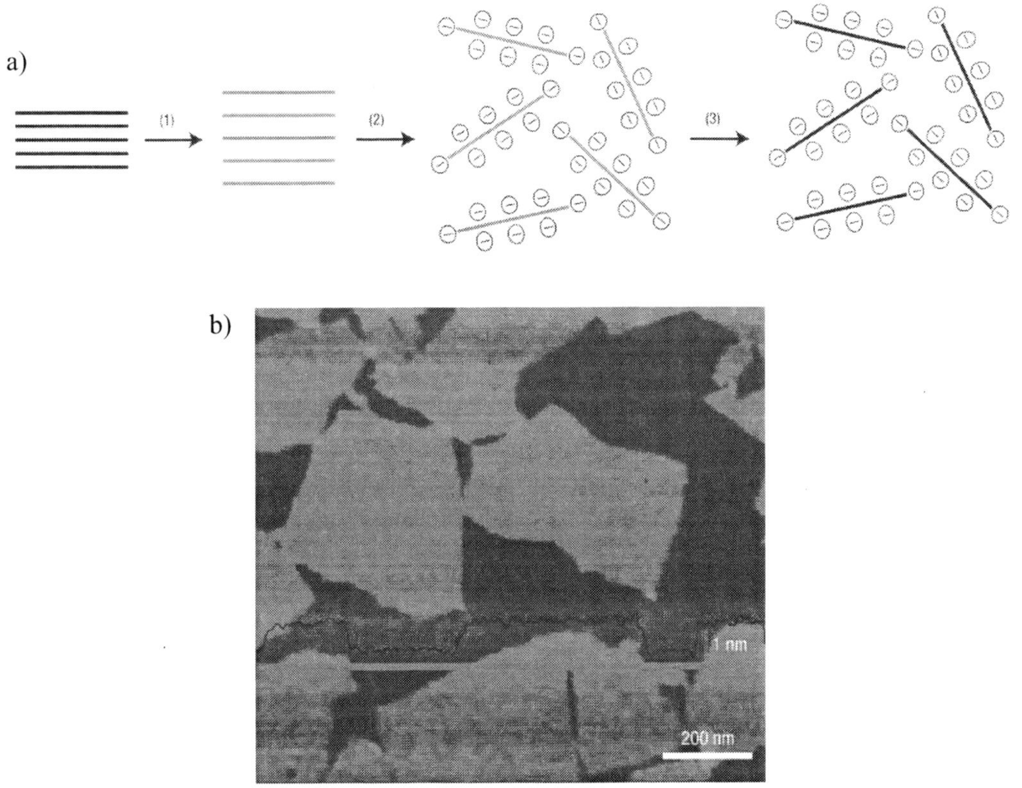

Figure 9. a) Schematic showing the chemical route to synthesis of aqueous graphene dispersions. (1) Oxidation of graphite (black blocks) to graphite oxide (lighter coloured blocks) with greater interlayer distance. (2) Exfoliation of graphite oxide in water by sonication to obtain GO colloids that are stabilized by electrostatic repulsion. (3) Controlled conversion of GO colloids to conducting graphene colloids through deoxygenation by hydrazine reduction. b) Tapping-mode AFM image of CCG sheets together with a cross-sectional profile (blue curve; scale bar, 1 nm) taken along the red line. The sample was prepared by drop-casting a dilute CCG dispersion onto a silicon wafer. Reproduced with permission from ref. [75] © 2008 NPG.

Biomolecules such as DNA and protein can be employed in the reduction process. Mann *et al.* [92] reduced GO suspension with hydrazine in the presence of freshly prepared single-stranded DNA (ssDNA) to produce a stable aqueous suspension of ssDNA-functionalized graphene sheets with concentration as high as 2.5 mg/mL. The negatively charged dispersions can be exploited for the spontaneous assembly of novel graphene-based bio-nanocomposites containing intercalated DNA molecules, or co-intercalated mixtures of DNA and the positively charged redox protein such as cytochrome c (see Figure 10).

Deng *et al.* [93] used bovine serum albumin (BSA) as both a reductant and a stabilizer to prepare BSA-GO/rGO conjugates. They further demonstrated that the incorporation of protein as a "universal glue" molecule successfully turns GO and rGO into general platforms toward the efficient assembly of nanoparticles with varying sizes, shapes, compositions, and surface properties.

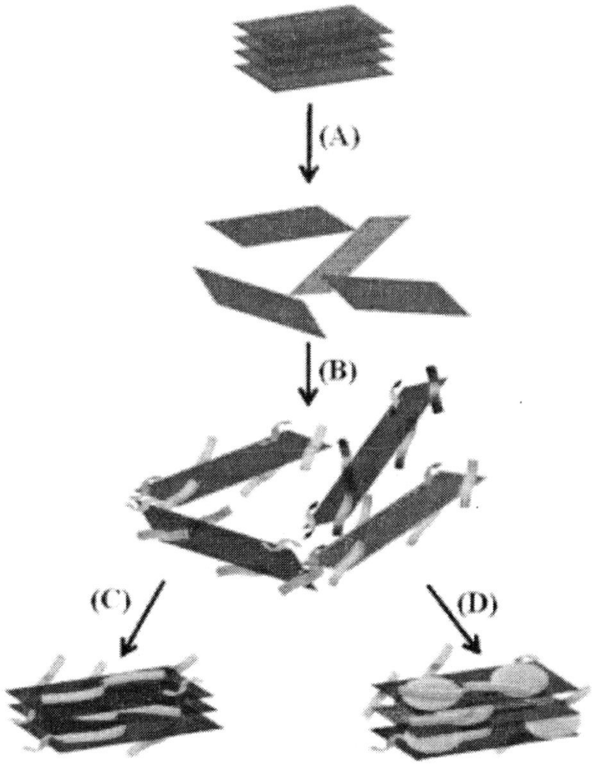

Figure 10. Schematic showing synthesis of DNA-stabilized graphene aqueous suspensions and fabrication of lamellar multifunctional nanocomposites. (A) Oxidative treatment of graphite (grey-black) yields delaminated nanometre-thick sheets of graphene oxide (GO) (brown). (B) Chemical reduction of GO sols with hydrazine in the presence of freshly prepared single-stranded DNA (ssDNA) produces a stable aqueous suspension of ssDNA-unctionalised graphene sheets (ssDNA-G). (C, D) Processing of ssDNA-G dispersions to produce ordered layered nanocomposites. (C) Evaporation induced deposition and self-assembly on flat substrates results in lamellar nanocomposite films with intercalated ssDNA molecules. (D) Co-assembly of negatively charged ssDNA-G sheets and positively charged cytochrome c produces co-intercalated multifunctional layered nanocomposites. Reproduced with permission from ref. [75] © 2009 Wiley.

Thermal treatment is another efficient route to turn insulating GO into electrical conductive rGO. Rapid heating of GO to approximately 1000℃ (heating rate > 2000 °Cmin^{-1}) splits and reduces the graphite oxide, or makes it burst like popcorn, to form a reduced product [51, 94]. Tour *et al.* [95] reported a one-step method to peel individual sheets from thermally converted graphene (TCG, < 5% oxygen content, obtained by the thermal expansion of GO) and concomitantly functionalize the predominantly graphene sheets with various aryl-addends. The expansion and reduction of the material is believed to be driven by the forceful (calculated to be as high as 130 MPa) release of small, gaseous molecules (principally CO and CO_2). Compared to reduction by chemical reductant, the thermal treatment can avoid the introduction of impurity. However, the thermal reaction process may lead to the creation of vacancies in the basal plane due to the evolution of carbon in the form of CO or CO_2 [51]. Thus, the thermal reduction cannot yield complete removal of oxygen, only resulting in partial restoration of sp^2 conjugated graphene network similar to chemical reduction. In other aspect, the measured surface areas, C/O ratios (atomic C/O ratio, 10.3), and conductivity ($1\text{-}2.3\times10^3$ Sm^{-1}) of the rGO by thermal treatment [51, 94] is often similar to that by chemical methods mentioned above and the rGO obtained by thermal approach can also be dispersed in some organic solvents [96].

Other reduction routes include electrochemical reaction [52], photocatalytic reduction (ultraviolet-assisted method) [81], and flash conversion [70]. For instance, Dong *et al.* [52] reported a simple, efficient, low-cost and environmentally friendly electrochemical method to produce electrochemically reduced graphene oxide films with an O/C ratio of less than 6.25 %. Huang *et al.* [97] reported a flash reduction process where GO can be photothermally reduced upon exposure to a pulsed Xenon flash at ambient conditions. In addition, this method could enable patterning of GO or GO/polymer films by flashing through a photomask.

Graphene from Liquid-Phase Exfoliation (LPE)

Liquid-phase exfoliation (LPE) is a method inspired by obtaining dispersions of graphite oxide to yield single layer graphene oxide. Similarly, the LPE process is to make dispersion of graphite in both aqueous solvents and organic solvents such as N,N-dimethylformamide, (DMF) or N-methylpyrrolidone (NMP), which is typically done by exposure of graphite powders in solvents to high intensity ultrasound [24]. LPE can be probably viewed as alternatives to the solid-phase micromechanical cleavage. In this micromechanical approach, solvent (such as DMF or NMP)-graphene interactions may lower the energy barrier of exfoliation by providing a better match of surface energies between the two components [47].

Graphite, graphite intercalation compounds or expandable graphites have been used as starting materials in LPE process. Compared to reduction from GO, the LPE method is believed to produce graphene suspensions with high quality approaching pristine graphene while it is more facile than solid-phase micromechanical cleavage. Coleman *et al.* [24] demonstrated graphene dispersions with concentration up to ~ 0.01 mg ml^{-1}, produced by dispersion and exfoliation of graphite in organic solvents such as N-methyl-pyrrolidone (NMP) (shown in Figure 11), though the concentration of monolayer yield was low(1-12 wt%). It was believed that the energy required to exfoliate graphene is balanced by the solvent-graphene interaction for solvents whose surface energy matches that of graphene. The thin films made of this material had conductivity of ~ 6, 500 S m^{-1} and optical transparency of ~ 42%. The low density of defects or the low concentration of oxides made this solution processing of graphene more favorable to applications from device and sensor fabrication. However, this process still has some drawbacks. These solvents are expensive and require special care when handling. In addition, they tend to have high boiling points, making it difficult to deposit individual monolayers on surfaces.

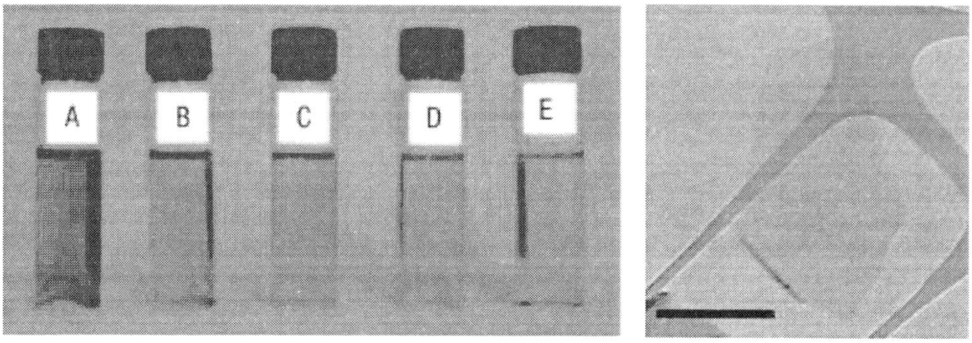

Figure 11. Left: Optical images of dispersions of graphite flakes in NMP, in the range of concentration from 6 mg/ml (A) to 4 mg/ml (E) after centrifugation. Right: Bright-field TEM images of monolayer graphene flakes deposited from NMP. Reproduced with permission from ref. [24] © 2008 NPG.

Dai *et al.* [98] reported that exfoliation-reintercalation-expansion of graphite can produce high quality monolayer graphene sheets in organic solvents. Expandable graphite was firstly exfoliated by rapidly heating to 1,000 °C in forming gas. Then the exfoliated graphite was reintercalated with oleum and inserted by tetrabutylammonium hydroxide (TBA) in DMF. Finally the TBA-inserted oleum-intercalated graphite in a DMF solution of 1,2-distearoyl-sn-glycero-3-phosphoethanolamine-N-[methoxy(polyethyleneglycol)-5000] (DSPE-mPEG) was sonicated for 60 min to form a homogeneous suspension (see Figure 12). This process could produce suspensions with ~90% single layer graphene sheets. The obtained graphene sheets exhibit high electrical conductance at room and cryogenic temperatures. Further, large transparent conducting films could be made by Langmuir-Blodgett assembly of the graphene sheets in a layer-by-layer manner.

Graphite can also be spontaneously exfoliated into single-layer graphene in chlorosulphonic acid, and dissolved at isotropic concentrations as high as ~ 2 mg ml^{-1}, which is an order of magnitude higher than other reported values.. [99] The acid was believed to protonate the graphene to induce repulsion between layers, which is similar to acid protonation of SWNTs. Moreover, chlorosulphonic acid disperses new forms of graphene such as graphene nanoribbons. High-concentration dispersions (~20-30 $mgml^{-1}$) of fully soluble material yield liquid-crystalline phases. Transparent and conducting films made from these dispersions showed 1,000Ω/sq and ~80% transparency.

Thc LPE processing can be conducted in aqueous solutions in a manner similar to surfactant aided carbon nanotube dispersion. Coleman *et al.* [25] exfoliated graphite to give rise of graphene suspended in water with the aid of ultrasonication, resulting in large quantities of multilayered graphene and smaller quantities of monolayer graphene (certificated by transmission electron microscopy (TEM) analysis). Atomic resolution TEM shows the monolayers to be well graphitized and largely defect free (see Figure 13). The exfoliated flakes are stabilized against aggregation by a relatively large potential barrier, which originates from the Coulomb repulsion between surfactant-coated sheets. Kar *et al.* [48] developed a scalable and facile technique for noncovalent functionalization of graphene with 1-pyrenecarboxylic acid (PCA) that exfoliated single-, few-, and multilayered graphene flakes that retain their high quality into stable aqueous dispersions. The exfoliation was achieved by the noncovalent functionalization of graphene with PCA in a polarity-controlled combination of media. The exfoliation mechanism was established using stringent control experiments and detailed characterization steps. Using the exfoliated graphene, they demonstrated highly sensitive and selective conductometric sensors (whose resistance rapidly changes >10 000% in saturated ethanol vapor), and ultracapacitors with extremely high specific capacitance (~ 120 F/g), power density (~ 105 kW/kg), and energy density (~ 9.2 Wh/kg).

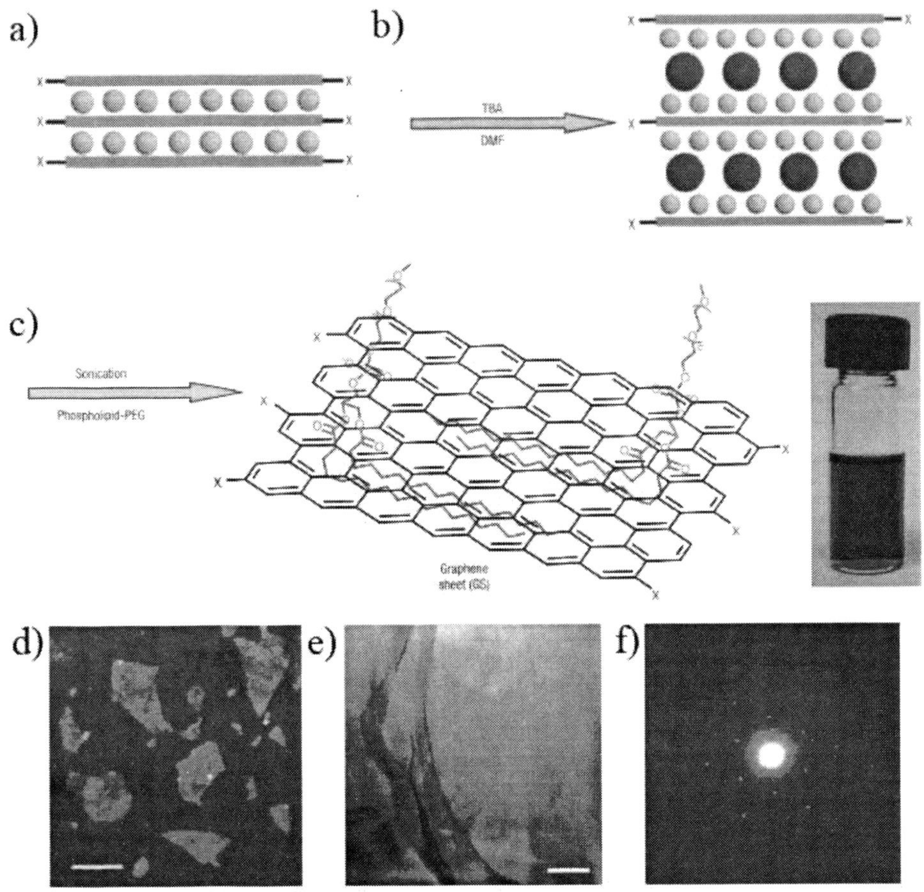

Figure 12. Chemically derived single-layer graphene sheets from solution phase. (a), Schematic of the exfoliated graphite reintercalated with sulphuric acid molecules (teal spheres) between the layers. (b) Schematic of TBA (blue spheres) insertion into the intercalated graphite. (c) Schematic of GS coated with DSPE-mPEG molecules and a photograph of a DSPE-mPEG/DMF solution of GS. (d) An AFM image of a typical graphene sheet several hundred nanometres in size and with a topographic height of ~1 nm. The scale bar is 300 nm. (e) Low- magnification TEM images of a typical graphene sheet several hundred nanometres in size. The scale bar is 100 nm. (f) Electron diffraction (ED) pattern of an as-made graphene sheet as in (e), showing excellent crystallization of the graphene sheet. Reproduced with permission from ref. [98] © 2008 NPG.

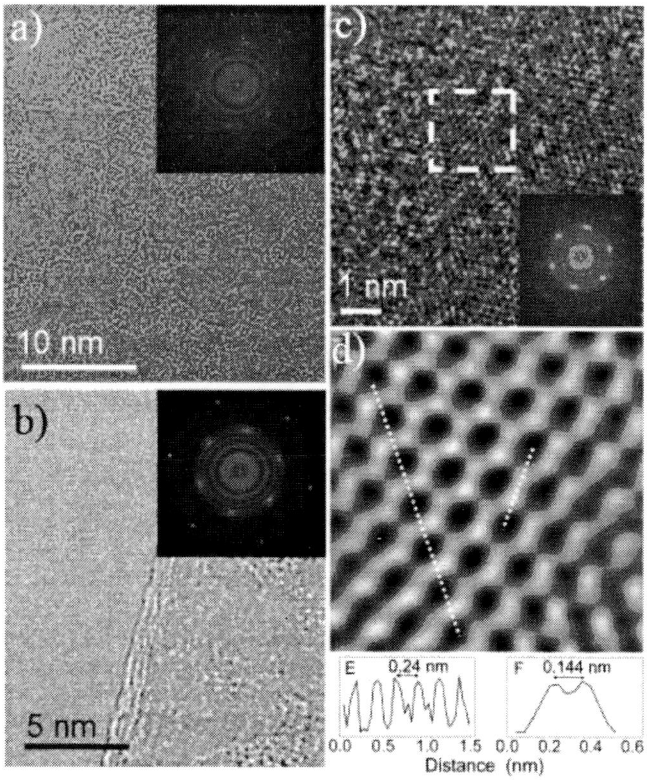

Figure 13. High-resolution TEM images of surfactant exfoliated graphene flakes. (a) A HRTEM image of a section of a graphene monolayer. Inset: Fast Fourier transform (equivalent to an electron diffraction pattern) of the image. (b) HRTEM image of a section of a trilayer. Inset: Fast Fourier transform of the image. (c) HRTEM image of part of a graphene monolayer. Inset: Fast Fourier transform of the region enclosed by the white square. The scale bar is 1 nm. (d) A filtered image of part of the region in the white square. (e) Intensity analysis along the left white dashed line shows a hexagon width of 2.4 Å. (f) Intensity analysis along the right white dashed line shows a C-C bond length of 1.44 Å. Reproduced with permission from ref. [98] © 2009 ACS.

CHALLENGES AND PERSPECTIVES

Graphene has already attracted widespread attention in the last several years. The unique properties of graphene make it a potential candidate for next-generation electronics and holds great promise for various applications such as ultrafast photodectors, transparent conductive electrodes, sensors, supercapacitors, and graphene-based composites. It is known that the realization of graphene-based technologies is highly dependent on the development of synthetic methods of this 2-dimensional material. Since the breakthrough report of graphene in 2004 by Novoselov and Geim, various methods have been developed and produced graphene sheets with tunable size and yield. However, fine thickness and crystalline structure controls over an entire graphene layer remain major challenges towards the realization of graphene-based technologies, in particular the applications related to electronics. Regarding the solution processing, the suspensions of CMGs hold great promise in the practical applications without requiring excellent electron transport property due to high yield and ready functionalization. Compared to CVD and surface segregation methods, this method generally does not involve high temperature process or complicated chemical post treatments.

However, the resulting rGO often contains a significant amount of oxygen species and significant numbers of defects, limiting the carrier transport abilities. Compared with reduction by reducing agent, thermal annealing can lead to enhanced transport properties though the elevated temperature is not desirable for the real applications. The liquid-phase exfoliation for producing graphene sheds a light on the synthesis of graphene from a solution process with fewer defects. But the yield of liquid-phase exfoliation is very low and less amenable to functionalization as the graphene sheets derived from graphene oxide. Certainly, the final choice of methods should depend on their use. For instance, the remained oxygen groups on rGO may be very useful when the applications focus on further chemical covalent functionalizations. In addition, the intermediate product, graphene oxide, is often referred to as a disordered material. But it is this inherent disorder induced by the presence of functional groups that provides opportunities for tailoring its chemical functionality as well as its optoelectronic properties. The insulating graphene oxide can even be viewed as semiconductors with wide energy gap with possibility to tailor it. Thus the energy gap opening of graphene might be realized by functionalization using graphene oxide as starting materials, which might be more efficient and much easier than preparation of graphene nanoribbons by physical methods with very low yield and difficulty to control the edges.

Despite significant progress has been made in understanding the structure, processing, and properties of chemical graphenes, the functionalization of the graphene layers, control of the number of graphene layers, and defect engineering remain to be explored towards the realization of the unique applications of graphene-based nanomaterials.

REFERENCES

[1] A. K. Geim and K. S. Novoselov, *Nature Mater.* **6**, 183 (2007).
[2] H. P. Boehm, A. Clauss, G. O. Fischer, U. Hofmann and Z. Naturforschung, **17**b, 150 (1962).
[3] J. W. May, *Surf. Sci.* **17**, 267 (1969).
[4] H. P. Boehm, R. Setton and E. Stumpp, *Carbon* **24**, 241(1986).
[5] K. S. Novoselov, A. K. Geim, S. V. Morozov, D. Jiang, Y. Zhang, S. V. Dubonos, I. V. Grigorieva and A. A. Firsov, *Science* **306**, 666 (2004).
[6] K. S. Novoselov, D. Jiang, F. Schedin, T. J. Booth, V. V. Khotkevich, S. V. Morozov and A. K. Geim, *Proc. Nat. Acad. Sci. U. S. A.* **102**, 10451 (2005).
[7] K. S. Novoselov, A. K. Geim, S. V. Morozov, D. Jiang, M. I. Katsnelson, I. V. Grigorieva, S. V. Dubonos and A. A. Firsov, *Nature* **438**, 197 (2005).
[8] Y. B. Zhang, Y. W. Tan, H. L. Stormer and P. Kim, *Nature* **438**, 201 (2005).
[9] S. Park and R. S. Ruoff, *Nat. Nanotechnol.* **4**, 217 (2009).
[10] A. K. Geim and P. Kim, *Sci. Am.* **298**, 68 (2008).
[11] M. D. Stoller, S. Park, Y. Zhu, J. An and R. S. Ruoff, *Nano Lett.* **8**, 3498 (2008).

[12] K. I. Bolotina, K. J. Sikesb, Z. Jiang, M. Klimac, G. Fudenberga, J. Honec, P. Kima, H. L. Stormer, *Solid State Commun.* **146**, 351 (2008).
[13] C. Lee, X. Wei, J. W. Kysar and J. Hone, *Science* **321**, 385 (2008).
[14] A. A. Balandin, S. Ghosh, W. Z. Bao, I. Calizo, D. Teweldebrhan, F. Miao and C. N. Lau, *Nano Lett.* **8**, 902 (2008).
[15] W. Cai, Y. Zhu, X. Li, R. D. Piner and R. S. Ruoff, *Appl. Phys. Lett.* **95**, 123115 (2009)
[16] F. Schwierz, *Nat. Nanotechnol.* **5**, 487 (2010).
[17] Y. Gao, H. L. Yip, S. K. Hau, K. M. O'Malley, N. C. Cho, H. Z. Chen and A. K. -Y. Jen, *Appl. Phys. Lett.* **97**, 203306 (2010).
[18] Y. Gao, H. L. Yip, K. S. Chen, K. M. O'Malley, O. Acton, Y. Sun, G. Ting, H. Z. Chen and A. K. -Y. Jen, *Adv. Mater.* DOI: 10.1002/adma.201100065.
[19] Y. Ohno, K. Maehashi and K. Matsumoto, *J. Am. Chem. Soc.* **132**, 18012 (2010).
[20] X. Yang, M. S. Xu, W. M. Qiu, X. Q. Chen, M. Deng, J. L. Zhang, H. Iwai, E. Watanabe and Hongzheng Chen, *J. Mater. Chem.* DOI: 10.1039/c1jm10697j.
[21] D. S. Yu and L. M. Dai, *J. Phys. Chem. Lett.* **1**, 467 (2010).
[22] D. C. Wei and Y. Q. Liu, *Adv. Mater.* **22**, 3225 (2010).
[23] X. Li, X. Wang, L. Zhang, S. Lee and H. J. Dai, *Science* **319**, 1229 (2008).
[24] Y. Hernandez, V. Nicolosi, M. Lotya, F. M. Blighe, Z. Y. Sun, S. De, I. T. McGovern, B. Holland, M. Byrne, Y. K. Gun'Ko, J. J. Boland, P. Niraj, G. Duesberg, S. Krishnamurthy, R. Goodhue, J. Hutchison, V. Scardaci, A. C. Ferrari and J. N. Coleman, *Nat. Nanotechnol.* **3**, 563 (2008).
[25] M. Lotya, Y. Hernandez, P. J. King, R. J. Smith, V. Nicolosi, L. S. Karlsson, F. M. Blighe, S. De, Z. M. Wang, I. T. McGovern, G. S. Duesberg and J. N. Coleman, *J. Am. Chem. Soc.* **131**, 3611 (2009).
[26] C. Vallés, C. Drummond, H. Saadaoui, C. A. Furtado, Maoshuai He, O. Roubeau, L. Ortolani, M. Monthioux and APénicaud, *J. Am. Chem. Soc.* **130**, 15802 (2008).
[27] S. Stankovich, D. A. Dikin, G. H. B. Dommett, K. M. Kohlhaas, E. J. Zimney, E. A. Stach, R. D. Piner, S. T. Nguyen and R. S. Ruoff, *Nature* **442**, 282 (2006).
[28] X. Li, H. Wang, J. T. Robinson, H. Sanchez, G. Diankov and H. Dai, *J. Am. Chem. Soc.* **131**, 15939 (2009).
[29] B. C. Brodie, *Ann. Chim. Phys.* **59**, 466 (1860).
[30] C. Mattevi, G. Eda, S. Agnoli, S. Miller, K. A. Mkhoyan, O. Celik, D. Mastrogiovanni, G. Granozzi, E. Garfunkel and M. Chhowalla, *Adv. Funct. Mater.* **19**, 2577 (2009).
[31] W. W. Cai, R. D. Piner, F. J. Stadermann, S. J. Park, M.A. Shaibat, Y. Ishii, D. X. Yang, A. Velamakanni, S. J. An, M. Stoller, J. An, D. M. Chen and R. S. Ruoff, *Science* **321**, 1815 (2008).
[32] G. Eda, G. Fanchini and M. Chhowalla, *Nat. Nanotechnol.* **3**, 270 (2008).
[33] A. E. Karu and M. Beer, *J. Appl. Phys.* **37**, 2179 (1966).
[34] A. N. Obraztsov, E. A. Obraztsova, A. V. Tyurnina and A. A. Zolotukhin, *Carbon* **45**, 2017 (2007).
[35] K. S. Kim, Y. Zhao, H. Jang, S. Y. Lee, J. M. Kim, K.S. Kim, J. -H. Ahn, P. Kim, J. -Y. Choi and B. H. Hong, *Nature* **457**, 706 (2009).
[36] A. Reina, X. T. Jia, J. Ho, D. Nezich, H. Son, V. Bulovic, M. S. Dresselhaus and J. Kong, *Nano Lett.* **9**, 30 (2009).

[37] S. Bae, H. Kim, Y. B. Lee, X. F. Xu, J. -S. Park, Y. Zheng, J. Balakrishnan, T. Lei, H. R. Kim, Y. Song, Y. -J. Kim, K. S. Kim, B. Özyilmaz, J. -H. Ahn, B. H. Hong and S. Iijima, *Nat. Nanotechnol.* **4**, 574 (2010).
[38] C. Berger, Z. M. Song, T. B. Li, X. B. Li, A. Y. Ogbazghi, R. Feng, Z. T. Dai, A. N. Marchenkov, E. H. Conrad, P. N. First and W. A. de Heer, *J. Phys. Chem. B.* **108**, 19912 (2004).
[39] P. W. Sutter, J. I. Flege and E. A. Sutter, *Nature Mater.* **7**, 406 (2008).
[40] K. V. Emtsev, A. Bostwick, K. Horn, J. Jobst, G. L. Kellogg, L. Ley, J. L. McChesney, T. Ohta, S. A. Reshanov, J. Röhrl, E. Rotenberg, A. K. Schmid, D. Waldmann, H. B. Weber and T. Seyller, *Nature Mater.* **8**, 203 (2009).
[41] M.Choucair, P. Thordarson and J. A. Stride, *Nature Nanotechnol.* **4**, 30 (2009).
[42] X. Wang, L. J. Zhi, N. Tsao, Ž. Tomović, J. L. Li, K. Müllen, *Angew. Chem.* **47**, 2990 (2008).
[43] J. Wu, W. Pisula and K. Mullen, *Chem. Rev.* **107**, 718 (2007).
[44] A. Reina, H. Son, L. Y. Jiao, B. Fan, M. S. Dresselhaus, Z. F. Liu and J. Kong, *J. Phys. Chem. C* **112**, 17741 (2008).
[45] A. Vijayaraghavan, C. Sciascia, S. Dehm, A. Lombardo, A. Bonetti, A. C. Ferrari and R. Krupke, *ACS Nano* **3**, 1729 (2009).
[46] M. S. Xu, D. Fujita, K. Sagisaka, E. Watanabe and N. Hanagata, *ACS Nano* **5**, 1522 (2011).
[47] J. N. Coleman, *Adv. Funct. Mater.* **19**, 3680 (2009).
[48] X. H. An, T. Simmons, R. Shah, C. Wolfe, K. M. Lewis, M. Washington, S. K. Nayak, S. Talapatra and S. Kar, *Nano Lett.* **10**, 4295 (2010).
[49] Y. T. Liang and M. C. Hersam, *J. Am. Chem. Soc.*, **132**, 17661 (2010).
[50] S. Stankovich, D. A. Dikin, R. D. Piner, K. A. Kohlhaas, A. Kleinhammes, Y. Jia, Y. Wu, S. T. Nguyen and R. S. Ruoff, *Carbon* **45**, 1558 (2007).
[51] H. C. Schniepp, J. -L. Li, M. J. McAllister, H. Sai, M. Herrera-Alonso, D. H. Adamson, R. K. Prud'homme, R. Car, D. A. Saville and I. A. Aksay, *J. Phys. Chem. B* **110**, 8535 (2006).
[52] Z. Ming, W. Yuling, Z. Yueming, Z. Junfeng, R. Wen, W. Fuan and D. Shaojun, *Chem. Eur. J.* **15**, 6116 (2009).
[53] C. N. Rao, A. K. Sood, R. Voggu and K. S. Subrahmanyam, *J. Phys. Chem. Lett.* **1**, 572 (2010).
[54] H. Ueta, M. Saida, C. Nakai, Y. Yamada, M. Sasaki and S. Yamamoto, *Surf. Sci.* **560**, 183 (2004).
[55] J. Coraux, A. T. Ndiaye, C. Busse and T. Michely, *Nano Lett.* **8**, 565 (2008).
[56] A. L. V. Parga, F. Calleja, B. Borca, M. C. G. Passeggi, J. J. Hinarejos, F. Guinea and R. Miranda, *Phys. Rev. Lett.* 2008, **100**, 056807 (2008).
[57] A. G. Starodubov, M. A. Medvetskii, A. M. Shikin and V. K. Adamchuk, *Phys. Solid State* **46**, 1340 (2004).
[58] X. S. Li, W. W. Cai, J. H. An, S. Kim, J. Nah, D. X. Yang, R. D. Piner, A. Velamakanni, I. Jung, E. Tutuc, S. K. Banerjee, L. Colombo and R. S. Ruoff, *Science* **324**, 1312 (2009).
[59] A. Dato, V. Radmilovic, Z. Lee, J. Phillips and M. Frenklach, *Nano Lett.* **8**, 2012 (2008).

[60] J. Wintterlin and M. L. Bocquet, *Surf. Sci.* **603**, 1841 (2009).
[61] F. Bonaccorso, Z. Sun, T. Hasan and A. C. Ferrari, *Nature Photonics* **4**, 611 (2010).
[62] Y. H. Hu, H. Wang and B. Hu, *ChemSusChem* **3**, 782 (2010).
[63] M. S. Xu, D. Fujita and N. Hanagata, *small* **5**, 2638 (2009).
[64] V. Yong and J. M. Tour, *small* **6**, No. 2, 313 (2010).
[65] D. R. Dreyer, R. S. Ruoff and C. W. Bielawski, *Angew. Chem. Int. Ed.* **49**, 9336 (2010).
[66] IUPAC in Compendium of Chemical Terminology (Eds.: A. D. McNaught, A. Wilkinson), 2nd ed., Blackwell Scientific, Oxford, 1997.
[67] S. Stankovich, R. D. Piner, X. Q. Chen, N. Q. Wu, S. T. Nguyen and R. S. Ruoff, *J. Mater. Chem.* **16**, 155 (2006).
[68] L. Staudenmaier, *Ber. Dtsch. Chem. Ges.* **31**, 1481(1898).
[69] W. S. Hummers and R. E. Offeman, *J. Am. Chem. Soc.* **80**, 1339 (1958).
[70] U. Hofmann, *Kolloid-Z.* 61, 297 (1932).
[71] A. Buchsteiner, A. Lerf and J. Pieper, *J. Phys. Chem. B* **110**, 22328 (2006).
[72] I. Jung, M. Pelton, R. Piner, D. A. Dikin, S. Stankovich, S. Watcharotone, M. Hausner and R. S. Ruoff, *Nano Lett.* **7**, 3569 (2007).
[73] H. He, J. Klinowski, M. Forster and A. Lerf, *Chem. Phys. Lett.* **287**, 53 (1998).
[74] G. Eda and M. Chhowalla, *Adv. Mater.* **22**, 2392 (2010).
[75] D. Li, M. Müller, S. Gilje, R. Kaner and G. Wallace, *Nat. Nanotechnol.* **3**, 101 (2008).
[76] J. P. Zhao, S. F. Pei, W. C. Ren, L. B. Gao and H. M. Cheng, *ACS Nano*, **4**, 5245 (2010).
[77] X. Sun, Z. Liu, K. Welsher, J. T. Robinson, A. Goodwin, S. Zaric and H. J. Dai, *Nano Res.* **1**, 203 (2008).
[78] H. P. Boehm, M. Eckel, W. Scholz and Z. Anorg, *Allgem. Chem.* **353**, 236 (1967)
[79] L. Zhang, J. J. Liang, Y. Huang, Y. F. Ma, Y. Wang, Y. S. Chen, *Carbon* **47**, 3365 (2009).
[80] J. Y. Luo, L. J. Cote, V. C. Tung, A. T. L. Tan, P. E. Goins, J. S. Wu, and J. X. Huang, *J. Am. Chem. Soc.* **132**, 17667 (2010).
[81] G. Williams, B. Serger and P. V. Kamat, *ACS Nano* **2**, 1487 (2008).
[82] V. C. Tung, M. J. Allen, Y. Yang and R. B. Kaner, *Nat. Nanotechnol.* **4**, 25 (2009).
[83] Y. Si and E. T. Samulski, *Nano Lett.* **8**, 1679 (2008).
[84] N. Mohanty, A. Nagaraja, J. Armesto and V. Berry, *Small* **6**, 226 (2009).
[85] A. B. Bourlinos, D. Gournis, D. Petridis, T. Szabo, A. Szeri and I. Dekany, *Langmuir* **19**, 6050 (2003).
[86] Y. Chen, X. Zhang, P. Yu and Y. Ma, *Chem. Commun.* 4527 (2009).
[87] D. W. Boukhvalov and M. I. Katsnelson, *J. Am. Chem. Soc.* **130**, 10697 (2008).
[88] W. Gao, L. B. Alemany, L. Ci and P. M. Ajayan, *Nature Chem.* **1**, 403 (2009).
[89] J. R. Lomeda, C. D. Doyle, D. V. Kosynkin, W.-F. Hwang and J. M. Tour, *J. Am. Chem. Soc.* **130**, 16201 (2008).
[90] Y. X. Xu, H. Bai, G. W. Lu, C. Li and G. Q. Shi, *J. Am. Chem. Soc.* **130**, 5856 (2008).
[91] S. J. Park, J. An, R. D. Piner, I. Jung, D. X. Yang, A. Velamakanni, S. T. Nguyen and R. S. Ruoff, *Chem. Mater.* **20**, 6592 (2008).
[92] A. J. Patil, J. L. Vickery, T. B. Scott and S. Mann, *Adv. Mater.* **21**, 3159 (2009).
[93] J. B. Liu, S. H. Fu, B. Yuan, Y. L. Li and Z. X. Deng, *J. Am. Chem. Soc.* **132**, 7279 (2010).

[94] M. J. McAllister, J. L. Li, D. H. Adamson, H. C. Schniepp, A. A. Abdala, J. Liu, M. Herrera-Alonso, D. L. Milius, R.Car, R. K. Prud'homme and I. A. Aksay, *Chem. Mater.* **19**, 4396 (2007).
[95] Z. Jin, J. R. Lomeda, B. K. Price, W. Lu, Y. Zhu and J. M. Tour, *Chem. Mater.* **21**, 3045 (2009).
[96] T. Ramanathan1, A. A. Abdala, S. Stankovich, D. A. Dikin, M. Herrera-Alonso, R. D. Piner, D. H. Adamson, H. C. Schniepp, X. Chen, R. S. Ruoff, S. T. Nguyen, I. A. Aksay, R. K. Prud'Homme and L. C. Brinso, *Nat. Nanotechnol.* **3**, 327 (2008).
[97] L. J. Cote, R. Cruz-Silva, J. Huang, *J. Am. Chem. Soc.* **131**, 11027 (2009).
[98] X. L. Li, G. Y. Zhang, X. D. Bai, X. M. Sun, X. R. Wang, E. G. Wang and H. J. Dai, *Nat. Nanotechnol.* **3**, 538 (2008).
[99] N. Behabtu, J. R. Lomeda, M. J. Green, A. L. Higginbotham, A. Sinitskii, D. V. Kosynkin, D. Tsentalovich, A. N. G. P. Vasquez, J. Schmidt, E. Kesselman, Y. Cohen, Y. Talmon, J. M. Tour and M. Pasquali, *Nat. Nanotechnol.* **5**, 406 (2010).

In: Graphene: Properties, Synthesis and Applications
Editor: Zhiping Xu
ISBN: 978-1-61470-949-7

Chapter 6

SYNTHESIS AND APPLICATION OF GRAPHENE IN SOLAR CELLS AND COMPOSITES

***Guifeng Fan*[1]*, Kunlin Wang*[1] *and Hongwei Zhu*[1,2*]**
[1]Key Laboratory for Advanced Materials Processing Technology and Department of Mechanical Engineering
Tsinghua University, Beijing, P. R. China
[2]Center for Nano and Micro Mechanics(CNMM), Tsinghua University, Beijing, P. R. China

Abstract

Graphene,a single atomic layer of two-dimensional carbon nanomaterials, owning excellent electrical, thermal and mechanical properties, has attracted intensive attentions since it was first separated by mechanical cleavage. However, this method can't provide large area graphene for research and application. To find a low cost preparation method that yields large scale and high quality graphene seems significantly important. Rapid progress has been made recently.Large area graphene films with thickness varying from single layer to few layers have been synthesized. These films, which show high conductance, transmittance and flexibility, have potential applicationin electronics and energy industry. However, some challenges must be overcome before industrial production. For example, there exists a gap between theoretical properties and practical properties due to the defects present in the film.Furthermore, some synthesis methods will raise pollution. Similarly, there are both opportunities and challenges for practical application.Solar cells composed of graphene show potential application prospects but required further improvement before practical in industry. Besides, graphene based composites display distinct properties, thanks to the unique structure of graphene.In this chapter, current status of graphene synthesis, including achievement and challenge, will be reviewed comprehensively. Oxidation reduction, liquid-phase exfoliation and chemical vapor deposition, which are promising for future research and application, will

[*]E-mail address: hongweizhu@tsinghua.edu.cn

be introduced in details. Besides, research on graphene based solar cells and composites will be reviewed as well.

Keywords: Graphene, Synthesis and transfer, Solar cells, Composites.

1. INTRODUCTION

Graphene, a single atomic layer of carbon hexagon, has aroused global interest since it was successfully separated in 2004. During the single year of 2008, there were about 1500 articles on graphene[1](or, equivalently, four papers per day). What makes graphene so attractive?

Graphene is known as the thinnest material,yet it owns the highest intensity at the same time[2]. With its conductivity of 10^6 S/m,graphene is the best conductive material at room temperature. Besides, the thermal conductivity of graphene reaches 5000 W/m·K[3], being five times higher than diamond. Researchers also find that graphene shows outstanding performances on gas isolation[2,4], hydrogen storage[5], photoelectricity[6-8], photocatalysis[9,10], semiconductors[11], etc. There are growing opportunities to discover the properties and application of this marvelous planar honeycomb lattice.

The electrical property of graphene is particularly eye-catching. For example, the electronic mobility of graphene can reach up to 2×10^5 $cm^2/V\cdot s$ after current annealing[12], while the electronic mobility of silicon is only 1500 $cm^2/V\cdot s$. That relatively low mobility limits furtherimprovement of semiconductor devices. From this point, substituting graphene for silicon will make a big progress in semiconductorindustry. Besides, the behavior of particle in graphene is special. Electron hole pairs travel in graphene with constant speed, masslessly, just like photon travels through space. This phenomenon involves special quantum electrodynamics combining with relativity theory[1]. Moreover, graphene is a semi-metal, with zero bandgap. These interesting propertiesattract dozens of physicists, for the possibilityof applying graphene to high energy physics and relativity physics research.

It is the unique structure of graphene that creates the wonder. Underhigh resolution transmission electron microscope (HRTEM)or scanning tunnelling microscope (STM), graphene is an atomically single layer structure, with carbon atoms closely arranging into a two-dimensionalhoneycomb lattice. In fact, the plane of graphene is different from graphite (Figure 1a).All carbon atom sites in graphene are equivalent (Figure 1b)[13]. In the past, both experiment and theory had demonstrated the impossibility of perfect two-dimensional crystal structure in a free state, due to the thermodynamic instability. J.C. Meyer et al demonstrated the structure of graphene was a rolling "quasi" plane while freestanding (Figure 1c & 1e)[14]. Through scanning electron microscope (SEM), graphene shows a smooth two-dimensional plane (in black) with many folds (in white) (Figure 1d).

For long, scientists have investigatedsufficiently on zero-dimensional (0D), one-dimensional (1D) and three-dimensional (2D) crystal structure materials, but only theory for 2Dmaterials. The discovery of graphene has undoubtedly opened the door to two-dimensional materials.Furthermore, graphene can be viewed as the foundation of carbon material family. While stacking graphene layer by layer, the result is three-dimensional graphite crystal; while rolling graphene into cylinder, the result is one-dimensional carbon nanotubes; while rolling graphene into sphere like a soccer ball, the result is zero-dimension fullerenes (Figure 2)[15].

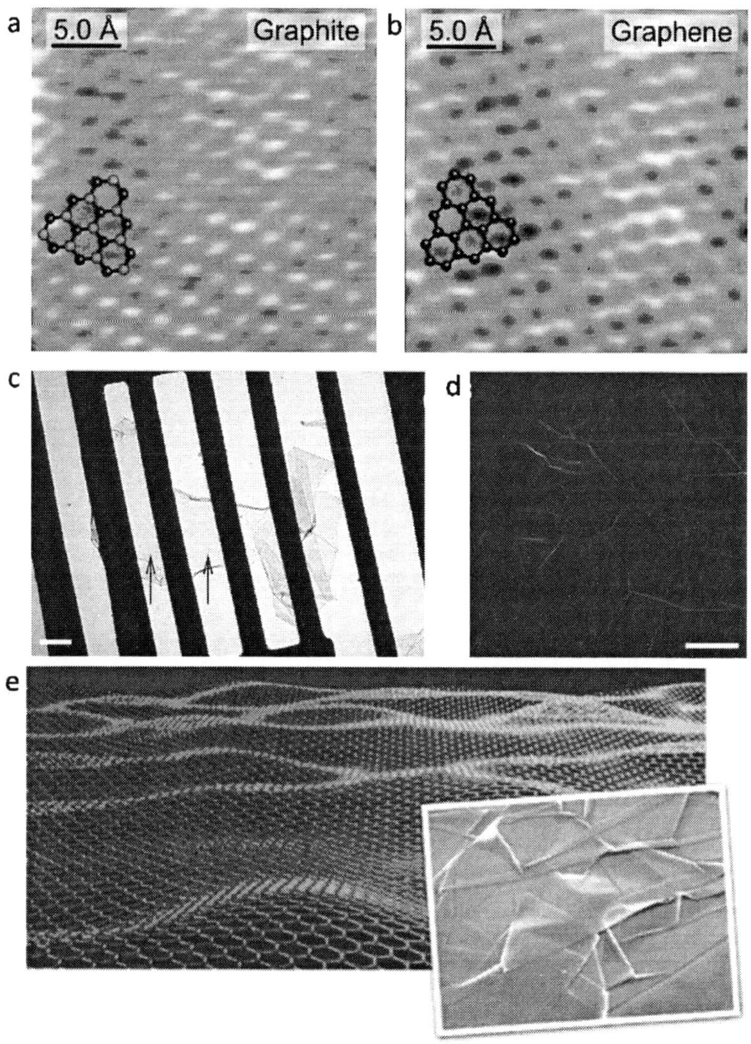

Figure 1.Structure and morphology of graphene.STM images of (a) graphite consisted of two non-equivalent atoms and (b) graphene with all carbon atom sites are equivalent. (c) Bright-field TEM image of a suspended graphene membrane. Its central part is homogeneous and its edges scrolled up. Scale bar, 500 nm. (d) SEM image of graphene on Si substrate. Scale bar, 2μm. (e) Simulated structure and topography of rolling quasi plane and microscope image at right.

It should be noted that there are defects and impurities in practical graphene. Performances of this graphene require further improvement. Besides, some nature characteristics of graphene become the obstacles of application. For example, graphene semiconductor devices are one of the most attractive applications. However, that is impossible before we change graphene from semi-metal to semiconductor, by opening its bandgap. There is still a long way to go.

Controllable synthesis and property improvement may be the firsttwo hot topics in ten years. In thischapter, attempts to prepare large scale, high quality graphene will be introduced. Besides, some promising applications, including graphene solar cells and graphene based composites, will be reviewed in a brief. Some representative devices, such as field emit transistors, however, will not be covered.

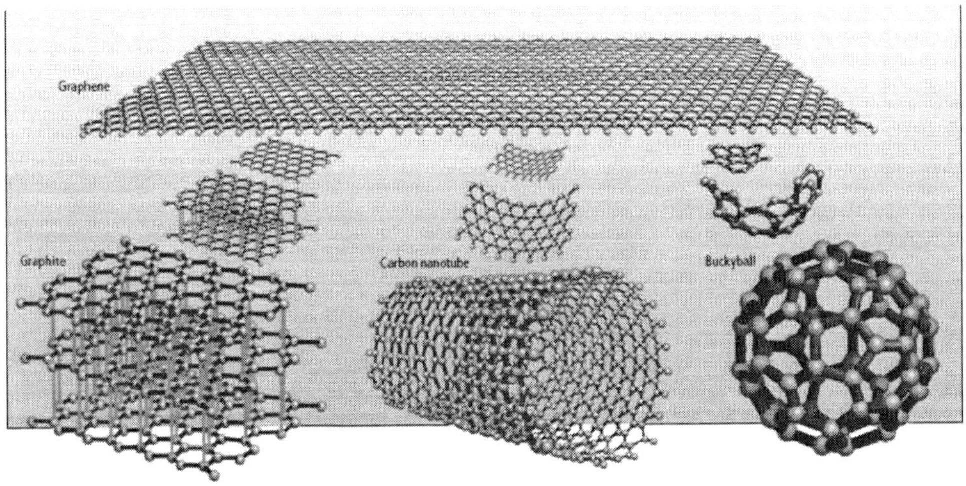

Figure 2.Graphene can be viewed as the foundation of carbon material family.

2. SYNTHESIS AND TRANSFER

Graphene was first exfoliatedfrom highly-oriented pyrolytic graphite (HOPG)by A.K. Geim and his team from Manchester University in 2004.It is a simple but ingenious method, named mechanical cleavage.First, 5 μm-deep mesas were prepared through dry etching platelets of HOPG with a thickness of 1 mm in oxygen plasma. By pressing and baking, the mesas were transferred to the photoresist layer. After cleaving them off the rest of HOPG, use scotch tape to peel flake of graphite off the mesas repeatedly.Finally,graphene is obtained[16]. Though simple, mechanical cleavage can't provide graphene large enough for research and application. Researchers have found other effective methods to prepare graphene. These methods can be classified into three groups by the synthesis principle. That is, exfoliating graphite method, direct growth method and carbon nanotube conversion method. Theexfoliating graphite method use graphite as raw material and adopt different approches, such as mechanical cleavage, oxidation and reduction, and chemical catalytic dispersed, to overcome the Van der Waals force between layers to obtain individual graphene sheets. The direct growth method makes carbon sourcesdecompose and compose under some certain conditions, such as high temperature, rapid cooling and specific gas atmosphere, to grow graphene on certain substrates. The carbon nanotube conversion method transforms nanotubes into graphene nanoribbons by cutting the nanotubes along the axis. So far, the oxidation and reduction method, liquid-phase exfoliation method and chemical vapor deposition (CVD) method have been considered to bemost effective. Besides, transforming nanotubes into graphene nanoribbons,which can display nonzero bandgap, could be promising for further application.

2.1. Oxidation and Reduction

The first step of this method is transforming graphite into oxidation graphite. Oxidation graphite is a layered covalent compound, the gap between layers differing from preparation methods. It can be regard as a stacking structure of graphene with oxygen functional groups on surfaces and edges[17]. Followed by expanding the gap, exfoliating, dispersing and deoxidizing in sequence, the result is graphene sheets dispersedin solution (Figure 3). In fact, Brodiefirst prepared exfoliated graphene oxides in 1859[18,19]. Back in 1950s, Hummers et al studied the oxidation of graphite and developed methods to prepare oxidation graphite. There are many effective methods to prepare oxidation graphite with various performances so far. In graphene preparation, stable dispersion and effective reduction should be the two key factors.

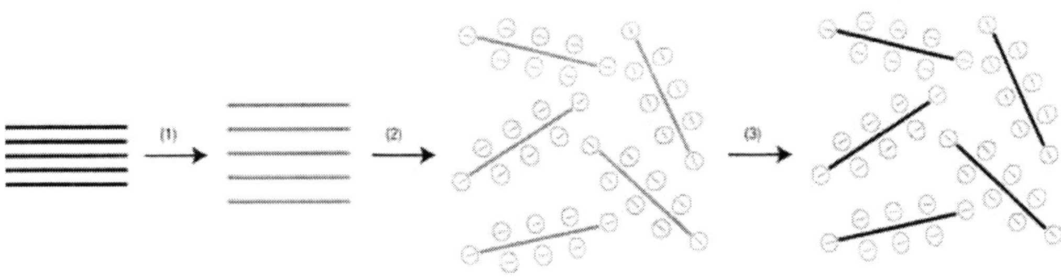

Figure 3.The process of oxidation and reduction method.Starting from graphite, graphene colloids obtained after (1)oxidation, (2)sonication and (3)deoxygenation.

In 2005, S.Stankovich et al first reported stable aqueous dispersions of graphitic nanoplatelets coated with polymer. This method made use of poly (sodium 4-styrenesulfonate) to keep graphitic nanoplatelets disperse stable in any kind of solution[20]. If these graphitic nanoplateletswere exfoliated completely, graphene could be obtained. Later the group achieved this goal and prepared a colloidal suspension ofgraphene oxide sheets (Figure4)[21]. Not only graphene composites can be obtained from this colloidal suspension, but also graphene oxide thin films.

Graphene membrane has great significance because it can be used as various functional layers, such as cover layers, chemical filter layers, battery electrodes, and supercapacitors. In 2007, D.A. Dikin et al prepared graphene oxide thin films by filtering graphene oxide colloid with inorganic membrane[17]. Control the concentration could control the film thickness. Preliminary characterization demonstrated that this film was superior to many other films in strength and stiffness.

Figure 4.Graphene oxide thin films on (a) filtration membrane, (b) glass and (c) plastic substrates.

On the basis of previous research work, G. Eda et al reported the reduction of graphene oxide film[22]. Graphene oxide film was prepared with the method mentioned above. After low-temperature annealing in hydrazine atmosphere, the oxide film could be effective reduced. The result was highly transparent and flexible graphene films from single layer to several layers. The membranes showed enhancing photoelectric properties, being potential for future semiconductive devices. In 2010, Z.J. Fan et al demonstrated an environmentally friendly route for the reduction of graphene oxide by adopting aluminum powder as reductant[23].

In 2009, K.K. Manga et al reported a self-assembly composite film of titania ($Ti_{0.91}O_2$) and graphene[24].In this structure, titania and graphene stacked alternately based on electrostatic interactions. By dip-coating a piece of quartz in polyethylene terephthalate(PEI) solution, TiO_2nanosheet colloid solution, PEI solution, and graphene oxide (GO) colloid solution in sequence, a kind of layered structure of($PEI/Ti_{0.91}O_2/PEI/GO)_n$ (Figure 5a) could finally emerged. After exposing to ultraviolet light for a certain time, graphene oxide could be made into graphene with PEI being completely decomposed. A new layered structure,with $Ti_{0.91}O_2$ and graphene (G) stacking alternately, $(Ti_{0.91}O_2/G)_n$could be obtained. In this expression, *n* could be controlled by the dip-coating times. It was photocatalysis between titania and graphene that realized the conversion. This method provides a new way for graphene oxide reduction. After that, H.B. Yao et al prepared this kind of composite nanosheet with pattern graphene (Figure 5b&5c)by adopting a mask when the film was exposed to UV irradiation[25].

So far, the oxidation and reduction method has been regarded as one of the most promising synthesis methods for its large production and low cost feature. However, there lie several obstacles. For example, how to guarantee the integrity of graphene after filter? Perfect graphene owns amazing strength as mentioned above, but most graphene films obtained at present will be broken at the snap of fingers. Besides, defects can bring degradation of conductivity, especially the hydroxyl and epoxide groups in graphene oxide[26]. Actually, graphene is an excellent conductor while graphene oxide is an insulator. Remove the oxide functional groups completely or control the amounts of these groups to achieve special graphene should be the two problems of this method.

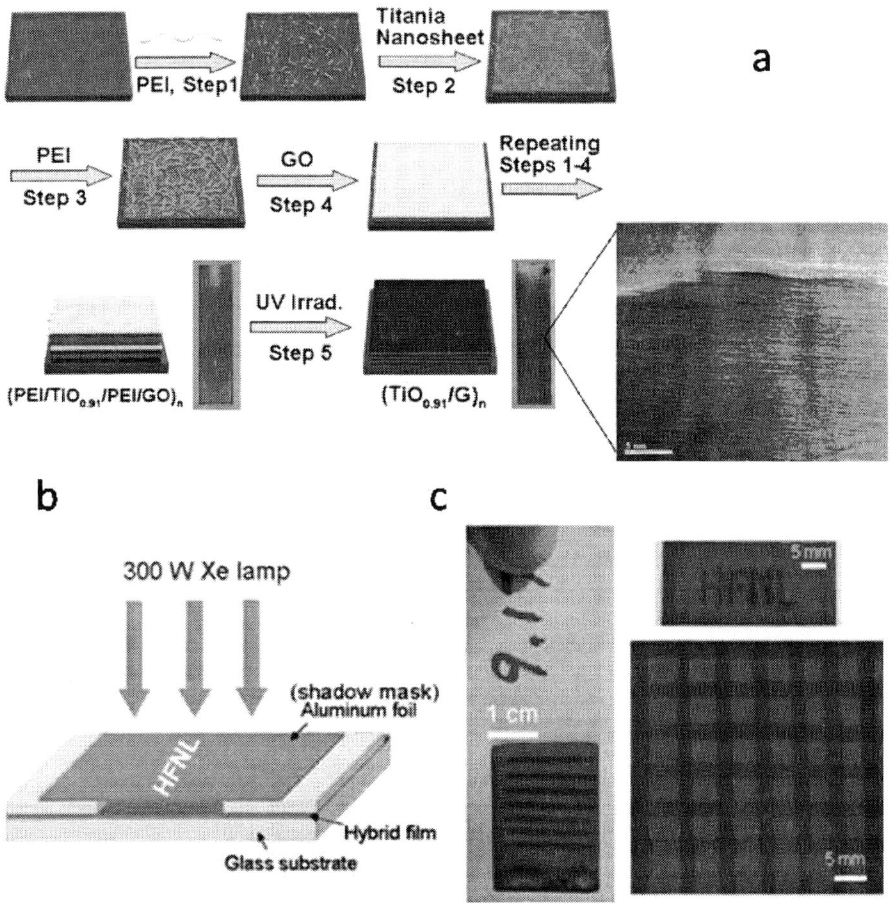

Figure 5.(a)Schematic illustration of the formation of $(PEI/Ti_{0.91}O_2/PEI/GO)_n$ and HRTEM image of $(Ti_{0.91}O_2/G)_n$multilayer membrane. (b) Adopting a shadow mask to fabricate (c) patterned graphene film.

2.2.Liquid-phase Exfoliation

Both the oxidation and reduction method and the liquid-phase exfoliation method belong to chemically dispersed method. The latter overcomes the Van der Waals force between graphite layers with chemical methods except oxidation, including chemical functionalization and intercalation. The process of liquid-phase exfoliation method does not include redox and this is the fundamental difference between oxidation and reduction method and liquid-phase exfoliation method. Without oxidation, liquid-phase exfoliation method can produce graphene of high quality.

Y. Hernandez et al successfully dispersed andexfoliated graphite (Figure 6a & 6b) in N-methyl-pyrrolidone (NMP) solvent[26]. Simply by ultrasonic dispersing graphite in NMP and followed by centrifugal separation, high quality graphene free of defects and epoxides could be obtained. Moreover, graphene produced by this method was less than five layers and there exited monolayer with a concentration about 1 wt%, which could be potentially improved to7-12 wt% with sediment recycling. Later experiment showed that other solvents, includingthe DMA, GBL and DMEU, could be qualified to the dispersion and exfoliation process (Figure 6c-e).It is the balance between surface energy of solvent and energy of exfoliate graphene that makes this simple method realize effective production of graphene. However, the solvents are difficult to remove. The worst of it is these solvents will do great harm to the environment.

In 2008, M. Lotya et al found a way to disperse and exfoliate graphite by adopting surfactant, resulting high quality graphene suspended in water[27]. In 2009, A.B. Bourlinos et al found new solvents to achieve the liquid phase exfoliation[28]. The method is developing fast.We believe that there are growing opportunities to produce graphene with structural integrity on a large scale by the liquid-phase exfoliation method.

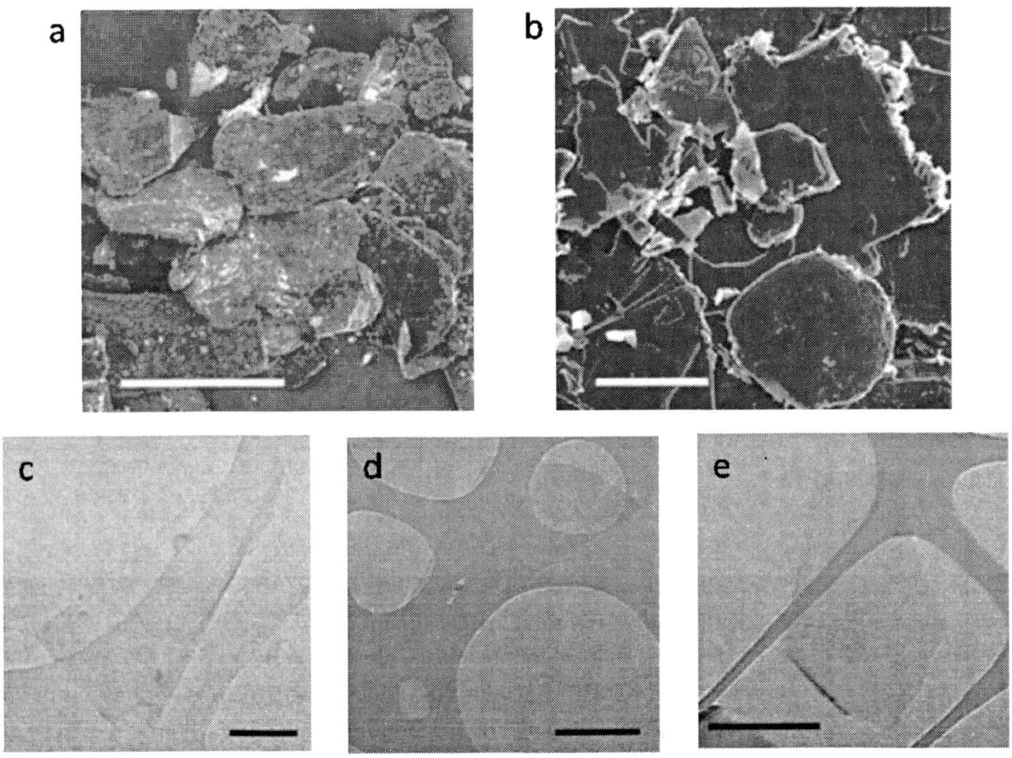

Figure 6.(a) SEM image of pristine graphite (scale bar, 500μm). (b) SEM image of sediment after centrifugation (scale bar, 25μm). TEM image of single-layer graphene from (c) GBL, (d) DMEU and (e) NMP.

2.3. CVD

This method seems most promising to produce graphene for industry. The CVD process will get a new material deposited after high temperature gas reaction. Actually, the CVD method is the most useful technology for deposition, surface treatment and material preparation in semiconductor industry. During graphene preparation,it is important to control the gas, temperature, growth time and catalysts. So far, adopting CVD methods to prepare graphene has produced some significant results.

In 2008, P.W. Sutter et al successfully grew arrays of macroscopic single-crystalline graphene on Ru (0001) substrate[29]. The first layer of this film showed relatively strong electron interaction with the substrate, while the second layer seemed almost completely independence, maintaining integrity of the electronic structure. Free standing graphene can be obtained after etching the substrate. Although the authors could only get small size graphene from noble metals, their works provide a new way for large area graphene synthesis.

In 2009, A. Reina et al realized a low cost and scalable CVD method by replacing the rubidium with polycrystalline Ni films[30], which is cheap and common. These continuous large area films that consisted of monolayer and few-layer graphene owned good transparency, conductivity, and ambipolar transfer characteristics. Later, K.S. Kim grew large area patterned graphene by adopting thin patterned nickel layers in the process[31]. At the same time, X.S. Li et al reported that copper foil could be a perfect substrate to grow graphene[32]. Differing from graphene on nickel substrate, the films deposited on copper foil were predominantly monolayer. Few layers only accounted for less than 5%. The group speculated that the difference of solubility of carbon between copper and nickel could be the reason for this result. So what is the truly mechanism of graphene growth in the CVD process?

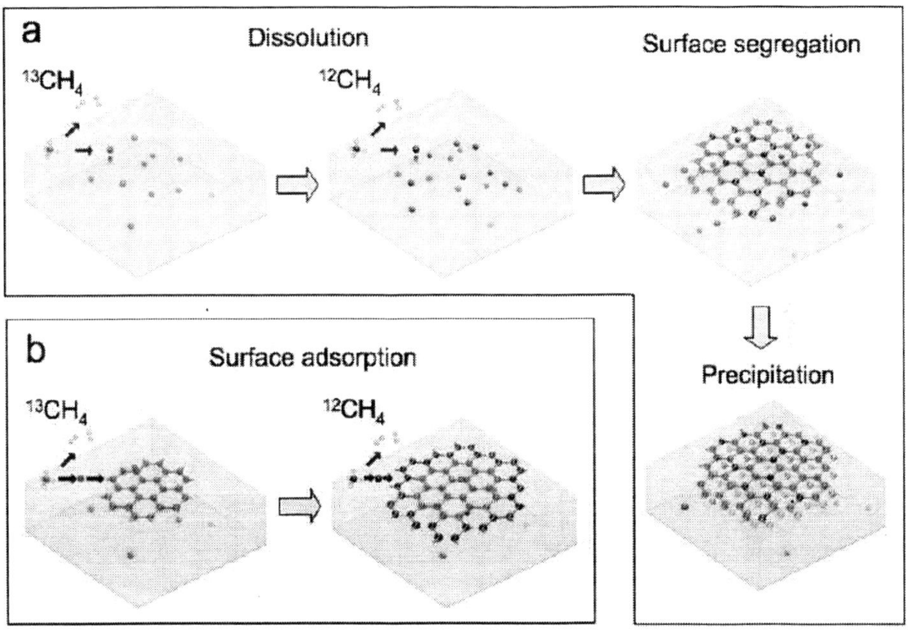

Figure 7.(a) Surface segregation and precipitation occurred on Ni and (b) surface adsorption on Cu.

In 2008, Q.K. Yu et al found an approach to prepare graphene on Ni by CVD[33]. Further, they pointed out that there may be a process of carbon dissolvant and segregation at metal surface. Under high temperature and inert gas protection, hydrocarbon decomposes and carbon atoms diffuse into the metal. By controlling the cooling rate, different segregation results can be obtained and at fast/medium cooling rate, graphene emerges at Ni surface. X.S. Li and co-workersfully described the mechanism of CVD process with substantial evidence[34]. By using carbon isotope labeling, they confirmed that surface segregation and precipitation occurred on Ni while surface adsorption on Cu (Figure 7). From this point, we can get monolayergraphene of high quality on copper foil, for the feasibility to control graphene growth.

The methods mentioned above are conventional CVD process based on methane or ethylene. In 2010, graphene with controllable thickness were grown from a modified CVD method[35]. Instead of gaseous materials, the method used different solid carbon sources as raw materials. Though hydrogen was still necessary as reducing gas and argon as protective gas, abandoning gaseous carbon sources pointed out a new direction to gas free "CVD" methods.

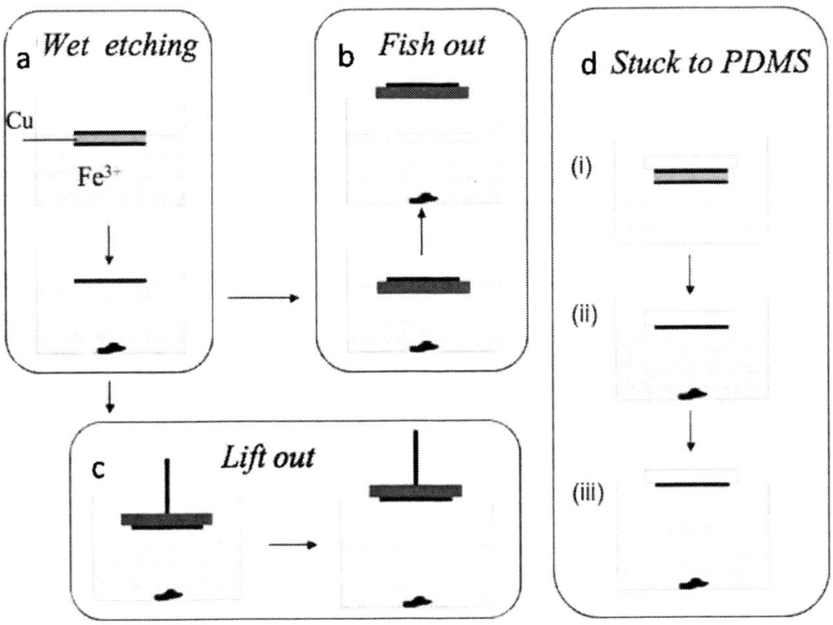

Figure 8.(a) The process of wet-etching Cu substrate and transfer to other substrates by (b) fishing out or (c) lifting out. (d) Stuck graphene to PDMS before etching and result in graphene on PDMS after etching.

The results above indicate the feasibility to prepare graphene on different substrates through CVD process. However, free standing graphene and graphene on other substrates are required in research and application. Therefore, shift graphene off the substrates is important. By adopting the common rule in chemistry, we can etch the substrates through substitution reaction to get free standing graphene. X.S. Li et al reported several effective methods to transfer graphene on other substrates by etching copper in ferric chloride ($FeCl_3$) solution (Figure 8)[36]. V.P. Verma et al realized large area graphene transfer from copper foil to PEI by hot press lamination (Figure 9)[37].It is worth mentioning that the graphene on PEI, with high transmittance and low resistance, can be used as flexible and transparent electrodes for future foldable electronics.

Dielectric substrate is one of the most important substrates for semiconductor industry. Epitaxial growth on silicon carbide can produce graphene on dielectric wafer[38-40]. However, thismethod of low yield requires high reaction temperature and expensive instruments. In 2010, A. Ismach et al found a direct deposition method to grow graphene on dielectric substrates[41]. A thin copper film was first deposited on dielectric substrates.After graphene synthesizing, the copper film was evaporated in the resistance furnace. This method gives new opportunities for graphene application in electron devices.

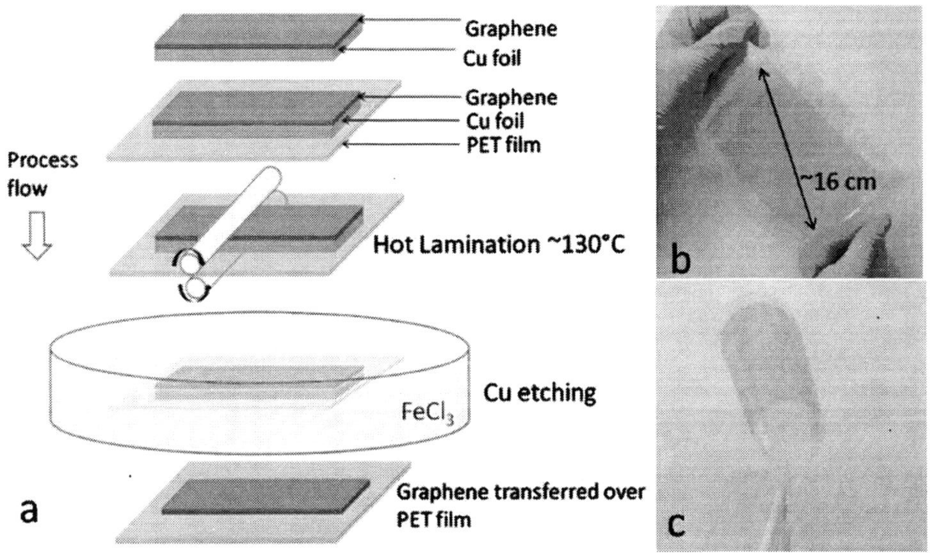

Figure 9.(a) Schematic process of hot press lamination and chemical etching. Large area (b) and flexible (c) graphene membrane on PET.

It is clear that CVD method is an effective method to grow large area continuous graphene with high conductivity and integrated structure. Moreover, the CVD method is a mature technology that has already taken hold in semiconductor industry. For this reason, research on graphene synthesis by CVD has been developed rapidly. So far, the size of graphene filmshas reached up to practical application. It is very prescient that CVD will take its place in graphene synthesis in future industry.

2.4. Carbon Nanotube Conversion

Graphene is a semi-metal material with zero bandgap. Before opening the energy gap of graphene, it is hard to adopt it to many useful semiconductor devices. So far, researchers have demonstrated several strategies to open the bandgap of graphene, including lithographic printing [42], oxidation[43-45] and hydrogenation[46,47], nanoparticles etching[48,49], etc. In addition, the carbon nanotube conversion method can provide graphene nanoribbons (GNRs) that show nonzero bandgap in certain conditions.

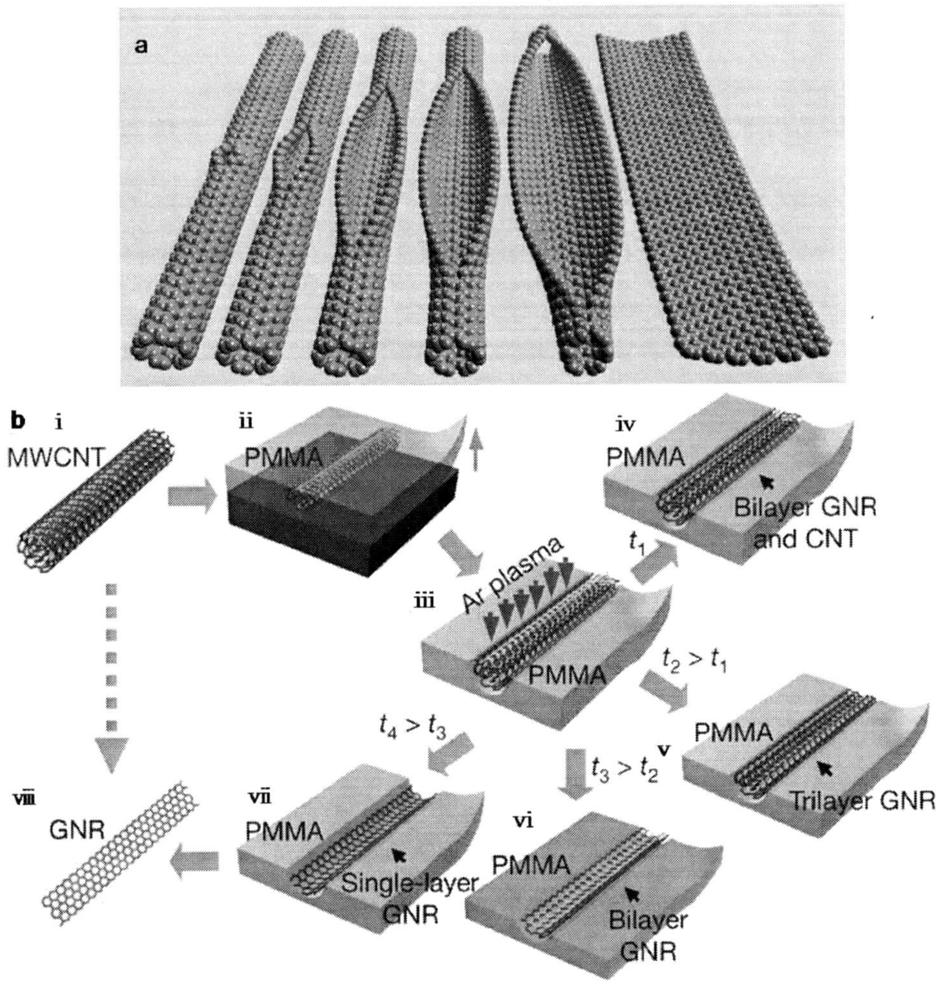

Figure 10.(a) Schematic of unzipping single wall CNT to form graphene nanoribbon. (b) The process of Ar plasma etching multiwall CNTs to form graphene nanoribbon. Different etching time will result in different graphene, including monolayer GNR, bilayer GNR and trilayer GNR.

In 2009, D.V. Kosynkin et al reported GNRs could be obtained from multiwalled carbon nanotubes (MWCNTs) by cutting them down on axial direction (Figure 10a)[50]. While suspending MWCNTs in sulfuric acid and potassium permanganate solution, oxidation takes place and manganate ester forms. As further oxidation, the tubes will be finally unzipped. At the same time, L.Y. Jiao et al found another way to gain GNRs with smooth edges from CNTs (Figure 10b)[51].MWCNTs coated with a PMMA substrate were exposed to argon plasma for certain times and monolayer GNRs could be obtained finally.

The carbon nanotube conversion method offers a new wayto prepare GNRs. However, it is difficult to control the widths of these GNRs on a large scope, for the limit of diameter of CNTS.

3. POTENTIAL APPLICATIONS

It's glamorous to adopt graphene to many application fields. Graphene owns many outstanding properties, such as high conductivity, high transparency, large specific surface area and diversechemical functions.To fabricate graphene devices is the current hot issue. There are numerous studies and reviews about graphene applications. To avoid reiteration, some promising applications, such as transistors, will not be covered in the chapter. Instead, graphenebased solar cells and composites that are rising in recent years will be introduced.

3.1. Solar Cells

To date, only silicon photovoltaic solar cells have been widely put into use. However, the high cost and relatively rapid degradation of silicon cells[52] push researchers to find alternative materials. Carbon materials, with highly chemical stability and structural diversity[52], are rising stars for solar cells. Recent studies have showed that CNTs of high conductivity and transparencycan substitute for traditional conductive materials, such as indium tin oxide (ITO). Y. Jia et al have demonstrated a model of CNTs-Si heterojunction solar cells with power conversion efficiency up to 7%[53,54]. In contrast with CNTs films, graphene membrane owns comparable conductivity and transparency while its two-dimensional continuous structure may outperform the porous structure of CNTs films. From these results, it is hopeful to adopt graphene to fabricate solar cells.

It is intuitionistic to make use of graphene as a transparent conductive electrode in thin film cells. In 2010, X.M. Li et al fabricated a Schottky junction solar cell by spreading graphene onto n-Si wafer (Figure 11a)[55]. The cell exhibited up to 1.5% power conversion efficiency and relatively high open circuit voltage (0.55-0.75V) (Figure 11b). The key element might be the smooth surface of graphene film with 100% coverage. In the process, the graphene film served as junction layer, charge transport path, and transparent electrode. The efficiency is relatively low due to the high resistance of the film. By increasing the number of the layers, the sheet resistance will reduce and charges will be harvested more effectively. However, stacking more graphene films will result in lowlight transparency. Moreover, K. Ihm et al demonstrated that the work function increased as layers increased, resulting in the decline of open circuit voltage[56]. From this result, choosing appropriate parameters of graphene can be conductive to improving the performance. Besides, more effective measures, such as chemical doping, should be introduced before this cell become practical. In our previous work, graphene/silicon nanowires Schottky junction solar cell was demonstrated. After chemical doping by thionyl chloride vapor, the efficiency of the cell increased from 0.68% to 2.86%. The great improvement could be attributed to the decrease of sheet resistance and enhancement of junction effect[57].

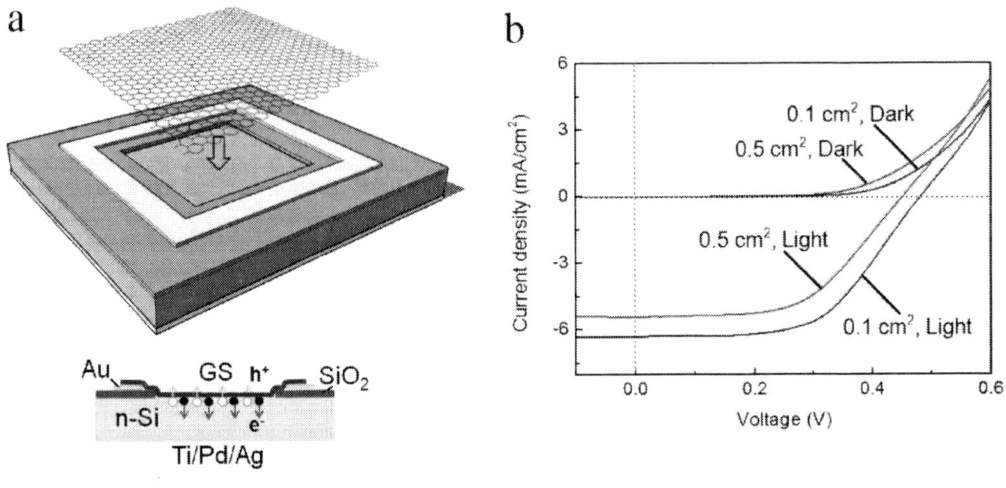

Figure 11.(a) Schematic picture of graphene/silicon Schottky junction solar cell. (b) The performance of the graphene based cells with different active area.

With special light absorption and carrier transportation mechanism, dye sensitized solar cells (DSSCs) are rising stars as the third generation photovoltaics. DSSCs own several advantages in contrast with Si based photovoltaics, including less sensitive to defects and much lower cost[58]. Recent research has demonstrated that graphene could play an important role in DSSCs, mainly in three parts.

Graphene as counter electrodes: Typical photoelectrochemical solar cell is generally composed of photoactive semiconductor working electrode and counter electrode[58]. DSSCs with platinum counter electrode had been demonstrated high power conversion efficiency[59]. H. Bonnemann and coworkers pointed out the important role of platinum nanoclusters in the iodide/triiodide redox system of DSSCs[60]. However, platinum is a precious metal and that prevents its mass application in DSSCs. Researchers found that graphene could perform comparably catalytic effect to platinum. DSSCs that applied functionalized graphene sheets (FGS) as counter electrode exhibited a conversionefficiency of 5.0% (Figure 12a)[61]. The group further utilized functionalized graphene sheets based ink, which was commercially available, to demonstrate DSSCs operation. W.J. Hong et al showed that graphene/PEDOT-PSS composite film could be an ideal alternative towards platinum counter electrode[62]. The composite film enhanced the electrocatalytic activity of ITO electrode while reserved the high transmittance as well.The performance of the cell was comparable to the cell with platinum electrode (Figure 12b). J.E. Trancik et al attributed the carbon catalytic mechanism to the high specific surface area and many chemical defects[63]. The high catalyzation activity of graphene may benefit from the similar mechanism.

Graphene as window electrodes or interfacial layer: More and more problems of indium tin oxide and fluorine tin oxide have been brought to our attention. X. Wang and coworkers successfully demonstrated that transparent, conductive, and ultrathin graphene films could be alternative window electrodes for DSSCs (Figure 12c)[64]. Due to the relatively high series resistance and low light transparency, the power conversion efficiency is relatively low at 0.26% (Figure 12d). By optimizing the key parameters of graphene films (here actually a kind of graphene constructed carbon thin film), G. Kalita et al enhanced the efficiency up to 0.68%[65]. The performance of graphene window electrodes can be further improved by upgrading the film, such as adopting graphene film prepared by CVD and chemical doping. On the other hand, graphene can be an interfacial layer to reduce the back-transport reactions in DSSCs,[66] which is a long-term problem in DSSCs research (Figure 12e).

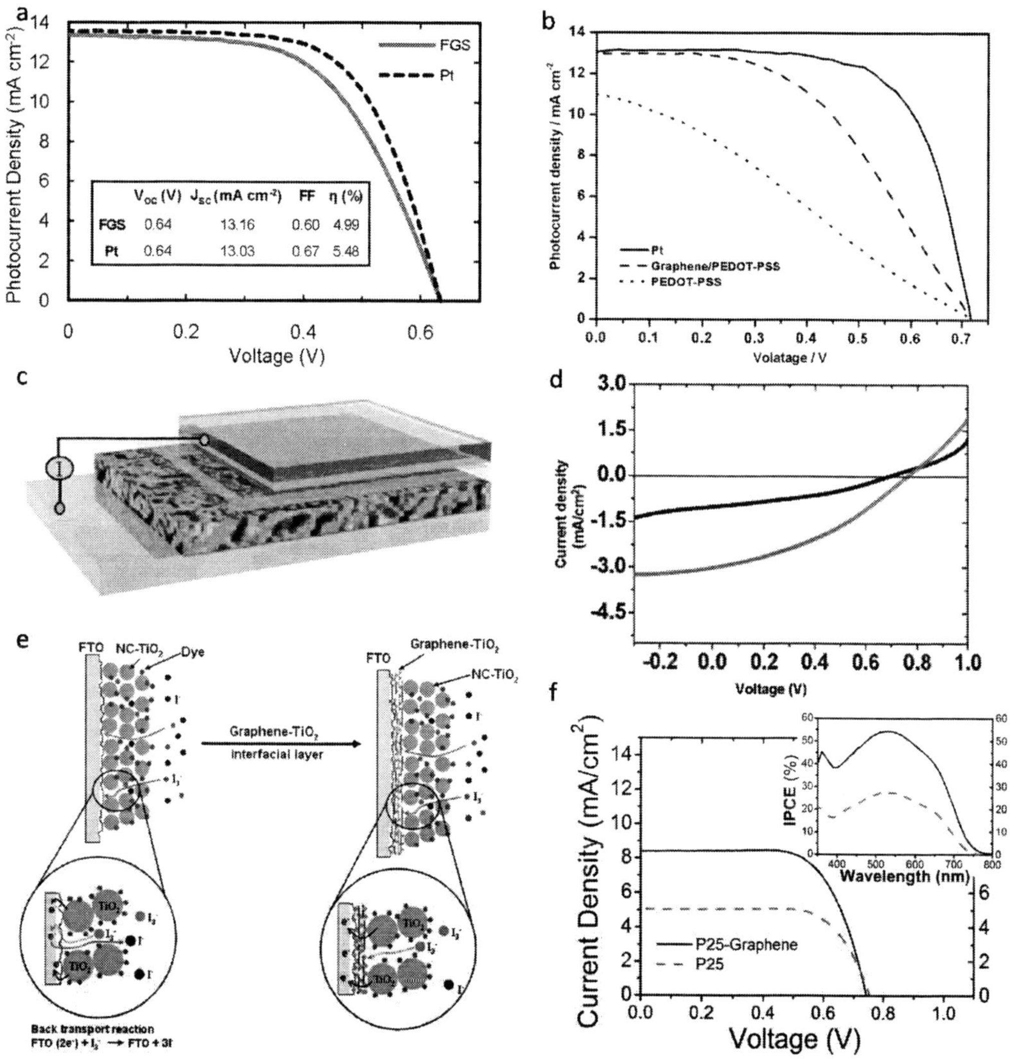

Figure 12.(a) FGS counter electrode exhibited comparable performance to platinum electrode. (b) Graphene/PEDOT-PSS counter electrode performed beyond pure PEDOR-PSS and closely to platinum electrode. (c) Illustration of DSSCs adopting graphene as window electrode and (d) its performance (black) compared to FTO-based cell (red). (e) Mechanism of graphene as interfacial layer to prevent back-transport reactions. (f) Photocurrent and IPCE (inset) characteristics showed great improvement of graphene/TiO_2 composites as photoelectrode.

Graphene asphotoelectrodes: H. Zhang et al demonstrated that TiO_2 cooperated with graphene would show obvious enhancement in dyes adsorptivity, light adsorption, and charge separation.[67]. It is the properties of high specific surface area and electronic mobilitythat make graphene an outstanding material for photoelectrodes. After using graphene TiO_2composites by heterogeneous coagulation,the performance of DSSCs significantly enhanced[68]. The graphene TiO_2 composites showed a tight interfacial binding between each others. After adopting the photoanode incorporating with 0.5 wt % graphene, J_{sc}increased from 5.04 mA/cm^{2}to 8.38 mA/cm^2 and η increased from 2.70% to 4.28% (Figure 12 f).A systematic investigation on graphene composites as photoelectrodes was reported by N.L. Yang et al[69,70]. In this work, the relation between graphene nanosheets concentration and performance of DSSCs were studied.The results showed that the power efficiency would be optimized using appropriate amount graphene. With redundant graphene loading, reduction of light harvesting in the dye and recombination would occur and result in total efficiency decrease.

3.2. Graphene Based Composites

Material science has greatly been enriched since the birth of graphene.Composite materials based on graphene and derivatives are already ahead. Synthesis, characterization, and application of graphene composites have been greatly exploared thanks to the development of graphene synthesis, especially the development of chemically dispersion method. Profiting fromthe special structure and properties of graphene,exciting potential application of graphene composites have been found. Beside transparent conductive film and composite electrodes for solar cells as mentioned above, graphene based composites found promising applications in lithium ion batteries[71-76], supercapacitors[77-81], magnetic materials[82-84], sensors[85,86], energy storage[86,87], etc. Graphene based composites for these applications and the behaviors of graphene in devices deserve a detailed discussion. These composites can be mainly classified as graphene/polymer, graphene/metal, graphene/metal oxides, graphene/non-graphene carbon, graphene/nitrides etc. Here, research status and remained challenges of graphene based polymer composites and graphene metal oxide composites will be reviewed in brief. Other composites, such as graphene/metal and graphene/CNTs, will not be covered in the chapter.

Graphene/polymer composites: Polymers can exhibit diversiform properties by combining with other plastics and mixing with talc and clay[88]. Graphene as additives for polymers offers new opportunities for application. This possibility was first demonstrated by S. Stankovich and coworkers[89]. They mixed polymer and isocyanate-derivatized graphite oxide and formed functionalized graphene polymer composite (Figure 13a).The graphene polystyrene composite synthesized by chemical mixing method exhibited an extraordinary low percolation threshold and sufficient conductivity for antistatic criterion in electrical application. Later, H. Pang and coworkers reported graphene polymer composites with enhanced electrically conductivities and lower percolation threshold of 0.07 vol.% (Figure13b)[90].

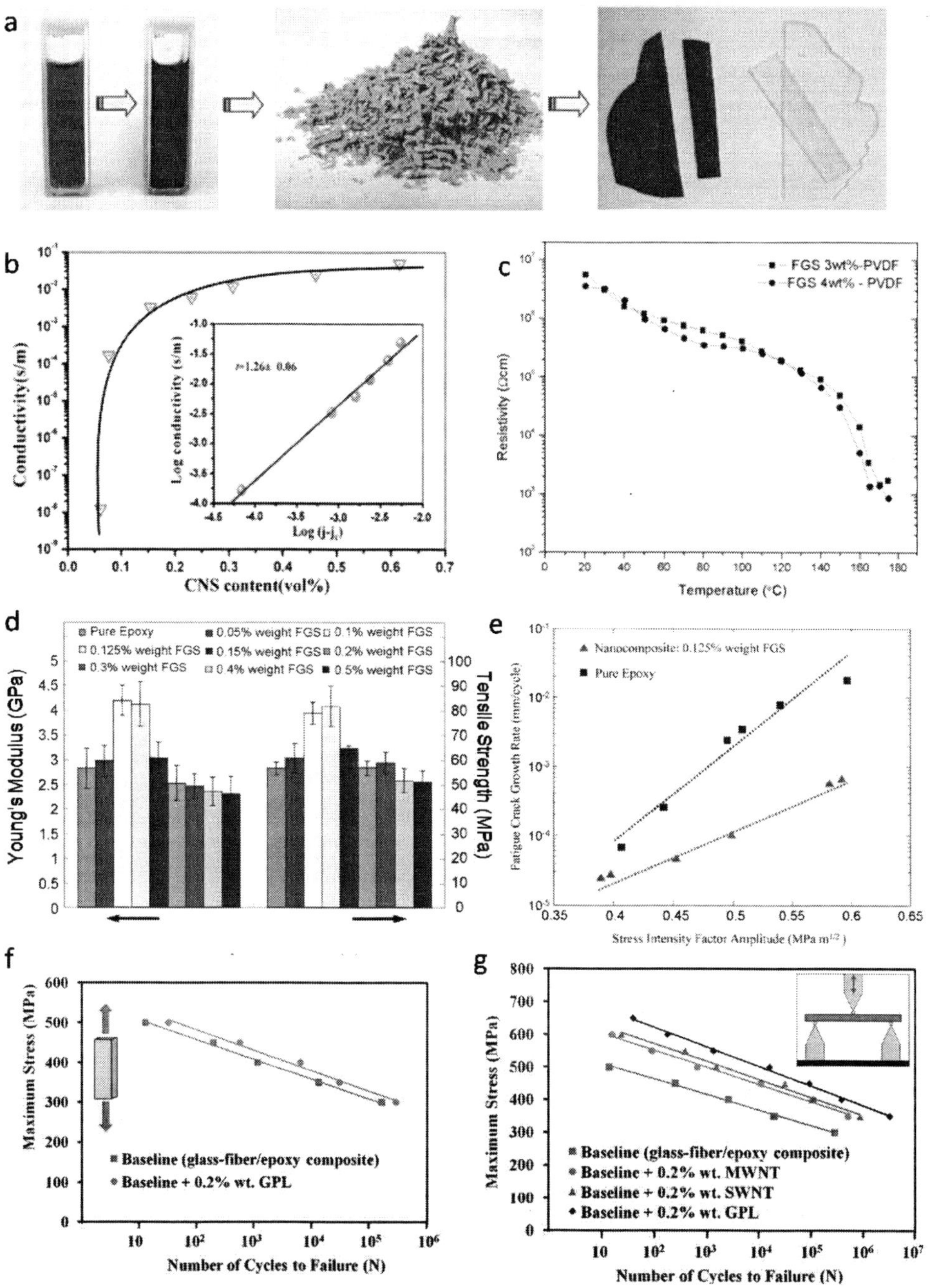

Figure 13.(a) Process of fabricating graphene/polymer composite. From left to right were suspension of graphene oxide and dissolved polystyrene in DMF, reduced by N,N-dimethylhydrazine, composite power after coagulation in methanol, and hot-press composite versus pure polystyrene. (b) Electrical conductivity enhancement and inset show a low percolation threshold of 0.07 vol.%. (c) The negative temperature coefficient behavior of FGS-PVDF. Exciting improvement of mechanical properties of polymer with graphene additives, including (d)Young' Modulus and tensile strength, (e)fatigue crack growth rate, (f)fatigue life under pure tension mode and (g) flexural bending mode. The reasons for these improvements may be the combination of high specific surface FGS/polymer composites.

The performance enhancement can attribute to the formation of a 2D conductive network. S. Ansari and E.P. Giannelis revealed the modified conductivity of poly (vinylidene fluoride) (PVDF) based on FGS prepared from graphite oxide, and exfoliated graphite (EG) prepared by solution processing, respectively[91].The structure and topography of FGS-PVDF and EG-PVDF are different. The most interesting phenomenon was the opposite resistance/temperature behavior between them (Figure 13c). The resistance of FGS-PVDF decreased with temperature, indicating a negative temperature coefficient behavior, while EG-PVDF showed a positive temperature coefficient behavior. This unusual behavior of FGS-PVDF was mainly caused by the transforming of resistance mechanism. The higher aspect ratio of FGS would lead to an increase of contacts between particles, thus contact resistance would predominate over tunneling resistance.

Systematic research on thermo mechanical properties of polymer with functionalized graphene sheets was done by the S. Stankovich's group later[92]. Exciting property improvements (Figure 13d~g), including Young's modulus (E), glass transition temperature (T_g), ultimate strength, and thermal degradation temperature, were revealed in functionalized graphene polymer composites. M.A. Rafiee et al reported that fracture and fatigue properties improved obviously by loading functionalized graphene sheets[93]. Later detailed study on hierarchical graphene fiberglass/epoxy composites demonstrated significantly increase in fatigue life under flexural bending or uniaxial tensile fatigue conditions[94]. From these results, we could catch a sight of the superior enhancement in polymer with graphene additives.

More detailed research has been done on graphene and derivatives based composites. The mechanical and electrical properties of polymer composites have been effectively highlighted by the special sp^2 carbon bonding. The detailed status of graphene based polymer composites was reviewed by P. Mukhopadhyay etc.[88]. Current graphene polymer composites were synthesized by solution phase mixing and melt blending.Due to the low surface energy of graphene, to ensure the effective dispersion of individual graphene sheets remain a challenge. Graphene owns high conductivity but low surface energy, while the situation is the opposite in graphene with defects[88]. Therefore, further development of graphene polymer composites should sit on the premise of enhancement of synthesis and dispersion of graphene.

Graphene/metal oxide composites: Metal oxides are one of the most popular materials in our daily life. The properties and application of metal oxides are various on the base of different element, valence and structure. However, more diversiform metal oxides are required to meet increasingly demands of industry. Meanwhile, practical applications of graphene call for an improvement of property. Combining graphene with metal oxides shows good synergistic effectsand promoted performance.

New functionalities wereintroducedby combining$Fe_3O_4$with graphene. G.X. Zhu et al reported a facile and robust route for graphene Fe_3O_4 composites fabricating[83]. Fe_3O_4nanoparticles and reduced graphene oxide were first synthesized respectively in colloidal dispersion. By mixing and stirring the two colloids, a two-dimensional membrane was formed with Fe_3O_4nanoparticles exclusively assembled on both sides of reduced graphene oxide sheet. The composite membrane exhibited considerable mechanical robustness, high magnetization moment, and superparamagnetic effect. K.F. Zhou et al reported a simplified method to assemble graphene Fe_3O_4 composites[95]. Instead of being pre-synthesized respectively, graphene and Fe_3O_4were synthesized at the same time during a chemical redox reaction (Figure 14a). In the process, graphene oxide was reduced to graphene and Fe_3O_4 nanoparticles grew on the graphene surface. High crystallinity of magnetite and considerable saturation magnetization were observed from the composites. Similar Fe_3O_4 graphene composites with special magnetic property were reported[96-98]. Besides, G.M. Zhou et al[99] and P.C. Lian et al[100] reported an improvement in lithium-ion batteries after adopting Fe_3O_4 graphene composites anode. The enhanced performance could be attributed to the stable dispersion of Fe_3O_4 nanoparticles which would otherwise aggregate during electrochemical cycling without graphene. At the same time, electron transport would be enhanced with the help of graphene.In addition, F.A.He et al demonstrated good adsorption capacity of Fe_3O_4 graphene oxide composite for methylene blue and neutral red cationic dyes[101].

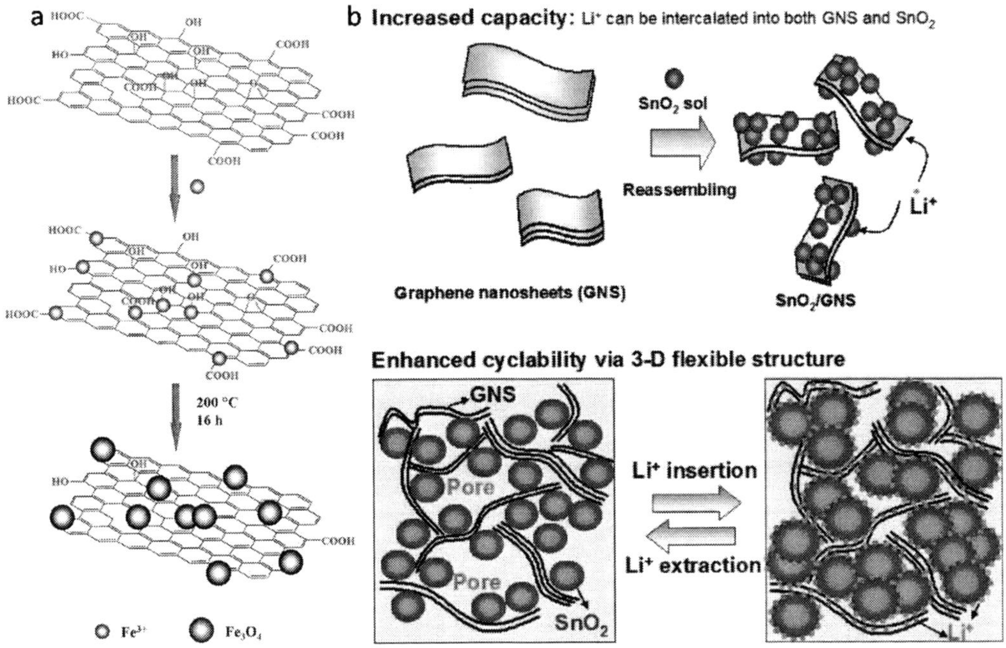

Figure 14.(a) Graphene and Fe_3O_4synthesized during a chemical redox reaction. Fe_3O_4 decorated on both sides of graphene. (b) Illustration of the synthesis and structure of graphene SnO_2 composite. The pores of the porous composite offered enough void spaces for volume change of SnO_2 during charge/discharge process.

Graphene SnO_2 composite is another hot issue for lithium-ion batteries and supercapacitors. SnO_2 shows high storage capacity but large volume change during charge/discharge process at the same time, which will result in aggregate effects and lead to a poor cyclic performance. SnO_2 based composite may be the solution to this question. S.M. Paek et al took the lead in graphene SnO_2 composite research[102]. Graphene nanosheets reassembled after adding rutile SnO_2 nanopraticles and formed a homogeneously nanoporous composite (Figure 14b). The composite electrode displayed a high reversible capacity and enhanced cycling performance. The charge capacity of graphene SnO_2 composite still remained 70% retention after 30 cycles while bare SnO_2nanopraticles rapidly droped from

550 mAh/g to 60 mAh/g only after 15 cycles. The great enhancement could be attributed to the fact that nanoporous composite offered enough void spaces for volume change of SnO_2 during charge/discharge process. Later Z.Y. Wang et al demonstrated the different performance between laterally confined graphene nanosheets, micro-scale graphene, andgraphene SnO_2 composites, respectively.The latter composites were prepared by arc-discharge evaporation of graphite, reduction of graphene oxide, and hydrolysis of $SnCl_4$ in sequence[103]. Results showed that laterally confined graphene with a size of ~100-200nm exhibited a higher reversible capacity than micro-scale graphene. Except for owning higher purity and better conductivity, laterally confined graphene offered a shorter transport path for lithium ions. However, both of the capacities of these graphene are not high enough for practically application. Graphene SnO_2 composites were examined with a better performance than these graphene forthe good cooperation of graphene and SnO_2. On the base of previous work, a simple method was developed to prepare graphene SnO_2 composites[104,105]. Instead of mixing the SnO_2 nanoparticles with graphene dispersions, graphene SnO_2 composites was synthesized by mixing graphene oxide with $SnCl_2 \cdot 2H_2O$ aqueous solution. In the process, graphene oxide was reduced to graphene and Sn^{2+} to Sn. SnO_2 was obtained after annealing in atmosphere. The morphology and performance of graphene SnO_2 composites had a close relation with the ratio of SnO_2 and graphene[106]. Apart from these two liquid phase method, a nonaqueous approach, atomic layer deposition, had developed to prepare amorphous or crystalline SnO_2 graphene nanosheet composites[107]. Graphene SnO_2 composites are also a promising material for supercapacitors[108,109]. The characteristics of ultracapacitor with graphene SnO_2 composites electrode indicated a good capacitive behavior. From the results above,we can image a variety of other application with these splendid electrodes.

Besides Fe_3O_4, SnO_2, and Ti_2O (in solar cells), various functional composite materials can be obtained from other metal oxides with graphene. These composites, including graphene/Cu_2O [110] or CuO[111,112], graphene/ZnO[108,113-115], graphene/Mn_3O_4[116,117], graphene/Co_3O_4[118-121], etc. show potential application and require detailed research.

CONCLUSION

We have reviewed several important aspects of graphene, including properties and structure, synthesis and transfer, and various potential applications. The single-atomic 2D honey structure gives graphene multiple of charming properties. Characteristics of high conductivity, transparency and mechanical strength are most eye-catching. One of the remaining challenges is to narrow the gap between properties of theoretical calculations and practical measurements. Finding a method to produce graphene with higher quality is significant important. The oxidation and reduction method, liquid-phase exfoliation method, CVD method, and carbon nanotube conversion method are introduced for their feasibility in future industrial production. Graphene film prepared by CVD method is superior in quality while graphene colloid by chemical dispersed is moderate. Graphenebased solar cells and composites are reviewed as representatives. In Schottky junction solar cells, graphene serves as transparent conductive electrodes. In DSSCs, graphene can play several important roles including counter electrodes, window electrodes, and photoelectrodes. Though graphene has positive impacts in these cells, relatively low energy conversion efficiency prevent them from practical applications. More efforts have to be made to improve the performance of graphene based solar cells. Graphene composites are another research emphasis. Polymer with graphene additives shows distinct enhancement in electrical and mechanical properties. Metal oxides cooperated with graphene exhibit more diverse functions. These fields deserve more attention and research. In conclusion, graphene, the unique 2D carbon material, together with other carbon materials, will find their indispensable positions in our industrial production and daily life.

REFERENCES

[1] Service, R. F. *Science*2009, *324*, 875-877.

[2] Geim, A. K. *Science*2009, *324*, 1530-1534.

[3] Balandin, A. A.; Ghosh, S.; Bao, W. Z.; Calizo, I.; Teweldebrhan, D.; Miao, F.; Lau, C. N. *Nano Letters*2008, *8*, 902-907.

[4] Jiang, D. E.; Cooper, V. R.; Dai, S. *Nano Letters*2009, *9*, 4019-4024.

[5] Wang, L.; Lee, K.; Sun, Y. Y.; Lucking, M.; Chen, Z. F.; Zhao, J. J.; Zhang, S. *Acs Nano*2009, *3*, 2995-3000.

[6] Park, H.; Rowehl, J.A.; Kim, K.K.; Bulovic, V.; Kong, J.*Nanotechnology*2010, *21*, 505204.

[7] Wei, D. *International Journal of Molecular Sciences*2010, *11*, 1103-1113.

[8] Hong, W. J.; Xu, Y. X.; Lu, G. W.; Li, C.; Shi, G. Q. *Electrochemistry Communications*2008, *10*, 1555-1558.

[9] Kamegawa, T.; Yamahana, D.; Yamashita, H. *Journal of Physical Chemistry C*2010, *114*, 15049-15053.

[10] Liang, Y. Y.; Wang, H. L.; Casalongue, H. S.; Chen, Z.; Dai, H. J. *Nano Research*2010, *3*, 701-705.

[11] Castro, E. V.; Novoselov, K. S.; Morozov, S. V.; Peres, N.; Dos Santos, J.; Nilsson, J.; Guinea, F.; Geim, A. K.; Neto, A. *Physical Review Letters*2007, *99,216802.*

[12] Bolotin, K. I.; Sikes, K. J.; Jiang, Z.; Klima, M.; Fudenberg, G.; Hone, J.; Kim, P.; Stormer, H. L. *Solid State Communications*2008, *146*, 351-355.

[13] Aristov, V. Y.; Urbanik, G.; Kummer, K.; Vyalikh, D. V.; Molodtsova, O. V.; Preobrajenski, A. B.; Zakharov, A. A.; Hess, C.; Hanke, T.; Buchner, B.; Vobornik, I.; Fujii, J.; Panaccione, G.; Ossipyan, Y. A.; Knupfer, M. *Nano Letters*2010, *10*, 992-995.

[14] Meyer, J. C.; Geim, A. K.; Katsnelson, M. I.; Novoselov, K. S.; Booth, T. J.; Roth, S. *Nature*2007, *446*, 60-63.

[15] Geim, A. K.; Kim, P. *Scientific American*2008, *298*, 90-97.

[16] Novoselov, K. S.; Geim, A. K.; Morozov, S. V.; Jiang, D.; Zhang, Y.; Dubonos, S. V.; Grigorieva, I. V.; Firsov, A. A. *Science*2004, *306*, 666-669.

[17] Dikin, D. A.; Stankovich, S.; Zimney, E. J.; Piner, R. D.; Dommett, G.; Evmenenko, G.; Nguyen, S. T.; Ruoff, R. S. *Nature*2007, *448*, 457-460.

[18] Gengler, R.; Spyrou, K.; Rudolf, P. *Journal of Physics D-Applied Physics*2010, *43,* 374015.

[19] Brodie, B. C. *Philosophical Transactions of the Royal Society of London*1859, *149.* 249-259.

[20] Stankovich, S.; Piner, R. D.; Chen, X. Q.; Wu, N. Q.; Nguyen, S. T.; Ruoff, R. S. *Journal of Materials Chemistry*2006, *16*, 155-158.
[21] Stankovich, S.; Dikin, D. A.; Piner, R. D.; Kohlhaas, K. A.; Kleinhammes, A.; Jia, Y.; Wu, Y.; Nguyen, S. T.; Ruoff, R. S. *Carbon*2007, *45*, 1558-1565.
[22] Eda, G.; Fanchini, G.; Chhowalla, M. *Nature Nanotechnology*2008, *3*, 270-274.
[23] Fan, Z. J.; Wang, K.; Wei, T.; Yan, J.; Song, L. P.; Shao, B. *Carbon*2010, *48*, 1686-1689.
[24] Manga, K. K.; Zhou, Y.; Yan, Y. L.; Loh, K. P. *Advanced Functional Materials*2009, *19*, 3638-3643.
[25] Yao, H. B.; Wu, L. H.; Cui, C. H.; Fang, H. Y.; Yu, S. H. *Journal of Materials Chemistry*2010, *20*, 5190-5195.
[26] Hernandez, Y.; Nicolosi, V.; Lotya, M.; Blighe, F. M.; Sun, Z. Y.; De, S.; McGovern, I. T.; Holland, B.; Byrne, M.; Gun'Ko, Y. K.; Boland, J. J.; Niraj, P.; Duesberg, G.; Krishnamurthy, S.; Goodhue, R.; Hutchison, J.; Scardaci, V.; Ferrari, A. C.; Coleman, J. N. *Nature Nanotechnology*2008, *3*, 563-568.
[27] Lotya, M.; Hernandez, Y.; King, P. J.; Smith, R. J.; Nicolosi, V.; Karlsson, L. S.; Blighe, F. M.; De, S.; Wang, Z. M.; McGovern, I. T.; Duesberg, G. S.; Coleman, J. N. *Journal of the American Chemical Society*2009, *131*, 3611-3620.
[28] Bourlinos, A. B.; Georgakilas, V.; Zboril, R.; Steriotis, T. A.; Stubos, A. K. *Small*2009, *5*, 1841-1845.
[29] Sutter, P. W.; Flege, J. I.; Sutter, E. A. *Nature Materials*2008, *7*, 406-411.
[30] Reina, A.; Jia, X. T.; Ho, J.; Nezich, D.; Son, H. B.; Bulovic, V.; Dresselhaus, M. S.; Kong, J. *Nano Letters*2009, *9*, 30-35.
[31] Kim, K. S.; Zhao, Y.; Jang, H.; Lee, S. Y.; Kim, J. M.; Kim, K. S.; Ahn, J. H.; Kim, P.; Choi, J. Y.; Hong, B. H. *Nature*2009, *457*, 706-710.
[32] Li, X. S.; Cai, W. W.; An, J. H.; Kim, S.; Nah, J.; Yang, D. X.; Piner, R.; Velamakanni, A.; Jung, I.; Tutuc, E.; Banerjee, S. K.; Colombo, L.; Ruoff, R. S. *Science*2009, *324*, 1312-1314.
[33] Yu, Q. K.; Lian, J.; Siriponglert, S.; Li, H.; Chen, Y. P.; Pei, S. S. *Applied Physics Letters*2008, *93,* 113103.
[34] Li, X. S.; Cai, W. W.; Colombo, L.; Ruoff, R. S. *Nano Letters*2009, *9*, 4268-4272.
[35] Sun, Z. Z.; Yan, Z.; Yao, J.; Beitler, E.; Zhu, Y.; Tour, J. M. *Nature*2010, *468*, 549-552.
[36] Li, X. S.; Cai, W. W.; Jung, I. H.; An, J. H.; Yang, D. X.; Velamakanni, A.; Piner, R.; Colombo, L.; Ruoff, R. S.*The Electrochemical Society Transactions*2009,*19*, 41-52.
[37] Verma, V. P.; Das, S.; Lahiri, I.; Choi, W. *Applied Physics Letters*2010, *96*, 203108.
[38] Rollings, E.; Gweon, G. H.; Zhou, S. Y.; Mun, B. S.; McChesney, J. L.; Hussain, B. S.; Fedorov, A.; First, P. N.; de Heer, W. A.; Lanzara, A. *Journal of Physics and Chemistry of Solids*2006, *67*, 2172-2177.
[39] Hass, J.; Feng, R.; Li, T.; Li, X.; Zong, Z.; de Heer, W. A.; First, P. N.; Conrad, E. H.; Jeffrey, C. A.; Berger, C. *Applied Physics Letters*2006, *89*, 143106.
[40] Aristov, V. Y.; Urbanik, G.; Kummer, K.; Vyalikh, D. V.; Molodtsova, O. V.; Preobrajenski, A. B.; Zakharov, A. A.; Hess, C.; Hanke, T.; Buchner, B.; Vobornik, I.; Fujii, J.; Panaccione, G.; Ossipyan, Y. A.; Knupfer, M. *Nano Letters*2010, *10*, 992-995.
[41] Ismach, A.; Druzgalski, C.; Penwell, S.; Schwartzberg, A.; Zheng, M.; Javey, A.; Bokor, J.; Zhang, Y. G. *Nano Letters*2010, *10*, 1542-1548.

[42] Han, M. Y.; Ozyilmaz, B.; Zhang, Y. B.; Kim, P. *Physical Review Letters*2007, *98*, 206805.
[43] Gilje, S.; Han, S.; Wang, M.; Wang, K. L.; Kaner, R. B. *Nano Letters*2007, *7*, 3394-3398.
[44] Kaiser, A. B.; Gomez-Navarro, C.; Sundaram, R. S.; Burghard, M.; Kern, K. *Nano Letters*2009, *9*, 1787-1792.
[45] Nourbakhsh, A.; Cantoro, M.; Vosch, T.; Pourtois, G.; Clemente, F.; van der Veen, M. H.; Hofkens, J.; Heyns, M. M.; De Gendt, S.; Sels, B. F. *Nanotechnology*2010, *21*, 435203.
[46] Sofo, J. O.; Chaudhari, A. S.; Barber, G. D. *Physical Review B*2007, *75*, 153401.
[47] Elias, D. C.; Nair, R. R.; Mohiuddin, T.; Morozov, S. V.; Blake, P.; Halsall, M. P.; Ferrari, A. C.; Boukhvalov, D. W.; Katsnelson, M. I.; Geim, A. K.; Novoselov, K. S. *Science*2009, *323*, 610-613.
[48] Datta, S. S.; Strachan, D. R.; Khamis, S. M.; Johnson, A. *Nano Letters*2008, *8*, 1912-1915.
[49] Ci, L.; Xu, Z. P.; Wang, L. L.; Gao, W.; Ding, F.; Kelly, K. F.; Yakobson, B. I.; Ajayan, P. M. *Nano Research*2008, *1*, 116-122.
[50] Kosynkin, D. V.; Higginbotham, A. L.; Sinitskii, A.; Lomeda, J. R.; Dimiev, A.; Price, B. K.; Tour, J. M. *Nature*2009, *458*, 872-875.
[51] Jiao, L. Y.; Zhang, L.; Wang, X. R.; Diankov, G.; Dai, H. J. *Nature*2009, *458*, 877-880.
[52] Zhu, H. W.; Wei, J. Q.; Wang, K. L.; Wu, D. H. *Solar Energy Materials and Solar Cells*2009, *93*, 1461-1470.
[53] Jia, Y.; Wei, J. Q.; Wang, K. L.; Cao, A. Y.; Shu, Q. K.; Gui, X. C.; Zhu, Y. Q.; Zhuang, D. M.; Zhang, G.; Ma, B. B.; Wang, L. D.; Liu, W. J.; Wang, Z. C.; Luo, J. B.; Wu, D. *Advanced Materials*2008, *20*, 4594-4598.
[54] Jia, Y.; Li, P. X.; Wei, J. Q.; Cao, A. Y.; Wang, K. L.; Li, C. L.; Zhuang, D. M.; Zhu, H. W.; Wu, D. H. *Materials Research Bulletin*2010, *45*, 1401-1405.
[55] Li, X. M.; Zhu, H. W.; Wang, K. L.; Cao, A. Y.; Wei, J. Q.; Li, C. Y.; Jia, Y.; Li, Z.; Li, X.; Wu, D. H. *Advanced Materials*2010, *22*, 2743-2748.
[56] Ihm, K.; Lim, J. T.; Lee, K. J.; Kwon, J. W.; Kang, T. H.; Chung, S.; Bae, S.; Kim, J. H.; Hong, B. H.; Yeom, G. Y. *Applied Physics Letters*2010, *97*, 032113.
[57] Fan, G. F.; Zhu, H. W.; Wang, K. L.; Wei, J. Q.; Li, X. M.; Shu, Q. K.; Guo, N.; Wu. D. H. *ACS Applied Materials &Interfaces*2011,*3*, 721-725.
[58] Wei, D. *International Journal of Molecular Sciences*2010, *11*, 1103-1113.
[59] Chiba, Y.; Islam, A.; Watanabe, Y.; Komiya, R.; Koide, N.; Han, L. Y. *Japanese Journal of Applied Physics Part 2-Letters & Express Letters*2006, *45*, L638-L640.
[60] Bonnemann, H.; Khelashvili, G.; Behrens, S.; Hinsch, A.; Skupien, K.; Dinjus, E. *Journal of Cluster Science*2007, *18*, 141-155.
[61] Roy-Mayhew, J. D.; Bozym, D. J.; Punckt, C.; Aksay, I. A. *Acs Nano*2010, *4*, 6203-6211.
[62] Hong, W. J.; Xu, Y. X.; Lu, G. W.; Li, C.; Shi, G. Q. *Electrochemistry Communications*2008, *10*, 1555-1558.
[63] Trancik, J. E.; Barton, S. C.; Hone, J. *Nano Letters*2008, *8*, 982-987.
[64] Wang, X.; Zhi, L. J.; Mullen, K. *Nano Letters*2008, *8*, 323-327.
[65] Kalita, G.; Matsushima, M.; Uchida, H.; Wakita, K.; Umeno, M. *Journal of Materials Chemistry*2010, *20*, 9713-9717.

[66] Kim, S. R.; Parvez, M. K.; Chhowalla, M. *Chemical Physics Letters*2009, *483*, 124-127.
[67] Zhang, H.; Lv, X. J.; Li, Y. M.; Wang, Y.; Li, J. H. *Acs Nano*2010, *4*, 380-386.
[68] Sun, S. R.; Gao, L.; Liu, Y. Q. *Applied Physics Letters*2010, *96*, 083113.
[69] Yang, N. L.; Zhai, J.; Wang, D.; Chen, Y. S.; Jiang, L. *Acs Nano*2010, *4*, 887-894.
[70] Tang, Y. B.; Lee, C. S.; Xu, J.; Liu, Z. T.; Chen, Z. H.; He, Z. B.; Cao, Y. L.; Yuan, G. D.; Song, H. S.; Chen, L. M.; Luo, L. B.; Cheng, H. M.; Zhang, W. J.; Bello, I.; Lee, S. T. *Acs Nano*2010, *4*, 3482-3488.
[71] He, Y. S.; Bai, D. W.; Yang, X. W.; Chen, J.; Liao, X. Z.; Ma, Z. F. *Electrochemistry Communications*2010, *12*, 570-573.
[72] Wang, X. Y.; Zhou, X. F.; Yao, K.; Zhang, J. G.; Liu, Z. P. *Carbon*2011, *49*, 133-139.
[73] Chou, S. L.; Wang, J. Z.; Choucair, M.; Liu, H. K.; Stride, J. A.; Dou, S. X. *Electrochemistry Communications*2010, *12*, 303-306.
[74] Wang, J. Z.; Zhong, C.; Chou, S. L.; Liu, H. K. *Electrochemistry Communications*2010, *12*, 1467-1470.
[75] Choi, D. W.; Wang, D. H.; Viswanathan, V. V.; Bae, I. T.; Wang, W.; Nie, Z. M.; Zhang, J. G.; Graff, G. L.; Liu, J.; Yang, Z. G.; Duong, T. *Electrochemistry Communications*2010, *12*, 378-381.
[76] Kim, H.; Kim, S. W.; Park, Y. U.; Gwon, H.; Seo, D. H.; Kim, Y.; Kang, K. *Nano Research*2010, *3*, 813-821.
[77] Zhang, Y. P.; Li, H. B.; Pan, L. K.; Lu, T.; Sun, Z. *Journal of Electroanalytical Chemistry*2009, *634*, 68-71.
[78] Yan, J.; Wei, T.; Shao, B.; Fan, Z. J.; Qian, W. Z.; Zhang, M. L.; Wei, F. *Carbon*2010, *48*, 487-493.
[79] Yan, J.; Wei, T.; Fan, Z. J.; Qian, W. Z.; Zhang, M. L.; Shen, X. D.; Wei, F. *Journal of Power Sources*2010, *195*, 3041-3045.
[80] Yan, J.; Wei, T.; Qiao, W. M.; Shao, B.; Zhao, Q. K.; Zhang, L. J.; Fan, Z. J. *Electrochimica Acta*2010, *55*, 6973-6978.
[81] Wu, Q.; Xu, Y. X.; Yao, Z. Y.; Liu, A. R.; Shi, G. Q. *Acs Nano*2010, *4*, 1963-1970.
[82] Semenov, Y. G.; Zavada, J. M.; Kim, K. W. *Journal of Applied Physics*2010, *107*, 064507.
[83] Zhu, G. X.; Liu, Y. J.; Xu, Z.; Jiang, T. A.; Zhang, C.; Li, X.; Qi, G. *Chemphyschem*2010, *11*, 2432-2437.
[84] Pisana, S.; Braganca, P. M.; Marinero, E. E.; Gurney, B. A. *Ieee Transactions On Magnetics*2010, *46*, 1910-1913.
[85] Shao, Y. Y.; Wang, J.; Wu, H.; Liu, J.; Aksay, I. A.; Lin, Y. H. *Electroanalysis*2010, *22*, 1027-1036.
[86] Pumera, M. *Chemical Record*2009, *9*, 211-223.
[87] Wang, D. H.; Kou, R.; Choi, D.; Yang, Z. G.; Nie, Z. M.; Li, J.; Saraf, L. V.; Hu, D. H.; Zhang, J. G.; Graff, G. L.; Liu, J.; Pope, M. A.; Aksay, I. A. *Acs Nano*2010, *4*, 1587-1595.
[88] Mukhopadhyay, P.; Gupta, R. K. *Plastics Engineering*2011, *67*, 32-42.
[89] Stankovich, S.; Dikin, D. A.; Dommett, G.; Kohlhaas, K. M.; Zimney, E. J.; Stach, E. A.; Piner, R. D.; Nguyen, S. T.; Ruoff, R. S. *Nature*2006, *442*, 282-286.
[90] Pang, H.; Chen, T.; Zhang, G. M.; Zeng, B. Q.; Li, Z. M. *Materials Letters*2010, *64*, 2226-2229.

[91] Ansari, S.; Giannelis, E. P. *Journal of Polymer Science Part B-Polymer Physics*2009, *47*, 888-897.
[92] Ramanathan, T.; Abdala, A. A.; Stankovich, S.; Dikin, D. A.; Herrera-Alonso, M.; Piner, R. D.; Adamson, D. H.; Schniepp, H. C.; Chen, X.; Ruoff, R. S.; Nguyen, S. T.; Aksay, I. A.; Prud'Homme, R. K.; Brinson, L. C. *Nature Nanotechnology*2008, *3*, 327-331.
[93] Rafiee, M. A.; Rafiee, J.; Srivastava, I.; Wang, Z.; Song, H. H.; Yu, Z. Z.; Koratkar, N. *Small*2010, *6*, 179-183.
[94] Yavari, F.; Rafiee, M. A.; Rafiee, J.; Yu, Z. Z.; Koratkar, N. *Acs Applied Materials & Interfaces*2010, *2*, 2738-2743.
[95] Zhou, K. F.; Zhu, Y. H.; Yang, X. L.; Li, C. Z. *New Journal of Chemistry*2010, *34*, 2950-2955.
[96] Liang, J. J.; Xu, Y. F.; Sui, D.; Zhang, L.; Huang, Y.; Ma, Y. F.; Li, F. F.; Chen, Y. S. *Journal of Physical Chemistry C*2010, *114*, 17465-17471.
[97] Yang, X. Y.; Zhang, X. Y.; Ma, Y. F.; Huang, Y.; Wang, Y. S.; Chen, Y. S. *Journal of Materials Chemistry*2009, *19*, 2710-2714.
[98] He, H. K.; Gao, C. *Acs Applied Materials & Interfaces*2010, *2*, 3201-3210.
[99] Zhou, G. M.; Wang, D. W.; Li, F.; Zhang, L. L.; Li, N.; Wu, Z. S.; Wen, L.; Lu, G. Q.; Cheng, H. M. *Chemistry of Materials*2010, *22*, 5306-5313.
[100] Lian, P. C.; Zhu, X. F.; Xiang, H. F.; Li, Z.; Yang, W. S.; Wang, H. H. *Electrochimica Acta*2010, *56*, 834-840.
[101] He, F. A.; Fan, J. T.; Ma, D.; Zhang, L. M.; Leung, C.; Chan, H. L. *Carbon*2010, *48*, 3139-3144.
[102] Paek, S. M.; Yoo, E.; Honma, I. *Nano Letters*2009, *9*, 72-75.
[103] Wang, Z. Y.; Zhang, H.; Li, N.; Shi, Z. J.; Gu, Z. N.; Cao, G. P. *Nano Research*2010, *3*, 748-756.
[104] Yao, J.; Shen, X. P.; Wang, B.; Liu, H. K.; Wang, G. X. *Electrochemistry Communications*2009, *11*, 1849-1852.
[105] Du, Z. F.; Yin, X. M.; Zhang, M.; Hao, Q. Y.; Wang, Y. G.; Wang, T. H. *Materials Letters*2010, *64*, 2076-2079.
[106] Wang, X. Y.; Zhou, X. F.; Yao, K.; Zhang, J. G.; Liu, Z. P. *Carbon*2011, *49*, 133-139.
[107] Meng, X. B.; Geng, D. S.; Liu, J. A.; Banis, M. N.; Zhang, Y.; Li, R. Y.; Sun, X. L. *Journal of Physical Chemistry C*2010, *114*, 18330-18337.
[108] Lu, T.; Zhang, Y. P.; Li, H. B.; Pan, L. K.; Li, Y. L.; Sun, Z. *Electrochimica Acta*2010, *55*, 4170-4173.
[109] Li, F. H.; Song, J. F.; Yang, H. F.; Gan, S. Y.; Zhang, Q. X.; Han, D. X.; Ivaska, A.; Niu, L. *Nanotechnology*2009, *20*, 455602.
[110] Xu, C.; Wang, X.; Yang, L. C.; Wu, Y. P. *Journal of Solid State Chemistry*2009, *182*, 2486-2490.
[111] Li, N.; Wang, Z. Y.; Zhao, K. K.; Shi, Z. J.; Xu, S. K.; Gu, Z. N. *Journal of Nanoscience and Nanotechnology*2010, *10*, 6690-6693.
[112] Wang, B.; Wu, X. L.; Shu, C. Y.; Guo, Y. G.; Wang, C. R. *Journal of Materials Chemistry*2010, *20*, 10661-10664.
[113] Lin, J. A.; Penchev, M.; Wang, G. P.; Paul, R. K.; Zhong, J. B.; Jing, X. Y.; Ozkan, M.; Ozkan, C. S. *Small*2010, *6*, 2448-2452.

[114] Hwang, S. W.; Shin, D. H.; Kim, C. O.; Hong, S. H.; Kim, M. C.; Kim, J.; Lim, K. Y.; Kim, S.; Choi, S. H.; Ahn, K. J.; Kim, G.; Sim, S. H.; Hong, B. H. *Physical Review Letters*2010, *105*, 127403.

[115] Cuong, T. V.; Pham, V. H.; Chung, J. S.; Shin, E. W.; Yoo, D. H.; Hahn, S. H.; Huh, J. S.; Rue, G. H.; Kim, E. J.; Hur, S. H.; Kohl, P. A. *Materials Letters*2010, *64*, 2479-2482.

[116] Wang, H. L.; Cui, L. F.; Yang, Y. A.; Casalongue, H. S.; Robinson, J. T.; Liang, Y. Y.; Cui, Y.; Dai, H. J. *Journal of the American Chemical Society*2010, *132*, 13978-13980.

[117] Wang, B.; Park, J.; Wang, C. Y.; Ahn, H.; Wang, G. X. *Electrochimica Acta*2010, *55*, 6812-6817.

[118] Kim, H.; Seo, D. H.; Kim, S. W.; Kim, J.; Kang, K. *Carbon*2011, *49*, 326-332.

[119] Chen, S. Q.; Wang, Y. *Journal of Materials Chemistry*2010, *20*, 9735-9739.

[120] Wu, Z. S.; Ren, W. C.; Wen, L.; Gao, L. B.; Zhao, J. P.; Chen, Z. P.; Zhou, G. M.; Li, F.; Cheng, H. M. *Acs Nano*2010, *4*, 3187-3194.

[121] Yan, J.; Wei, T.; Qiao, W. M.; Shao, B.; Zhao, Q. K.; Zhang, L. J.; Fan, Z. J. *Electrochimica Acta*2010, *55*, 6973-6978.

In: Graphene: Properties, Synthesis and Applications
Editor: Zhiping Xu
ISBN: 978-1-61470-949-7

Chapter 7

Electrochemical Performance of Graphene

Dale A. C. Brownson and Craig E. Banks*
Faculty of Science and Engineering,
School of Science and the Environment,
Division of Chemistry and Environmental Science,
Manchester Metropolitan University, Manchester, Lancs, UK

Abstract

The electrochemical performance and utilisation of graphene is explored and overviewed. Starting with graphene's use as a sensor substrate through to its utilisation in energy production and storage devices, this chapter details the reported enhancements that graphene has imposed upon various electrochemical applications. A comprehensive approach is employed in which the possible limitations of graphene's exploitation within electrochemistry are highlighted and insight into the future prospects of graphene electrochemistry is sought at each stage.

Introduction

Graphene, a single atomic layer thick two-dimensional sp2 bonded carbon nanostructure, has captured the attention of scientists worldwide since its discovery in 2004. [1-4] Its reported unique physical, chemical, thermal and electrical properties have led to its exploitation in a variety of scientific disciplines, [1-4] however, one area of particular interest is electrochemistry where graphitic forms of carbon have widespread applicability as disposable electrode materials due to their non-toxic and highly conductive attributes. 5 Graphene is a key derivative of carbon and originates from a large family of fullerene nanomaterials where it is the essential 'building block' for many of the allotropic dimensionalities that have significant and widespread use as electrode materials, [2] for example, graphene can be 'folded' into zero-dimensional spherical buckyballs, 'rolled' into

* E-mail address: c.banks@mmu.ac.uk

one-dimensional carbon nanotubes (CNTs), or 'stacked' into three-dimensional graphite; [2-4, 6] a conceptual depiction of graphene is shown in Figure 1.

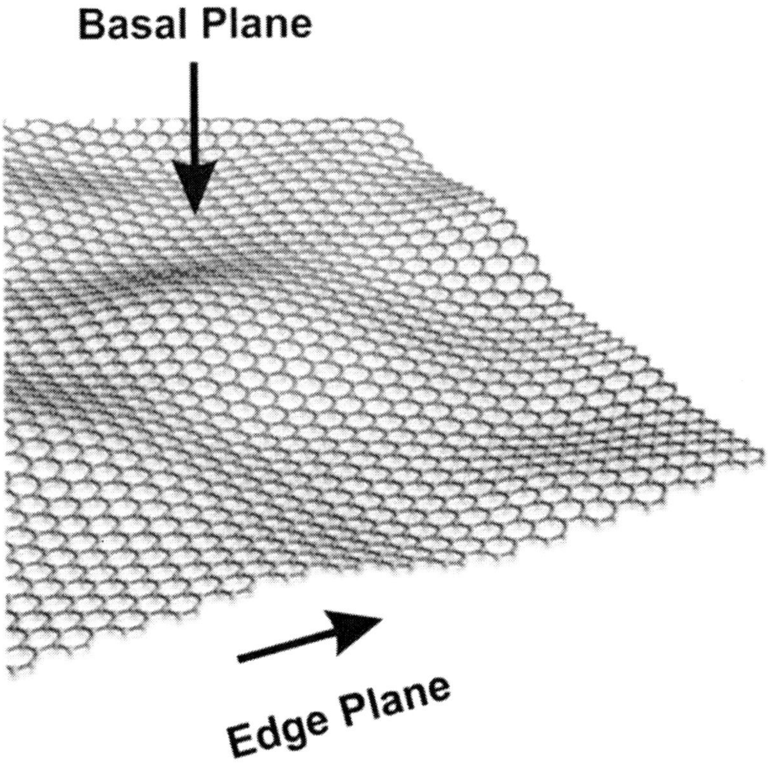

Figure 1. A conceptual model depicting the structure of graphene; also shown are the sites of electron transfer, basal and edge plane like- sites.

Electrochemistry has implications within a plethora of industrial, medical and academic applications. [7] In addition to aiding the understanding and visualisation of many scientific principals and theories, [8] electrochemistry is utilised for the fabrication of sensing devices that have global ramifications for the detection of substances harmful to human health and the environment, [2, 7] and has further relevance in both energy production and storage devices that are of vital significance given the increasing demand for greener energy related devices with improved power output and efficiency. [9] Importantly, it is the properties of the electrode material itself that are most significant when influencing the performance obtained from fabricated sensing or energy related devices. [2] It is therefore vital that research be conducted into the utilisation of various electrode materials (such as graphene), where research has the potential to result in the fabrication of sufficiently enhanced renewable energy devices, high capacitive durable batteries, and the quick, cheap, sensitive, and portable detection apparatus of the future. [2]

In this chapter we explore the electrochemical applicability of graphene and highlight key areas where the impact of this novel nanomaterial is of elevated significance. We begin by detailing the advancements made in sensing, covering the electro-catalysis of graphene, sensing of gaseous species, trace metal analysis, bio-sensing, and other interesting areas of sensing, after which we extend our overview into energy storage and production devices, including graphene based capacitors, batteries, and fuel cells. The possible limitations of utilising graphene for electrochemical applications as well as an insight into the future prospects of graphene and a direct comparison of graphene towards existing and competing electrode materials (such as CNTs) is carefully considered throughout.

FABRICATION FOR ELECTROCHEMISTRY

Many fabrication methods exist for the synthesis of graphene, however, not all are suitable for each of the applications graphene has to offer and no 'single' method exists for the pertinent production of graphene for use within all of these potential areas, consequently, because it is the fabrication method that defines the resultant properties and suitability of graphene for use within various applications (for example its electrochemical performance), fabrication routes are heavily researched. [2, 3, 6, 10]

The enormous selection of synthesis methods for graphene include: dry mechanical exfoliation, suited ideally for the investigation of graphene's physical properties; [1, 3, 11, 12] the chemical exfoliation of graphite, where bulk quantities of graphene sheets are created through an intercalation process where small molecules (surfactants such as tetrabutyl ammonium hydroxide) are incorporated between the graphene layers in graphite, consequently leading to the separation of the graphene sheets; [3, 4] the '*in*' or '*ex*' *–situ* reduction of graphene oxide following the exfoliation of oxidised graphite; [10, 13, 14] the unzipping of CNTs through a variety of methods such as electrochemical, chemical, or physical processes; [12] the epitaxial growth of graphene on various substrates (for example utilising chemical vapour deposition (CVD)); [15, 16] and finally the reduction of sugar (such as glucose, fructose, or sucrose), [17] this last method is leading the innovation for the cheap-large-scale production of graphene for applications in various fields. [17]

Note, although the cheap, rapid and large-scale production of 'high quality, large-area' graphene is obtainable through the implementation of many of the above fabrication methods (for example, reduction of graphene oxide or epitaxial growth methodologies) [10] there are a number of limitations that must be addressed: to elevate the mistaken use of multi-layered graphene (graphite) where it is assumed single layer graphene is present, appropriate characterisation needs to be performed prior to graphenes use in electrochemical applications where the control of layer thickness and the manipulation of its morphology are variables (see later), [2, 9, 10] further, adequate characterisation is required where the reduction of graphene oxide is employed as some reduction methodologies may only be partial, leading to incomplete reduction; [10] additionally, caution towards the electrochemical activity of surfactants must be considered when incorporated into fabrication processes (see later), [18-20] and although CVD synthesis of graphene generally does not incorporate the same metallic impurities that are commonly the 'hidden' origin of electrochemical activity in CNTs, [16, 21] appropriate control measures need to be performed to ensure a sufficient lack of interferents (as is the case for each of the limitations noted above – see later). Note, as with CNTs the reproducibility of fabrication and control of surface defects are likely to be commonly encountered problems associated with graphene. [2]

GRAPHENE AS AN ELECTRODE MATERIAL

Carbon derivatives are widely utilised as electrode materials in both analytical and industrial electrochemistry, where in numerous areas they have out-performed more traditional 'noble metal' alternatives.[2] This success and diversity originates largely from carbons structural polymorphism, chemical stability, low cost, wide potential window, rich surface chemistry, and electro-catalytic activity towards a variety of analytes.[2, 5] Most recently the classical carbon materials (graphite, carbon black, glassy carbon, diamond and so on) have been out-performed by the distinctive properties of micro-fabricated carbon structures, such as CNTs, facilitating novel applications in energy and sensing related devices.[2]

When reviewing the essential characteristics of an electrode material for widespread applicability within electrochemistry, it is apparent that graphene is in many ways superior over rival materials. A large surface area is of high importance for both sensing and energy related applications and graphene exhibits a higher theoretical specific surface area (2630 m^2 g^{-1}) than both graphite and CNTs (~10 and 1315 m^2 g^{-1} respectively), [3] with additional benefits including graphene's high flexibility (as opposed to graphite's brittleness), [9] attractive $\pi - \pi$ interactions, and its strong absorptive capability suggesting a viable future in energy and molecule storage as well as bio-sensing. [22] Of further significance for electrochemical applications is the electrical conductivity of graphene (~64 mS cm^{-1}) resulting from its extensive conjugated sp^2 carbon network, [9, 23] which is calculated to be ~60 times higher than that of CNTs and has been shown to remain stable over a vast range of temperatures. [4, 24] Furthermore, electrons on the surface of graphene move ballistically without scattering or colliding; with mobilities as high as 10,000 cm^2 V^{-1} s^{-1} at room temperature graphene exhibits the fastest heterogeneous electron mobilities and transfer rates when compared to other possible materials, thus 'theoretically' graphene based sensor electrodes can react much faster and efficiently. [1, 2, 4, 25]

Another advantage of graphene that corresponds to determining its electrochemical performance is the availability for addition of various moieties (for example oxygen-containing groups) at its edges or surface. These moieties will greatly influence the electrochemical performance of graphene in terms of heterogeneous electron transfer rates as is observed for graphite and CNT electrodes – which can be either advantageous or detrimental towards the sensing of a target analyte. [2, 26, 27] Furthermore, where controlled attachment of functional elements is required, such as in the development of nano-architectonics or the attachment of bio-molecules (which are expected to be similar to that observed for CNTs), [28] these sites provide convenient attachment sites where specific groups can be introduced that play vital roles in electrochemical sensing and battery applications, hence the electrochemical properties of graphene-based electrodes can be modified or tuned by chemical modification and tailored to suit appropriate applications. [2]

Table 1. A table depicting the relative limits of detection and linear ranges of graphene based electrodes when utilised towards various analyte.

Analyte	Analytical parameter		Electro-catalytic w.r.t (graphitic form)	Underlying electrode / complex	Comments	Ref.
	LOD	Linear Range				
4-Aminophenol	0.057 µM	0.2 – 550 µM	G	GCE and CH/GCE	G synthesised *via* RGO	[103]
Adenine	0.75 µM	5 – 200 µM	G	G-NAF complex on a GCE	Explored DNA samples in milk powder, urine and herring sperm	[53]
Ascorbic Acid	0.12 mM	0.4 – 6.0 mM	G	PPFE	G synthesised *via* DMF exfoliation	[36]
Cadmium	0.18 nM	13.3 – 266 nM	G	G-NAF complex on a GCE and bare NAF/GCE	G synthesised *via* RGO	[41]
Caffeine	0.12 µM	0.4 – 40 µM	G	G-NAF complex on a GCE	N/A	[104]
Chlorpromazine	6 nM	0.01 – 9 µM	GPE	N/A	N/A	[105]
Dopamine	2.64 µM	4 – 100 µM	G	GCE	G synthesised *via* RGO. Tested in the presence of ascorbic acid	[52]
Ethanol	5 µM	5 – 200 µM	ADH/IL-functionalised G	GCE	G synthesised *via* RGO. Explored in diluted wine and beer samples	[106]
Glucose	0.02 mM	0.08 – 12 mM	GO	GCE and CH/GCE	GOD/GO/CH Nanocomposite film	[50]
Guanine	0.58 µM	2 – 200 µM	G	G-NAF complex on a GCE	Explored DNA samples in milk powder, urine and herring sperm	[53]
Guanosine	0.75 µM	2.0 µM – 0.35 mM	G	CH/GCE, nano-Fe_3O_4/GCE and G/CH/GCE	G synthesised *via* RGO	[107]

Analyte	Analytical parameter		Electro-catalytic w.r.t (graphitic form)	Underlying electrode / complex	Comments	Ref.
	LOD	Linear Range				
Hydrazine	1 mM	1 – 25 mM	G	GCE	G synthesised *via* RGO. 0.1 M KOH electrolyte	[29]
Hydrogen Peroxide	1.7 μM	5.0 μM – 5.13 mM	G	Au/HRP/GS/CH complex on a GCE	G synthesised *via* RGO	[108]
Lead	0.096 nM	7.2 – 14.5 nM	G	G-NAF complex on a GCE and bare NAF/GCE	G synthesised *via* RGO	[41]
NADH	0.25 mM	0.25 – 2 mM	ADH/IL-functionalised G	GCE and SWCNTs	G synthesised *via* RGO	[106]
NADH	1.9 μM	45 – 360 μM	G	PPFE	G synthesised *via* DMF exfoliation	[35]
Nitric Oxide	3.6 μM	3.6 – 43.2 μM	G	GCE	G synthesised *via* RGO	[109]
Paracetamol	0.032 μM	0.1 – 20 μM	G	GCE	G synthesised *via* RGO. Explored in real human blood, plasma, and pharmaceutical preparations	[22]
TNT	1 ppm	1 – 15 ppm	G	GCE	G synthesised *via* SC exfoliation	[49]

Key: ADH – Alcohol dehydrogenase; Au – Gold; CH – Chitosan; DMF – Dimethylformamide; G – Graphene; GCE – Glassy carbon electrode; GO – Graphene Oxide; GOD – Glucose oxidase; GPE – Graphene paste electrode; GS – Graphene sheet; HRP – Horseradish peroxidise; IL – Ionic liquid; N/A – Not applicable; NADH – β-Nicotinamide adenine dinucleotide; NAF – Nafion; PB – Prussian Blue; PPFE – Pyrolysed photoresist film electrode; RGO – Reduced graphene oxide; SC – Sodium cholate; SWCNTs – Single walled carbon nanotubes; TNT – 2,4,6-trinitrotoluene; w.r.t – With respect to.

As highlighted above graphene's inimitable electrochemical properties are superior over various other carbon allotropes, theoretically making graphene the ideal replacement electrode material for electrochemical applications, where until now CNTs have fore-fronted innovation and dominated the field. Graphene is reported to offer many advantageous properties and is likely to exhibit fewer of the weaknesses that have plagued CNTs, thus relevant research can be built on the wealth of techniques and knowledge accumulated from decades of research on graphite and CNTs, where graphene has the opportunity to outperform at each stage. [2, 5]

SENSING APPLICATIONS

The detection of various substances including heavy metals, bio-molecules, gases, and miscellaneous organic and inorganic compounds is of high importance due to the many human and environmental health implications, thus the development of improved analytical sensors is of great significance. Table 1 overviews a selection of literature reports where graphene has been utilised in numerous areas of sensing.

Electro-Catalysis

The 'electro-catalytic' behaviour and enhanced analytical performance of graphene and related structures has been widely reported. For example, Wang *et al.* [29] have reported the electrochemical sensing of hydrazine, greatly significant in fuel cells, rocket propellants, insecticides and explosives. [2] On graphite surfaces hydrazine usually exhibits irreversible kinetics, such that the electrochemical signal occurs at high potentials close to the end of the accessible voltammetric window, but is more electrochemically reversible on metallic surfaces. [30, 31] As evident in Figure 2A a graphene (synthesised *via* the reduction of graphene oxide) modified glassy carbon electrode (GCE) has been shown to possess excellent 'electro-catalytic' activity towards the sensing of hydrazine as compared to the underling GCE substrate. [29] While the analytical utility was not examined in much depth, the smallest addition was 1 mM which was linear up to 25 mM. [29] Furthermore, another notable example is the sensing of paracetamol using graphene sheets functionalised with both hydroxyl and carboxylic groups (synthesised from graphite oxide). [22] Figure 2B depicts the voltammetric profiles where a quasi-reversible redox process of paracetamol at graphene-modified GCE was obtained showing that the over-potential of paracetamol was significantly decreased when compared to that of the same analysis upon an unadulterated GCE, thus the graphene-modified electrode was reported to exhibit excellent 'electro-catalytic' activity towards paracetamol. The graphene modified GCE outperformed the unadulterated GCE consistently where at low concentrations the graphene-modified GCE exhibited a pair of well-defined redox waves, and in contrast the unadulterated GCE showed an irreversible behaviour with relatively weak redox current peaks. [22] The enhancement of the graphene based electrode compared to the unadulterated GCE is claimed to be due to the nanocomposite film of graphene 'accelerating' the electrochemical reaction, [22] and the large surface area of graphene can be attributed to the large background current exhibited by the graphene modified electrode in Figure 2B. Thus, the graphene based electrochemical sensor shows excellent performance for detecting paracetamol with a detection limit of 3.2 x 10^{-8} M, a linear range of 0.1 – 20 μM, and a reproducibility of 5.2% RSD. [22] This sensor demonstrated that graphene based electrodes can be utilised for the simple, sensitive, and quantitative detection and screening of paracetamol. [22]

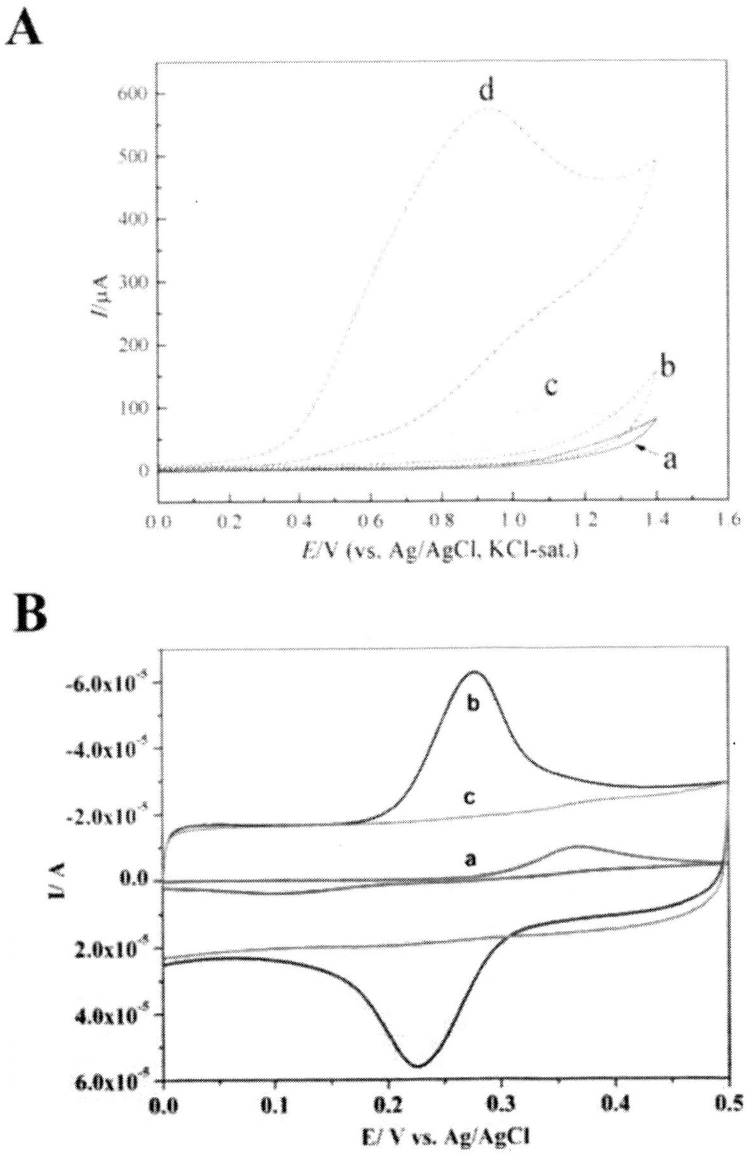

Figure 2. **A.** Cyclic voltammetric responses recorded at (a and c) the bare glassy carbon and (b and d) the graphene modified glassy carbon electrode (a and b) in the absence and (c and d) in the presence of 10 mM hydrazine in 0.1 M KOH solution at a scan rate of 100 mV s^{-1}. Reproduced from Ref. [29] with permission from Elsevier. **B.** The electrochemical sensing of 100 μM paracetamol at a bare glassy carbon electrode (a); and the sensing of 20 μM paracetamol at a graphene modified glassy carbon electrode (b), and without paracetamol (c) in the buffer of 0.1 M NH3•H2O–NH4Cl, pH 9.3, scan rate: 50 mV s^{-1}. Reproduced from Ref. [22] with permission from Elsevier.

It is interesting to note that in the above examples (and many others in Table 1) the electro-catalysis of graphene is compared *only* to the underlying electrode and not to other relevant carbon materials, such as graphite or graphene oxide. A similar issue occurred when CNTs were explored in electrochemistry and 'electro-catalysis' was claimed when compared only to the underlying substrates (usually GCEs), and consequently was proven to be false. [32] In an attempt to gain insight into the electro-catalytic performance of graphene Mao and co-workers [33] compared the electrochemical activity of four kinds of carbon materials, namely single-walled CNTs (SWCNTs), pristine graphene oxide nanosheets (GONs), chemically reduced GONs, and electrochemically reduced GONs, using potassium ferricyanide, β-nicotinamide adenine dinucleotide (NADH) and ascorbic acid (AA) as redox probes. Mao and co-workers [33] found that the electron transfer kinetics of the analytical probes were dependant on the kind of material used, of which the redox processes of the probes at SWCNTs and electrochemically reduced GONs were faster than those at the pristine and chemically reduced GONs. The authors inferred that synergetic effects of the surface chemistry, such as the surface charge, C/O ratio, presence of quinone-like groups, surface cleanliness, and the conductivity of the materials utilised contribute greatly. [33] In essence, Mao's paper highlights that the electron transfer kinetics of the redox probes employed are essentially dependent on the type of carbon-based materials utilised. [2, 33]

Further important work towards understanding the electrochemical performance of graphene is from Lin *et al.* [[34]] who compared graphene 'film' modified basal and edge plane pyrolytic graphite (BPPG and EPPG respectively) electrodes for the electro-catalytic oxidation of hydrogen peroxide and NADH with that of bare BPPG and EPPG electrodes. They observed that the application of graphene to the electrode surface has the ability to lower the electro-oxidation potentials of hydrogen peroxide and NADH in comparison with unmodified BPPG and EPPG electrodes, thus indicating that graphene appeared to exhibit an enhanced "electro-catalytic" activity towards both analytes. [34] However, a problem with their study [34] is that the graphene utilised contained on average 4-6+ layers, which is approaching that of graphite! Elegant work by Keeley *et al.* [35, 36] has attempted to preclude the issues that have plagued earlier attempts at utilising graphene, where towards the sensing of NADH and AA they used dimethylformamide (DMF) exfoliated graphene immobilised on pyrolysed photoresist film electrodes (PPFEs): the DMF fabrication was chosen in an attempt to avoid a high oxygen content (an issue with reduced graphene oxide fabrication methodologies) where for the stable and sensitive detection of NADH a low oxygen content is vital; PPFEs were chosen opposed to a GCE substrate as they are less electrochemically active and thus permit the de-convolution of kinetic data. Although the authors report favourable results (see Table 1) appropriate controls were not performed and thus the electro-catalytic influence of DMF is unclear, and in addition to this characterisation of the graphene utilised indicated a thickness of on average 5 layers, which again is approaching that of graphite.

Interestingly, we have also shown that surfactants, which are routinely incorporated into graphene's fabrication to reduce the coalescing of graphene layers (thus causing the separation of graphene layers – see above), [37] play an important role on the electrochemical response either beneficially or detrimentally. [19] Using NADH and paracetamol (see Figure 3) as analytical probes we demonstrated that where surfactants are incorporated the electro-catalytic response of graphene should not be assumed to be solely due to the graphene itself, *viz* edge plane like sites, but rather the surfactant (sodium cholate) in this case, which is evident in Figure 3, the electro-catalytic behaviour exhibited from a surfactant modified electrode (in the absence of graphene) is similar to that of a graphene modified electrode; thus the 'observed' electro-catalytic response from graphene will reside truly from a contributory effect of both graphene and the surfactant. [19]

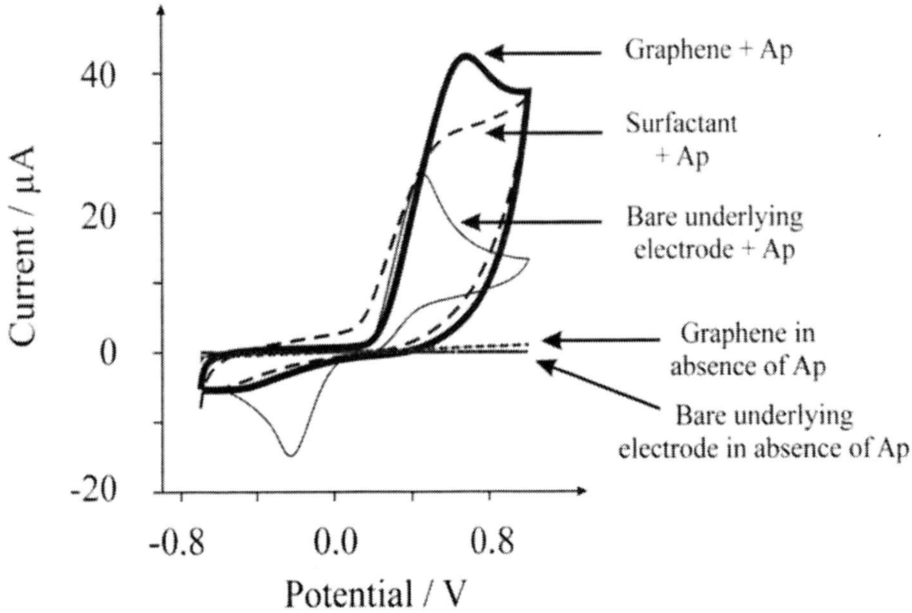

Figure 3. Cyclic voltammetric profiles recorded for 1 mM paracetamol (Ap) in a pH 7 phosphate buffer solution using a 0.5 µg graphene modified electrode (thick line) and in the absence of Ap (dotted line) and the response of an unmodified basal plane like- screen printed electrode in the presence (thin line) and absence (solid line) of Ap. Also shown is the response of a surfactant modified (dashed line) electrode in the presence of Ap. Scan rate: 200 mV s^{-1} (vs. Ag/AgCl). Reproduced from Ref. [19] with permission from Wiley.

We have recently demonstrated, with the use of EPPG and BPPG of highly ordered pyrolytic graphite (HOPG) electrodes, that the electro-catalytic nature of graphene resides in electron transfer from the edge of the graphene structure, which resembles the behaviour of edge plane (rather than basal plane) of HOPG; [38] Figure 1 shows a conceptual picture of graphene's electron transfer sites where the rate of electron transfer at its edge is anomalously faster than that of its side. Additionally, we have demonstrated the importance of control experiments and offered further insight into the electro-catalytic performance of graphene [39] through the comparison of graphene and graphite modified screen printed graphite electrodes (SPEs) towards the amperometric sensing of hydrogen peroxide. We have demonstrated that graphite exhibits a superior electrochemical response towards hydrogen peroxide due to its enhanced percentage of edge plane sites when compared to graphene, where sensitivities were measured at 32.8 and 49.0 µA/mM for graphene and graphite modified electrodes respectively. [39] Furthermore, we offered insight into the fabrication of graphene based amperometric sensors, demonstrating that through the addition of Nafion, the orientation of graphene upon the electrode surface can be altered leading to an increase in available edge plane sites and thus an increase in sensitivity from 32.8 to 42.2 μA/mM, which coincidently outperformed a Nafion-graphite modified electrode which exhibited a decrease in sensitivity from 49.0 to 32.0 μA/mM, attributed to the blocking of edge plane sites. [39] Note however that the original sensitivity of the graphite modified electrode remains higher than that of graphene. In addition to these findings, work by Pumera *et al.* [40] has emerged which demonstrates that single-, few-, and multi-layer graphene does not provide a significant advantage over graphite micro-particles in terms of sensitivity, linearity, and repeatability towards the electro-analytical detection of uric acid, where sensitivities of 4.65 and 5.11 μA/mM were recorded for graphene and graphite modified electrodes respectively. [40]

Although there are many encouraging examples of the reported 'electro-catalysis' of graphene where improved sensitivity and detection limits have been obtained, it is clear that control experiments with EPPG and BPPG electrodes need to be conducted to gain a clearer understanding of the observed 'electro-catalysis' at graphene modified electrodes, as well as with surfactant modified electrodes, to ensure that the observed response is truly due to the presence of graphene. [2, 19, 38, 39]

Metals

Given the widespread exploration of graphene in many electrochemical areas, the use of graphene for target heavy metals is surprisingly limited. Recently however graphene (dispersed in Nafion (Nafion-G) solution) was used in combination with an 'in-situ' plated bismuth film electrode for the sensing of lead and cadmium *via* differential pulse anodic stripping voltammetry (DPASV). [41] It was found that the prepared Nafion-G composite film not only exhibited improved sensitivity for metal ion detection, but also alleviated the interferences due to the synergistic effect of graphene nanosheets and Nafion. The detection limits were estimated to be around 0.02 ppb for Pb^{2+} and Cd^{2+}, and the linear calibration curves ranged from 0.5 to 50 ppb for Pb^{2+} and 1.5 to 30 ppb for Cd^{2+} respectively. [41] As is evident in Figure 4A the authors report sharper and higher peak currents for the target metal ions using the Nafion-G modified electrode, an improvement of 79% in comparison to a Nafion coated electrode. Additionally, using a CNT/Nafion modified electrode, a 69% improvement was observed, indicating a similar behaviour to graphene. Shown in Figure 4B are SEM images of the Nafion and Nafion-G which are distinctively different. [41] The Nafion modified electrode shows a flat and homogeneous film while the Nafion-graphene composite film exhibits a rough and stratified structure – allowing a more effective active area for the nucleation of the bismuth due to this unique structure. [41] Impedance changes of the electrode surfaces were conducted where it was found that when compared with that of the pure Nafion modified electrode, the electron-transfer resistance of the Nafion-G modified electrode was greatly reduced, suggesting the interfusion of the graphene into the Nafion film greatly reduced electron-transfer-limited process; due to the high conductivity resulting from the extensive conjugated sp^2 carbon network, the change in the morphologies and the structure, and the interfacial electron-transfer properties of the Nafion-G modified electrode. [41] Note, as with CNTs, care should be taken in identifying what enhancements are simply due to increased surface area. [2]

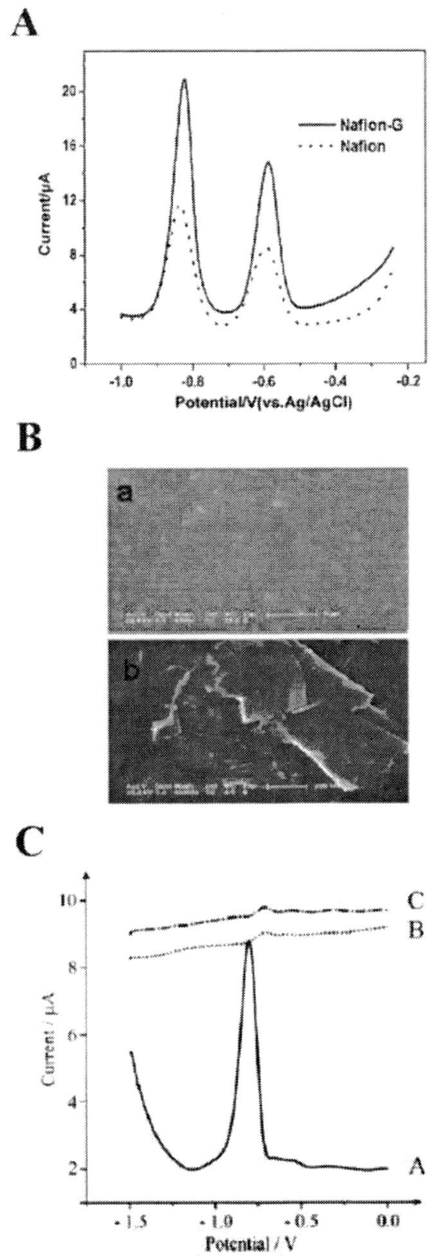

Figure 4. **A.** DPASVs for 20 μg L^{-1} each of Cd^{2+} and Pb^{2+} on an in situ plated Nafion-BFE, Nafion-G-BFE in solution containing 1 mg L^{-1} Bi^{3+}. **B.** SEM image of bismuth film deposited on the Nafion (a) and Nafion-G (b) modified GCE. Supporting electrolyte: 0.1 M acetate buffer (pH 4.5); deposition potential: −1.2 V; deposition time: 120 s; amplitude: 80 mV; increment potential: 5 mV; quiet time: 10 s. **A** and **B** reproduced from Ref. [41] with permission from Elsevier. **C.** Typical square-wave voltammetric profiles for detection of 200 μgL^{-1} cadmium in pH 2.5 buffer solution obtained at a bare screen printed graphite electrode (SPE) (A), a 0.5 μg graphene modified SPE (B), and a surfactant modified SPE (C). A deposition potential of −1.3 V (vs. SCE) for 120 seconds was utilised. Reproduced from Ref. [18] with permission from Elsevier.

The reported detection limits of around 0.02 ppb for lead and cadmium, appear to be more sensitive than previously reported Nafion film modified bismuth electrodes; 0.1 ppb for lead and cadmium [42] and yet comparable to CNT/Nafion coated bismuth film electrodes; 0.025 ppb for lead and 0.04 ppb for cadmium. [43] Recently this has been extended for graphene oxide-Nafion nanocomposite film modified GCEs for the sensing of cadmium; with a reported detection limit of 0.005 ppb. [44] Clearly graphene exhibits excellent potential for use in trace metal detection, and appears to provide state-of-the-art analytical responses.

Once again however, care must be taken to avoid the misinterpretation of experimental data when surfactants have been incorporated into graphene's fabrication process. [18] Utilising graphene modified SPEs we have recently shown that the surfactant (sodium cholate) used in graphene's fabrication actually inhibits the electro-analytical sensing of cadmium (II) *via* anodic stripping voltammetry (ASV). [18] As is evident in Figure 4C, inspection of the stripping step reveals that the surfactant inhibits the transition of cadmium metal to cadmium ions, where a poor voltammetric signature is exhibited for both the graphene (in the presence of surfactant) and the surfactant (no graphene) modified electrodes; in direct comparison the bare underlying electrode exhibits a large and quantifiable signal. [18]

Gaseous and miscellaneous sensing

Interesting work has been reported by Schedin and co-workers, [37] whom have demonstrated that graphene sensors are capable of detecting *individual* gas molecules attaching or detaching from the surface of graphene. The process of molecules attaching/detaching changes the local carrier concentration in graphene, allowing the resistance to be used as a convenient means of measurement. Furthermore, work using density functional theory has shown that molecules trapped between graphene electrode plates can be used to sense molecules *via* vibrational fluctuations. This was experimentally shown to be the case using water, trapped between two graphene molecules connected to a potential difference, where electrical current fluctuations were generated; corresponding to the fluctuations of the vibrational modes. [45] Other work has demonstrated that graphene oxide sheets deposited on gold interdigitated electrodes allow the sensing of NO_2, NH_3, and 2,4-dinitriotoluene *via* resistance measurements. [46, 47]

Additional uses of graphene have been for the rapid detection (after 10 seconds of exposure) of an explosive and three main classes of chemical-warfare agents at ppb concentrations, which were found to be superior to that previously reported using CNTs. [48] Furthermore, Pumera *et al.* [49] has recently utilised various graphitic forms towards the detection of 2,4,6-trinitrotoluene (TNT) in seawater where a detection limit of 1 ppm (1 – 15 ppm linear range) for single layer graphene was obtained, however the authors elegantly compared the responses of single-, few-, and multi-layer graphene nanoribbons to that of graphite microparticle-based electrodes where no significant difference in the performance was noted. Note however, surfactants are used in the fabrication of the graphitic forms utilised and (as stated earlier) could be dominating the observed electrochemical response – thus no observable difference would be evident. [19]

Bio-Sensing

As we have shown above, graphene based electrode materials are beginning to revolutionise certain sensing applications, and have been shown to exhibit improved selectivity and sensitivity, display wider dynamic ranges, and exhibit lower detection limits than other analytical methods available today; [22] the application of graphene within bio-sensing where electron transfer between redox enzymes is concerned is no exception to this trend. Improvements in graphene's conduction and electrode kinetics, its larger surface area, thermal stability, flexibility, and the ability to act as a nano connector for electrical connection with electrode substrates [2] mean that graphene exhibits a high enzyme surface load capacity, and in addition to this graphene offers a truly unique opportunity within bio-sensing because unlike in most semiconductor systems, its two-dimensional electronic states are not buried deep under the surface, [50] and can be accessed directly by tunnelling and other local probes, meaning that its enzyme catalytic properties can truly be realised; [4] because of this, graphene is an ideal material for electrochemical bio-sensing.

A novel and promising electrochemical sensing platform for detecting bio-molecules has been demonstrated by Zhu *et al.* [17] whereby they have shown that graphene sheets (also known as graphene nano-sheets (GNSs)) (of which were synthesised through reducing sugar) exhibit electro-catalytic activity towards the most abundant catecholamines (dopamine, epinephrine, and norepinephrine), which act as neuromodulators in nervous systems or as hormones in the circulatory system. [17] It was shown that the utilisation of a GNS modified GCE for the electro-catalysis detection of cholamines achieved greatly increased current densities and exhibited an enhanced electrochemical activity over that of an unadulterated GCE. The improved electrochemical response was ascribed to graphene's capability of accumulation for various analytes, improved conductivity and greater surface area, and the $\pi - \pi$ interactions between the analytes and the basal planes of graphene were also proposed to play an important role in electro-catalysis. [17]

Furthermore, graphene modified electrodes have also been used within bio-sensing as a glucose sensor in which Kang *et al.* [50] have reported a study upon the direct electrochemistry of a glucose oxidase-graphene-chitosan nanocomposite (loaded upon a GCE), claiming that the immobilised enzyme retained its bioactivity and exhibited a surface confined, reversible two-proton and two-electron transfer reaction, exhibiting activity, stability, and fast heterogeneous electron transfer rates. The authors commented that the use of a graphene based material obtained a greatly improved enzyme loading capability than glassy carbon surfaces. [50] When compared to other nanostructured supports (including: SWCNTs, multi-walled CNTs (MWCNTs), and an unadulterated GCE) the fabricated bio-sensor exhibited a wide linear range of 0.08 to 12 mM, and had a lower glucose detection limit of 0.02 mM in addition to a greatly enhanced sensitivity, demonstrating that a glucose oxidase-graphene-chitosan nanocomposite film can be used as a sensitive and effective glucose detection strategy. Other work [51] on glucose sensing has reported that graphene oxide exhibits high sensitivity, their linear range was up to 28 mM for glucose, with potential for graphene based glucose bio-sensors for clinical diagnosis in the future. [51]

Recently, Kim *et al.* [52] have developed a sensitive electrochemical sensor for the determination of dopamine without the interference of ascorbic acid (the oxidation potentials of the two analytes are very similar – thus poor reproducibility and selectivity plagued earlier attempts where the signals overlap). Modifying a GCE with graphene and comparing its performance to unmodified GCEs, using cyclic voltammetry, the authors found that the difference between the two peak potentials of dopamine and ascorbic acid oxidation was well over 200 mV with use of the graphene modified electrode and observed a linear range for the detection of dopamine from 4 to 100 μM, with a detection limit of 2.64 μM. [52] This detection was superior to conventional electrodes such as Au, Pt, and bare GCE, where the interference from ascorbic acid results in an overlap of signals. [52] Furthermore, graphene modified electrodes have shown superior bio-sensing performance over SWCNTs toward dopamine detection in the presence of another common interfering agent, serotonin, [23] and it has also been shown that adenine and guanine can be simultaneously detected using a Nafion-G composite film modified GCE with a separation between the two oxidation peaks of 364 mV and a low μM limit of detection for both analytes being reported. [53]

Finally, of critical importance in many areas of modern biomedical and biochemical research is the sensitive detection of disease-related proteins (immunoproteins). [54] Research into graphene has already produced some exciting prospects in this area, where reliable and quick diagnosis of, for example, cancer or tumour biomarkers could lead to the utilisation of pro-active treatment long before the disease has taken hold – increasing the patients chances of recovery. [2] Zhong *et al.* [55] have developed a highly sensitive electrochemical immunosensor that was designed to quantify carcinoembryotic antigens (CEA) as a tumour marker using nanogold-enwrapped graphene nanocomposites (NGGNs) as trace labels in clinical immunoassays. The device consisted of a GCE coated in Prussian Blue (PB), on whose surface the NGGNs were electrochemically deposited which was then further modified with the specific analyte-capturing molecule, anti-CEA antibodies. Their results indicated that the method using the anti-CEA-NGGN-PB-GCE as detection antibodies showed high signal amplification and exhibited a dynamic working range of 0.05 – 350 ng mL^{-1} with a low limit of detection of 0.01 ng mL^{-1}. In comparison, performance of an identical analysis using an anti-CEA-nanogold-PB-GCE for detection exhibited less sensitivity and lower signal values. Further work by Du and co-workers [54] has reported a novel electrochemical immunosensor for the sensitive detection of the cancer biomarker α-fetoprotein (AFP) utilising a GNS sensor platform and functionalised carbon nanosphere (CNS) labelling. The developed immunosensor showed a 7-fold increase in detection signal where graphene greatly enhanced the sensitivity for the cancer biomarker compared to the immunosensor without graphene modification and CNS labelling. The author's method could respond to 0.02 ng mL^{-1} AFP with a linear range of 0.05 – 6 ng mL^{-1}. Thus graphene appears to be a promising platform on which future enhancements will no doubt develop providing ultrasensitive assay sensors for clinical applications in the future.

Energy Storage Applications

Energy generation and storage devices are inundated with both environmental and performance related issues, [56-58] where increasing demands for the improved portability, energy efficiency, and power output of capacitors, batteries, and fuel cells constitute immense research interest and consequently the exploration and exploitation of new electrode materials is of great significance and importance. [59-61]

Table 2. A table depicting the specific capacitance and power outputs for a variety of graphene based materials and various other comparable materials for use in super-capacitor devices

Electrode material	Performance parameter		Cyclic Ability	Comments	Ref.
	Specific capacitance / F g^{-1}	Power density / kW kg^{-1}			
CNT/PANI	780	NT	Capacitance decreased by 67 % after 1000 cycles	Obtained from CV measurement at a SR of 1 mVs^{-1}	[63]
Fe_3O_4	104	NT	Remaining stable after 1000 cycles of charge/discharge	Obtained at a specific charge/discharge current of 5 A g^{-1}	[72]
Fe_3O_4/RGO	480	NT	Remaining stable after 1000 cycles of charge/discharge	Obtained at a specific charge/discharge current of 5 A g^{-1}	[72]
GNS	150	NT	Specific capacitance was maintained with the specific current of 0.1 A g^{-1} for 500 cycles of charge/discharge	N/A	[64]
GNS	38.9	2.5	NT	Synthesised *via* screen-printing and ultrasonic spray pyrolysis. Data obtained from CV measurement at a SR of 50 mVs^{-1}	[68]
GNS	205	10 @ 28.5 W h kg^{-1}	~ 90% specific capacitance remaining after 1200 cycles	Graphene was prepared *via* graphite oxide	[66]
GNS	62.6	NT	NT	Obtained at a discharge current of 67 nA	[20]
GNS/CNT/PANI	1035	NT	Capacitance decreased by 6 % after 1000 cycles	Obtained from CV measurement at a SR of 1 mVs^{-1}	[63]
GNS/PANI	1046	70 @ 39 W h kg^{-1}	NT	Synthesised using *in-situ* polymerisation. Capacitance obtained from CV measurement at a SR of 1 mVs^{-1}	[67]
GNS-Cobalt (II) hydroxide nano-composite	972.5	NT	NT	N/A	[75]
GNS-Nickel foam	164	NT	Specific capacitance remained 61% of the maximum capacitance after 700 cycles	Obtained from CV measurement at a SR of 10 mVs^{-1}	[70]

Electrode material	Performance parameter		Cyclic Ability	Comments	Ref.
	Specific capacitance / F g^{-1}	Power density / kW kg^{-1}			
GNS-SnO_2	42.7	3.9	NT	Synthesised *via* screen-printing and ultrasonic spray pyrolysis. Data obtained from CV measurement at a SR of 50 mVs^{-1}	[68]
GNS-ZnO	61.7	4.8	NT	Synthesised *via* screen-printing and ultrasonic spray pyrolysis. Data obtained from CV measurement at a SR of 50 mVs^{-1}	[68]
GOS	8.1	NT	NT	Obtained at a discharge current of 67 nA	[20]
Graphite	5.4	NT	NT	Obtained at a discharge current of 67 nA	[20]
MWCNT/PANI	463	NT	NT	Synthesised using *in-situ* polymerisation. Capacitance obtained from CV measurement at a SR of 1 mVs^{-1}	[67]
NIHYD/GNS	1335	NT	NT	Obtained at a charge/discharge density of 2.8 A g^{-1}	[73]
PANI	115	NT	NT	Synthesised using *in-situ* polymerisation. Capacitance obtained from CV measurement at a SR of 1 mVs^{-1}	[67]
PANI/GOS	531	NT	NT	Nanocomposite with a mass ratio of PANI/graphene, 100:1. Capacitance obtained by charge-discharge analysis	[56]
RGO	139	NT	Remaining stable after 1000 cycles of charge/discharge	Obtained at a specific charge/discharge current of 5 A g^{-1}	[72]
RuO_2/GNS	570	10 @ 20.1 W h kg^{-1}	~97.9% specific capacitance remaining after 1000 cycles	N/A	[74]
SC	156	NT	NT	Obtained at a discharge current of 1.3 μA	[20]
SC integrated GNSs	148.6	NT	NT	GNS in the presence of surfactant due to the fabrication process. Data obtained at a discharge current of 1.3 μA	[20]

Key: CNT – Carbon nanotube; CV – Cyclic voltammetry; GNS – Graphene nano-sheet; GOS – Graphene oxide sheets; MWCNT – Multi-walled carbon nanotube; N/A – Not applicable; NIHYD – Nickel (II) hydroxide nano-crystals; NT – Not tested; PANI – Polyaniline; RGO – Reduced graphene oxide; RuO_2 – Hydrous ruthenium oxide; SC – Sodium cholate (surfactant); SR – Scan rate.

Super-Capacitor

Table 2 overviews a selection of literature reports where graphene has been utilised as a super-capacitor material in comparison to existing materials, for example graphite and CNTs.

Electrochemical super-capacitors are passive and static electrical energy storage devices that find application in portable electronics, memory back-up systems, and hybrid cars, where extremely fast charging is a valuable feature. [9, 62, 63] In super-capacitor devices energy is stored either due to the formation of an electrical double layer at the interface of the electrode and the electrolyte (electrical double layer capacitors) [64] or due to electron transfer between the electrolyte and the electrode through fast Faradiac redox reactions (pseudo-capacitors). [65] For super-capacitors the ability to utilise both of the previous energy storage capabilities with high efficiency, where these two mechanisms can function simultaneously depending on the nature of the electrode material, is often the capacitance determining factor. [62]

Current research on electrochemical capacitors is focused on increasing the power and energy densities of fabricated devices, as well as lowering fabrication costs whilst incorporating environmentally friendly materials. [9] The core materials studied for super-capacitor electrodes are carbons, metal oxides, and conducting polymers, [56] with recent innovation having focused on CNTs. However, graphene based materials have shown great practical and theoretical advantages, such as excellent conductivity, a high surface area and relatively low production costs (mass production). Thus unsurprisingly the use of graphene as a super-capacitor material has been extensively investigated, where it appears that many studies report graphene as a far superior super-capacitor material than existing carbon and polymer based materials.

Numerous reports have emerged where graphene has been utilised as an independent super-capacitor material. For example, Du *et al.* [64] mass-produced graphene nano sheets (GNSs) with a narrow mesopore distribution of ~4 nm from natural graphite *via* oxidation and rapid heating processes finding them to maintain a stable specific capacitance of 150 F g^{-1} under the specific current of 0.1 A g^{-1} for 500 cycles of charge/discharge. [64] Furthermore, Wang *et al.* [66] report a maximum specific capacitance of 205 F g^{-1} with a power density of 10 kW kg^{-1} at a energy density of 28.5 Wh/kg whilst utilising graphene, excellent cyclic ability was obtained also with ~90% specific capacitance remaining after 1200 cycles. Further exciting prospects are apparent when considering work by Vivekchand *et al.* [65] who have shown graphene (prepared *via* graphitic oxide whilst utilising an ionic liquid – thus enabling the operating voltage to be extended up to 3.5 V) to exhibit a specific capacitance and energy density of ~75 F g^{-1} and 31.9 Wh kg^{-1} respectively, that surpasses both SW- and MWCNTs, 64 and 14 F g^{-1} respectively, when tested in an identical manner. It is claimed that the performance characteristics of graphene are directly related to its quality in terms of numbers of graphene layers, inherent surface area, and available edge plane sites. [65] These comparable results suggest potential for graphene based capacitors to be fabricated, but scalability and reproducibility might be common issues.

It is clear however, that the general specific capacitance of graphene is not as high as expected, and thus it is notable that many researchers have turned to fabricating graphene based hybrid materials in the pursuit of improved capacitance performance. [9] In one notable example Wang and colleagues [56] report a novel 'high performance electrode material' based on fibrillar polyaniline (PANI) doped graphene oxide sheets. The authors obtained a nanocomposite with a mass ratio of PANI/graphene, 100 : 1, which exhibited a high specific capacitance of 531 F g^{-1} (obtained by charge-discharge analysis), that when compared to individual PANI (216 F g^{-1}) it was clear doping has a profound effect on the electrochemical capacitative performance of the nanocomposites. Further work on graphene/PANI composites has been reported by Yan *et al.*, [67] where their in-situ polymerisation synthesised composite obtained a high specific capacitance of 1046 F g^{-1} (examined using cyclic voltammetry (CV) and galvanostatic charge/discharge analysis), which compared to 115 F g^{-1} for pure PANI, 463 F g^{-1} for SWCNT/PANI and 500 F g^{-1} for MWCNT/PANI. Note also, the energy density of the GNS/PANI composite could reach 39 W h kg^{-1} at a power density of 70 kW kg^{-1}. [67] It is apparent that GNS/PANI modifications offer a highly conductive support material where the well-dispersed depositions of nanoscale PANI particles are attributable to the GNSs flexibility and large surface area. Other work on this topic [63] has investigated the effect of a GNS/CNT/PANI composite, claiming responses similar to the composites mentioned above, however excellent cyclic performance was also noted, where after 1000 cycles the capacitance decreased by only 6 % of the initial compared to 52 and 67 % for GNS/PANI and CNT/PANI composites respectively. It is evident that a hybrid-graphene material may exhibit the ultimately desired properties required for superior energy related devices to be realised. [59, 63]

In addition to these reports, further work has indicated that graphene hybrid materials have the potential to outperform counterparts as super-capacitors. For example it has been highlighted that graphene-ZnO composites exhibit enhanced capacitative performance's when compared to CNTs, as well as an enhanced reversible charge/discharging ability. [57, 66, 68] In one notable example [68] a screen-printing approach was employed to deposit a graphene film onto a graphite substrate while ZnO or SnO_2 was deposited onto the graphene film *via* ultrasonic spray pyrolysis. The performances of these electrodes were then tested through electrochemical impedance spectroscopy, CV, and chronopotentiometry, and results demonstrated that the inclusion of either ZnO or SnO_2 improved the capacitive performance of the graphene electrode. [68] When compared to the GNS-SnO_2 and pure graphene electrodes, that exhibited capacitances of 42.7 and 38.9 F g^{-1} and power densities of 3.9 and 2.5 kW kg^{-1} respectively, the GNS-ZnO composite electrode exhibited a higher capacitance value of 61.7 F g^{-1} and maximum power density of 4.8 kW kg^{-1}. One explanation for this superior capacitive behaviour, besides graphene's enhanced surface area (and other qualities discussed earlier), has been ascribed to the increase in lattice defect density exhibited by graphene (such as the fracture of graphene layers and the ratio of edge to basal carbon surface), where it is believed that increased edge plane increases the capacitance of a material. [9, 64] However, work by Stoller and co-workers [69] stated that on chemically modified graphene electrodes the super-capacitor performance does not depend on a rigid porous structure to deliver its large surface area but on graphene's flexibility, where graphene sheets adjust their position depending on the electrolyte used. Stoller *et al.* [69] found the weight specific capacitance of chemically modified graphene to be up to 135 F g^{-1}.

It appears that hybrid materials are the way forward for enhanced super-capacitor devices and numerous other materials have been utilised in combination with graphene to form these potential hybrid materials, for example, the electrophoretic deposition of graphene onto nickel foams, and the deposition of silver nanoparticles onto graphene sheets have both been reported to improve specific capacitance values. [70, 71]

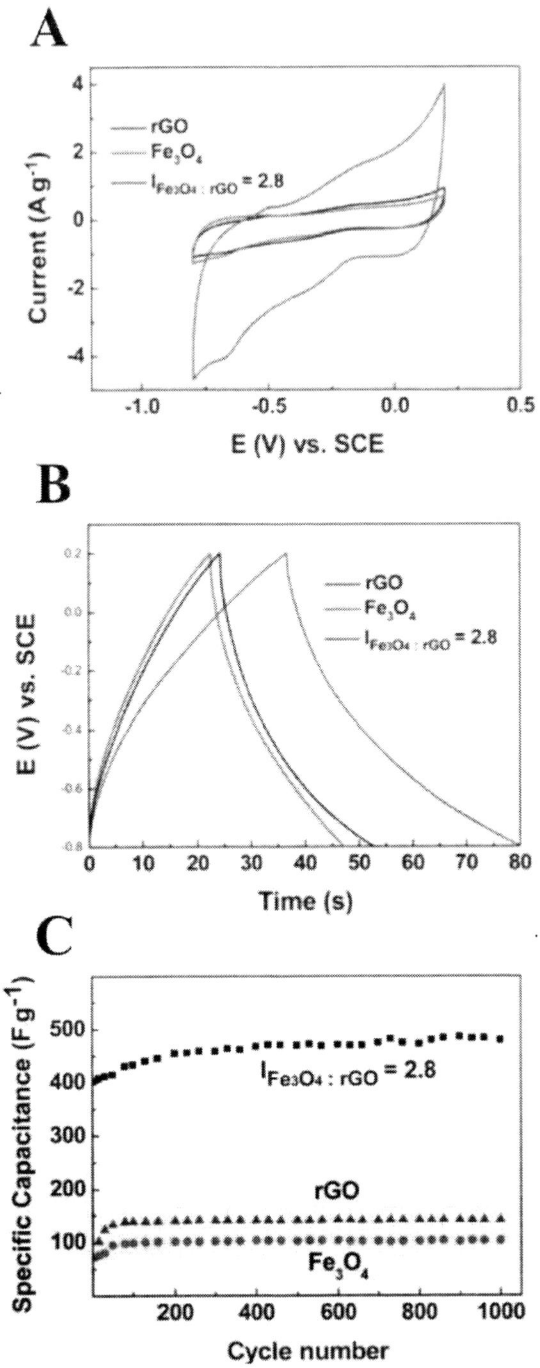

Figure 5. **A.** Cyclic voltammogram curves of Fe3O4 nanoparticles (NPs), reduced graphene oxide (rGO) and nanocomposites with $I_{Fe3O4\ :\ rGO}$ = 2.8 at a scan rate of 5 mV s^{-1} in 1 M KOH. **B.** Galvanostatic charge/discharge curves of rGO, Fe_3O_4 and nanocomposite (as earlier) in 1 M KOH solution at 5 A g^{-1}. **C.** The cycling performance of Fe_3O_4 NPs, rGO and the nanocomposite electrode (as earlier) at a current density of 5 A g^{-1} in 1 M KOH. Reproduced from Ref. [72] by permission of The Royal Society of Chemistry.

In one notable example Shi *et al.* [72] synthesised nanocomposites with Fe_3O_4 nanoparticles attached onto reduced graphene oxide (rGO) sheets (*via* a single step solvothermal process), where it was shown that (at the optimum Fe_3O_4 : rGO weight ratio of 2.8) the Fe_3O_4/rGO nanocomposite exhibited a greater specific capacitance (480 F g^{-1}) than both the individual components (139 and 104 F g^{-1} for rGO and Fe_3O_4 respectively) at a specific discharge current of 5 A g^{-1}; Figure 5 depicts cyclic voltammetric and galvanostatic charge/discharge data in addition to cycling data where it is evident the fabricated nano-composite exhibits stable cycling performance (exhibiting no decrease in capacitance after 1000 charge/discharge cycles).

Other work has been explored where inorganic nanomaterials have been anchored onto graphene's surface, resulting in high performance super-capacitors. [73-75] Wang *et al.* [73] have grown nickel (II) hydroxide nano-crystals on graphene sheets, finding that as well as exhibiting excellent cyclic ability the resultant material exhibited a high specific capacitance of ~1335 F g^{-1} at a charge/discharge density of 2.8 A g^{-1} and ~953 F g^{-1} at 45.7 A g^{-1} respectively; when comparison was sought, a physical mixture of pre-synthesized nickel (II) hydroxide nano-plates and graphene exhibited a lower specific capacitance, highlighting the importance of the direct growth of these nanomaterials on graphene to enhance the intimate interactions and charge transport between the conducting underlying network and the active nanomaterials. [73] Further work by Wu and co-workers [74] has demonstrated that hydrous ruthenium oxide/graphene composites exhibit a high specific capacitance of ~570 F g^{-1} (for 38.3% ruthenium weight loading) with enhanced rate capability and excellent electrochemical stability (~97.9% retention after 1000 cycles). Interestingly, it was found that the total specific capacitance of the composite material was higher than the sum of specific capacitances of pure graphene and pure ruthenium oxide in their relative ratios, which was ascribed to be indicative of a positive synergistic effect of graphene and ruthenium oxide on the improvement of the electrochemical performance. [74] Moreover, Chen *et al.* [75] have reported on the utilisation of a graphene-cobalt (II) hydroxide nano-composite and shown the electrochemical specific capacitance of the graphene composite to reach 972.5 F g^{-1}, leading to a significant improvement in relation to each individual component (137.6 and 726.1 F g^{-1} for graphene and cobalt·hydroxide respectively). Thus these findings suggest the importance of rational design and the synthesis of graphene-based nano-composite materials for high-performance energy-storage systems. [73-75]

It is important to note that work by Pumera *et al.* [76] has shown it is not always beneficial to exfoliate graphitic structures to single layer graphene in order to achieve maximum electrochemical performance and capacitance. The authors demonstrated, using electrochemical impedance spectroscopy, that multilayer (>10 layers) graphene nanoribbons with cross sections of 100 x 100 nm provide a larger capacitance (15.6 F g^{-1}) than few-layer (3 to 9 layers) graphene nanoribbons (14.9 F g^{-1}) and a far greater capacitance than single layer graphene sheets (10.9 F g^{-1}) with the same cross section; [76] likely caused by graphene layers laying flat on top of one another blocking/concealing the edge plane sites and available surface area – as has been shown previously where basal plane based electrodes have greatly reduced charge storage when compared to edge plane surfaces, likely due to a larger proportion of available edge plane sites, [20, 77, 78] thus Pumera *et al.* [76] state that vertically orientated GNSs are a near-ideal structure due to larger inter-layer spaces allowing for improved capacitance due to the complete utilisation of available surface area and edge plane sites. [76, 77] Interestingly however, we have recently studied the capacitance of graphene and graphite modified electrodes, where our findings are in direct contrast to those noted above. [20] Galvanostatic charge/discharge analysis was performed on both graphene and graphite modified SPEs (under identical conditions) where the specific capacitances of 62.6 F g^{-1} and 5.4 F g^{-1} were obtained respectively at a discharge current of 67 nA. [20] It appears that although graphite has a larger portion of edge plane sites when compared to graphene, it is a 'bulk' material where tightly stacked layering leads to the reduced accessibility of available surface area. [20] We infer that the larger capacitance of graphene is attributed to a combination of capacitance due to available edge plane sites and surface area, where enhanced layer spacing results in the superior accessibility of these available components in flexible graphene sheets as opposed to bulk graphite, consequently leading to the higher capacitance values observed. [20] We also highlighted the impact of surfactants that, as within sensing applications, when present, require appropriate control experiments to be performed to de-convolute the true origin the responses observed. [20] The specific capacitance of an alternatively available commercially produced graphene (in the presence of the surfactant sodium cholate – due to the fabrication process) was found to be 148.6 F g^{-1} at a discharge current of 1.3 μA, which compared to 156.0 F g^{-1} for the surfactant alone (no graphene) under identical conditions. [20] It is clear that the surfactant dominates the capacitive behaviour of graphene and thus the observed response is likely from the surfactant and not the graphene, where graphene's influence (the difference) is diminutive. [20] Thus, without performing appropriate control experiments the misinterpretation of experimental data can potentially lead to false claims of graphene's superior capacitive behaviour. [20] The observed response of Pumera *et al.* [76] (see above) could therefore be attributed to the presence of surfactants, where due to the increased interlayer spacing caused by the presence of surfactants (see earlier) in addition to the electro-activity of the surfactant itself, the effects of increasing graphene layer numbers in the presence of surfactant is not truly representative of the same comparison performed in the absence of surfactant. This observation is further supported when considering the use of hybrid materials, where the combination of graphene with various other materials leads to the increased availability of the graphene's edge-plane sites and surface area, as well as influencing other factors such as inter-layer spacing leading to numerous advantageous prospects in capacitance and energy storage applications. [20]

It is clear that graphene has made a significant impact and revealed itself to be a promising material for application within future research concerning the fabrication of super-capacitors, where the possibility of hybrid graphene materials may lead to the fabrication of super-capacitor composites that are able to acquire currently unthinkable capacitance levels and greatly improved cyclic abilities; of course a major limitation is the presence of interfering compounds (such as surfactants), the current cost of graphene, reproducibility, scalability, and characterisation, where drawing from the latter many reports are emerging where claims of the utilisation of graphene are made without adequate characterisation and control experiments being performed (where results are compared to other appropriate carbon forms such as graphite or activated carbons) and in such cases perhaps graphene is not present, rather multiple layer or greater. [20]

Batteries (Li-ion storage)

Table 3 overviews a selection of various graphene based electrode materials reported in the literature for use in lithium based batteries, where comparison to other electrode materials (namely graphite and CNTs) is also highlighted.

As with super-capacitors, there is an increasing worldwide demand for advanced lithium (Li) -ion batteries with longer cycle lifetimes and higher energy capacities, which are promising with regards to their application in electric vehicles. [11, 79, 80] Li-ion batteries store and supply electricity over long periods of time, where the electrode material (i.e. the anode/cathode) plays a dominant role in the batteries performance. [81, 82] Currently the anode material employed for lithium based batteries is usually graphite because of its high Coulombic efficiency (the ratio of the extracted Li to the inserted Li) [81] where it can be reversibly charged and discharged under intercalation potentials with a reasonable specific capacity. [9, 60] To improve battery performance the relatively low theoretical capacity associated with graphite batteries (372 mA h g-1) and the long diffusion distances of the Li-ions need to be overcome. [9, 81] Graphene's reportedly superior physical attributes suggest that it should constitute a beneficial replacement electrode material for lithium based rechargeable batteries.

In one interesting example Yoo *et al.* [83] fabricated a graphene based electrode using CNTs as spacers to prevent the restacking of graphene sheets upon the electrode surface, where the modified electrode was demonstrated to possess a specific capacitance (towards lithium) of 730 mA h g^{-1}, [83] which is greatly improved when compared to existing graphite (372 mA h g^{-1}) [3] and an alternate graphene modified electrode without the use of spacers (540 mA h g^{-1}). Note also, the use of graphene without spacers exhibits a larger capacitance than that of comparable graphite. [3, 11, 60, 79, 84] When Yoo and co-workers utilised C_{60} molecules in addition to spacers between the graphene sheets, the capacitance increased further to 784 mA h g^{-1}. [83] In another interesting example Lian *et al.* [79] reported a large reversible capacity where high quality graphene sheets were used as an anode material in coin-type cells for Li-ion batteries, showing that the GNSs possess a curled morphology with a large specific surface area. The first reversible specific capacity of the prepared GNS was as high as 1264 mA h g^{-1} at a current density of 100 mA g^{-1}, where even at high current densities of 500 mA g^{-1} the reversible capacity remained at 718 mA h g^{-1}. Further on this subject Pan *et al.* [11] have observed that 'few layer' graphene samples exhibit superior characteristics when they are disordered in nature. The authors reported the fabrication of highly disordered graphene nanosheets which exhibit promising possibilities in high-capacity Li-ion batteries because of their high reversible capacity (1054 mA h g^{-1}) that is far superior when compared to highly ordered GNSs (540 mA h g^{-1}) and graphite samples (372 mA h g^{-1}) (as above), thus the observable capacity of graphene based materials was attributed (as with super-capacitors) by Pan and colleagues to larger inter-layer spacing and increased edge plane sites and accessible surface area – made available due to the disorder of the graphene sheets. [11] The disorder of GNSs can be achieved *via* a variety of methods including the introduction of hybrid materials or for example by electron-beam irradiation as utilised in the work highlighted here. [11] Graphene's advantageous application in lithium storage devices has also been partly explained by Takamura *et al.* [85] whom linked the high rate discharge capability of Li-ion batteries to nano-sized holes identified within the graphene sheets. However, this is not the only issue when looking at the performance of batteries, as cyclic abilities also need to be considered. [9] Thus Bhardwaj and colleagues [86] have reported on the electrochemical Li intake capacity of carbonaceous one-dimensional GNSs obtained by unzipping pristine MWCNTs. The authors showed that oxidised-GNSs (ox-GNSs) outperformed all other materials tested (including GNSs and MWCNTs) in terms of energy density, obtaining a first charge capacity of ~1400 mA h g^{-1} with a low Coulombic efficiency for the first cycle, ~53%, and ~95% for subsequent cycles, which was shown to be notably superior to MWCNTs and GNSs and similar to that of graphite. [86] The cyclic capacity of the ox-GNSs was within the range of 800 mA h g^{-1}, with an early capacity loss per cycle of ~3% (decreasing upon subsequent cycles). [86] The cyclability is the capacity loss per cycle, and is a useful parameter to evaluate battery performance, where here, MWCNTs and GNSs perform better than the proposed ox-GNSs (capacity losses of ~1.4% and ~2.6% respectively), however, due to a higher initial capacitance of the ox-GNSs the performance of this material can still be regarded as superior over its counterparts as the 'stable capacitance' does not fall below that of the other materials. [86]

As is the case for super-capacitors, research regarding Li-ion batteries has focused strongly towards the fabrication of hybrid graphene composite materials when looking for enhanced battery performance. For example, previous work by Paek *et al.* [60] involved the fabrication of a graphene/SnO_2 based nano-porous electrode, that was shown to exhibit a reversible capacity of 810 mA h g^{-1} and a cyclic performance that was drastically enhanced when compared to bare SnO_2 nanoparticles, where after 30 cycles the charge capacity of the graphene/SnO_2 based electrode remained at 570 mA h g^{-1} (70% retention of the reversibile capacity) and in comparison the bare SnO_2 nanoparticles first charge capacity was 550 mA h g^{-1} which dropped rapidly to 60 mA h g^{-1} after only 15 cycles at 50 mA g^{-1}. Paek and co-workers [60] also reported superior cyclic performances over graphite modified nanoparticles, where the initial capacitance was ~500 mA h g^{-1}, which dropped slightly during cycling – the bare graphene electrode resided only slightly above this level. Further work on SnO_2-graphene hybrid based Li-ion batteries has yielded similar results with regards to capacitance and obtained good cyclic abilities, retaining 520 mA h g^{-1} after 100 cycles. [87, 88] In other work Wang and co-workers [89] have developed a two-step solution-phase reaction to form Mn_3O_4 nanoparticles on rGO sheets. The authors claim their selective growth of Mn_3O_4 nanoparticles onto the rGO sheets (in contrast to free particle growth in solution) allowed for the electrically insulating Mn_3O_4 nanoparticles to be 'wired-up' to a current collector through the underlying conducting graphene network. [89] The Mn_3O_4/rGO hybrid material demonstrated a high specific capacity up to ~900 mA h g^{-1} (near the theoretical capacity of Mn_3O_4 – ~936 mA h g^{-1}) with a good rate capability and cycling stability (a capacity of ~730 mA h g^{-1} at 400 mA g^{-1} was retained after 40 cycles) owing to the intimate interactions between the two components. [89] Even at a high current density of 1600 mA g^{-1} the specific capacity was shown to be ~390 mA h g^{-1}, which is higher than the theoretical capacity of graphite – commonly used within Li-ion batteries (372 mA h g^{-1}). [89] Control experiments involving free Mn_3O_4 nanoparticles (synthesised *via* the same process) (without the presence of graphene) showed them to exhibit poor responses; at a low current density of 40 mA g^{-1} the free Mn_3O_4 nanoparticles exhibited a capacity lower than 300 mA h g^{-1}, which further decreased to ~115 mA h g^{-1} after only 10 cycles. [89] Consequently Wang and co-workers claim that the above hybrid material could be a promising candidate for the low-cost, high-capacity, and environmentally friendly anode materials (for Li-ion batteries) of the future, since the observed performance surpasses that of graphite alternatives. [89]

Table 3. Overview of the specific capacitance and cyclic stabilities for a variety of graphene based materials and various other comparable materials for utilisation as a Lithium-ion battery electrode

Electrode compound	Specific capacitance / mA h g^{-1}	Cyclic stability	Comments	Ref.
GNS	540	300 after 30 cycles at 50 mA g^{-1}	N/A	[60, 83]
GNS	1264	848 after 40 cycles at 100 mA g^{-1}	GNSs in coin-type cells	[79]
GNS	1233	502 after 30 cycles at the current density of 0.2 mA cm^{-2}	GNSs were prepared from artificial graphite by oxidation, rapid expansion and ultrasonic treatment.	[80]
GNS/Fe_3O_4	1026	580 after 100 cycles at 700 mA g^{-1}	N/A	[82]
GNS/SnO_2	860	570 after 30 cycles at 50 mA g^{-1}	GNSs were homogeneously distributed between the loosely packed SnO_2 nanoparticles.	[60]
GNS/SnO_2	840	590 after 50 cycles at the current density of 400 mA g^{-1}	The optimum molar ratio of SnO_2/graphene was 3.2:1.	[88]
GNS/with C_{60} spacer molecules	784	NT	N/A	[83]
GNSs with CNT spacers	730	NT	N/A	[83]
Graphite	372	240 after 30 cycles at 50 mA g^{-1}	N/A	[60, 83]
Mn_3O_4/RGO	~900	~730 after 40 cycles at 400 mA g^{-1}	N/A	[89]
Ox-GNSs	~1400	Cyclic stability within the range of 800 cycles	Capacity loss per cycle of ~3% for early cycles; this decreasing for subsequent cycles.	[86]

Key: C_{60} – Carbon 60; CNT – Carbon nanotube; GNS – Graphene nanosheet; N/A – Not applicable; NT – Not tested; Ox – Oxidised; RGO – Reduced graphene oxide.

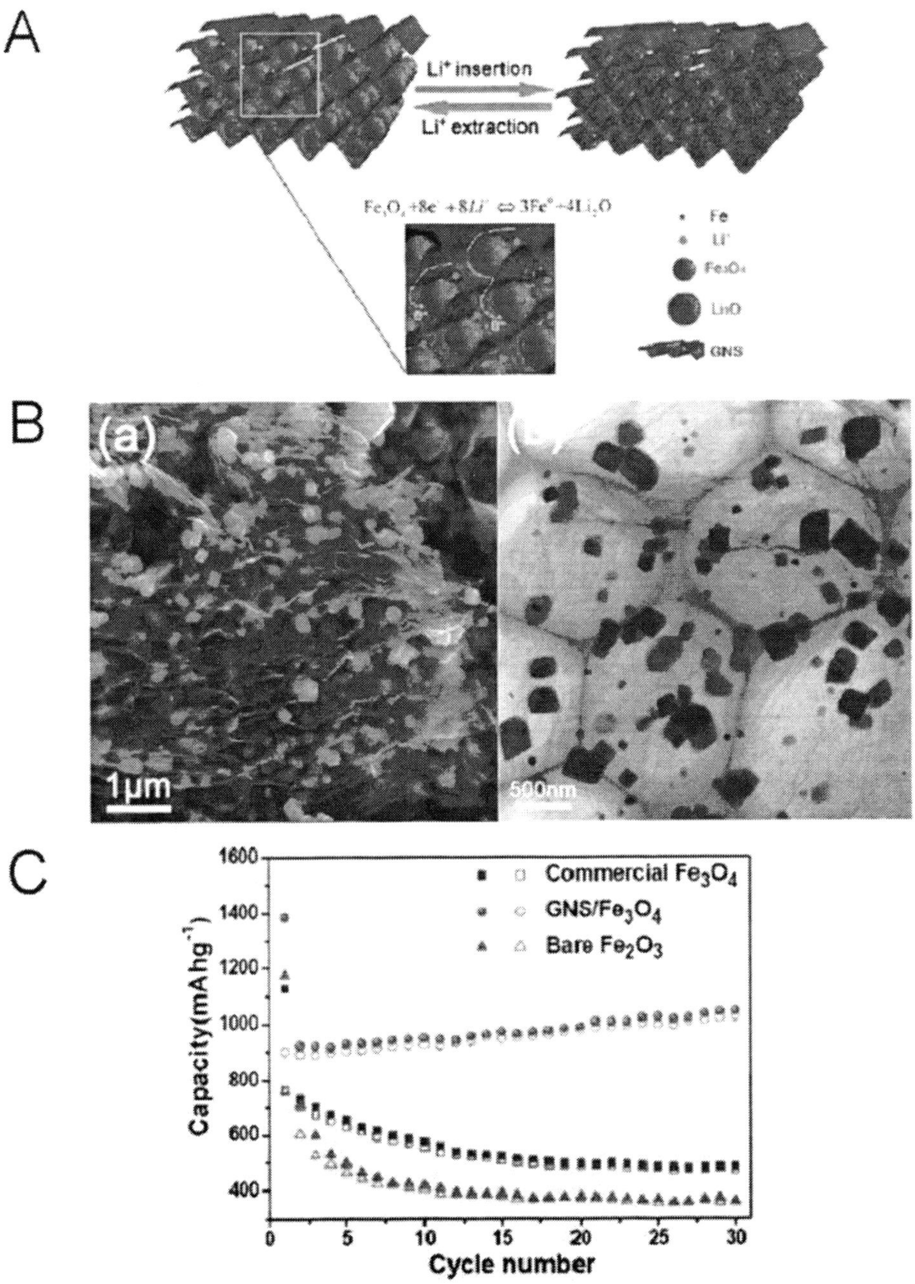

Figure 6. **A.** Schematic of a flexible interleaved structure consisting of graphene nanosheets (GNS) and Fe_3O_4 particles. **B.** (a) SEM image of the cross-section of GNS/Fe_3O_4 composite, and (b) TEM. **C.** Cycling performance of commercial Fe_3O_4 particles, GNS/Fe_3O_4 composite, and bare Fe_3O_3 particles at a current density of 35 mA g^{-1} – solid symbols, discharge; hollow symbols, charge. Reproduced with permission from Ref. [82]. Copyright (2010) American Chemical Society.

Further on the topic of hybrid materials, Fe3O4 has been explored for potential use as an anode material with low cost, high capacity, eco-friendliness, and natural abundance. [82] However, it has attracted the attention of scientists for use in the creation of hybrid materials because of a problem with rapid capacity fading during cycling. [82] Cause for excitement comes from elegant work by Zhou *et al.* [82] who fabricated a well-organised flexible interleaved composite of graphene decorated with Fe_3O_4 particles through the in-situ reduction of iron hydroxide between the graphene sheets; the interleaved network of GNSs produce a pathway for electron transport as shown in Figure 6. The GNS/Fe_3O_4 composite exhibits a reversible specific capacity approaching 1026 mA h g^{-1} after 30 cycles at 35 mA g^{-1} and 580 mAh g^{-1} after 100 cycles at 700 mA g^{-1}, as well as improved cyclic stability and an excellent rate capability when comparison is sought towards the capacities of bare Fe_2O_3 and commercial Fe_2O_4 particles after 30 cycles. [82] The authors inferred the multifunctional features of the GNS/Fe_3O_4 composite to be as follows: (i) the cycling life of the electrode was extended due to the GNSs exhibiting a "flexible confinement" function to enwrap Fe_3O_4 particles, which can compensate for the volume change of Fe_3O_4 and prevent the detachment and agglomeration of pulverized Fe_3O_4; (ii) the accessible capacity was improved as GNSs provide a large contact surface for individual dispersion of well-adhered Fe_3O_4 particles and act as an excellent conductive agent to provide a highway for electron transport; (iii) the adsorption and immersion of electrolyte on the surface of electro-active material was improved due to the Fe_3O_4 particles separating the GNSs and preventing their restacking; and (iv) ion transportation was facilitated due to the porosity formed by lateral GNSs and Fe_3O_4 particles. [82] As a result, this unique laterally confined GNS/Fe_3O_4 composite can dramatically improve the cycling stability and the rate capability of Fe_3O_4 as an anode material for lithium ion batteries. [9, 82]

Intriguingly graphene composites have been utilised as graphene/$LiFePO_4$ [90] and graphene/$Li_4Ti_5O_{12}$ [91] hybrid anode and cathode materials, where their initial application was promising for Li-ion secondary batteries, however, such outputs are not as high as previous reports, questioning their suitability into this field. [9]

In light of the literature discussed above current research regarding graphene as a Li-ion storage device indicates it to be beneficial over graphite based electrodes, exhibiting higher capacitance and improved cyclic performances for applications within Li-ion batteries which can be attributed to the easily accessible surface area and edge plane sites of graphene and thus lithium storage sites. Again, exciting future developments in this area are expected and there is no doubt that among the greatest performers to materialise within the near future, a hybrid graphene based material will reside. Although once again caution is advised towards limitations concerning the reproducibility, scalability, and characterisation of the fabricated graphene materials.

ENERGY PRODUCTION APPLICATIONS

With the depletion of non-renewable energy sources ever-nearing the search for replacements has resulted in great interest in fuel cell applications and consequentially the examination and development of new electrode materials is of high importance. [9] Table 4 summarises the various applications of graphene in fuel cell devices.

Table 4. Overview of the power and current densities obtained through the utilisation of a range of graphene based materials and various other comparable materials as an electrode material in fuel cells

Electrode complex	Fuel/Oxidant	Current/Power density at maximum power	Power density vs. time	Comments	Ref.
Au/GNS/GOx	Glucose	156.6 ± 25 μA cm^{-2}	24.3 ± 4 μW at 0.38 V (load 15 kΩ)	Tested daily with a 15 kΩ external load. Power output: 6.2 % loss after 24 hours; 50 % loss after 7 days.	[102]
Au/SWCNT/GOx	Glucose	86.8 ± 13 μA cm^{-2}	7.8 ± 1.1 μW at 0.25 V (load 15 kΩ)	N/A	[102]
GMS-*E. coli*	Glucose	142 mW m^{-2}	NT	N/A	[101]
GNS-Pt	H_2/O_2	390 mW cm^{-2}	NT	PEMFC	[93]
GO-Pt	H_2/O_2	161 mW cm^{-2}	NT	N/A	[110]
Nitrogen doped-GNS-Pt	H_2/O_2	440 mW cm^{-2}	NT	PEMFC	[93]
Pt	H_2/O_2	96 mW cm^{-2}	NT	N/A	[110]
SSM-*E. coli*	Glucose	2668 mW m^{-2}	NT	N/A	[101]

Key: Au – Gold; GMS – Graphene modified 'SSM'; GNS – Graphene nanosheet; GO – Graphene oxide; GOx – Glucose oxidase; N/A – Not applicable; NT – Not tested; PEMFC – Proton exchange membrane fuel cell; Pt – Platinum; SSM – Stainless steel mesh; SWCNT – Single walled carbon nanotube.

Fuel cells

Graphene is an attractive electrode material for use within fuel cells because of its large surface area, improved enzymatic binding ability, unique electrical conductivity, widely applicable electro-catalytic activity, and low production costs. [2-4, 81, 92] One notable contribution has been from Qu *et al.* [58] who explored graphene and nitrogen doped graphene formed *via* CVD towards the electrochemical reduction of oxygen (oxygen reduction reaction (ORR)), comparing it with commercially available platinum (Pt) loaded carbon – Pt has long been regarded as one of the best catalysts for the electrochemical ORR, and is consequently used in fuel cells; however, it exhibits time-dependent drift due to the poisoning/deactivation of the Pt surface with carbon monoxide and also has the drawback of being expensive. [58] Qu *et al.* have showed that nitrogen-doped graphene acts as a metal-free electrode, greatly enhancing electro-catalytic activity, long-term operation stability, and tolerance to crossover effect when compared to existing Pt materials towards the ORR (*via* a four-electron pathway in alkaline solutions producing water as a product). [58]

In proton exchange membrane fuel cells (PEMFCs) Pt based electro-catalysts are widely used as anode and cathode materials for hydrogen oxidation and for ORRs respectively. [9] In recent work Jafri *et al.* [93] utilised GNSs and nitrogen doped-GNSs as the catalyst support for Pt nanoparticles for ORRs in PEMFCs, where the authors constructed a fuel cell with Pt loaded GNSs exhibiting power densities of 440 and 390 mW cm^{-2} for nitrogen doped-GNS-Pt and GNS-Pt respectively. [93] Jafri and co-workers commented on the improved performance inferring that the nitrogen doping process creates pyrrolic nitrogen defects acting as anchoring sites for the deposition of Pt nanoparticles, and is also likely to increase electrical conductivity or improve carbon-catalyst binding. [93] However, contrasting earlier work by Zhang *et al.* [94] has demonstrated that the fabrication of low-cost graphite submicronparticles (GSP) can be employed as a possible support for polymer electrolyte membrane (PEM) fuel cells where Pt nanoparticles were deposited on GSP in addition to carbon black and CNTs (*via* a ethylene glycol reduction method). Zhang and co-workers report that the Pt/GSP composite exhibited the highest electro-catalytic activity towards the ORR and that a durability study indicated the Pt/GSP was 2-3 times more durable than the CNT and carbon black alternatives. [94]

Direct methanol fuel cells have drawn great attention recently due to their high energy densities, low pollutant emission, ease of handling (the liquid 'fuel'), and low operating temperatures (60-100 ^oC), however, low electro-catalytic activity towards methanol oxidation is hindering exploitation. [95] Xin *et al.* [95] have recently linked graphene to the enhanced electro-catalytic activity of catalysts for fuel cell applications, by demonstrating that the utilisation of a Pt/GNS catalyst exhibits a high catalytic activity for both methanol oxidation and the ORR when compared to Pt supported on carbon black (C). The authors deposited Pt nanoparticles onto GNSs *via* simultaneous reduction of H_2PtCl_6 and graphene oxide suspension using $NaBH_4$, where *via* CV analysis towards methanol electro-oxidation it is apparent that the current density of the Pt/GNS catalyst (182.6 mA mg^{-1}) outperforms the response of Pt/C (77.9 mA mg^{-1}), as evident in Figure 7. [95] However, note that prior heat treatment of the Pt/GNS catalyst improved the performance further by ~3.5 times over that of the Pt/C response; a response supported by previous literature. [95, 96] Moreover, chronoamperometric analysis was performed to analyse the stability of the fabricated catalysis towards methanol oxidation, and it was shown that (after all of the catalysts decayed rapidly in the initial stages) a steady state response was observed where the greatest activity and stability was achieved by the heat treated Pt/GNS with an ending current of 55.5 mA mg^{-1}, followed by Pt/GNS and Pt/C respectively – suggesting that the heat treatment of a Pt/GNS catalyst provides greatly improved methanol oxidation than current materials. [95] Other similar work [97] has demonstrated that GNS supported Pt-Ruthenium (Ru) nano-composites, improved catalytic activity towards the oxidation of methanol when contrasted to a Pt-Ru/Vulcan base alternative. Moreover, work by Dong *et al.* [61] has demonstrated that Pt and Pt-Ru nanoparticles synthesised onto GNSs exhibit high electro-catalytic activity towards methanol and ethanol when contrasted against graphite alternatives, leading to a greatly reduced over-potential and increased reversibility, thus these findings favour the use of graphene sheets as catalyst supports for both direct methanol and ethanol fuel cells. [61] Additionally, Shang *et al.* [98] have investigated the application of uniform and porous GNSs as a support for catalytic Pt nanoclusters in direct-methanol electro-oxidation, where again, as observed throughout the energy related field, it appears that hybrid materials exhibit distinctly superior electrochemical characteristics than the individual standard components, particularly in the fabrication of fuel cells, as this novel graphene supported Pt based nanostructure has the potential to serve as a low-cost and highly efficient electrode material for methanol fuel cells. [9]

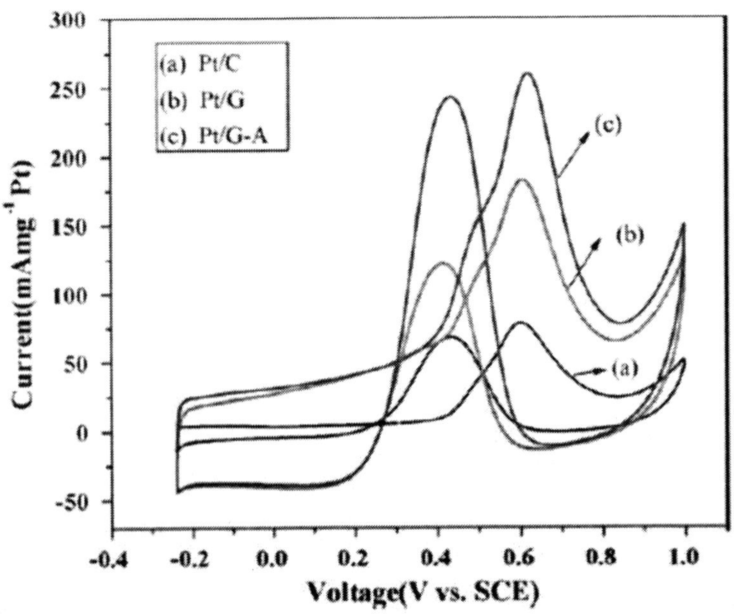

Figure 7. Cyclic voltammograms of methanol oxidation on (a) Pt/Carbon black, (b) Pt/Graphene, and (c) Pt/Graphene-heat treated modified composites. Scan rate: 20 mV s^{-1}, 0.5 M H_2SO_4 and 0.5 M CH_3OH. Reproduced from Ref. [95] with permission from Elsevier.

Graphene is a good candidate for potential applications in the fabrication of high-energy 'greener' solutions to current issues surrounding fuel cells, particularly for use as a supporting material in high-loading metal catalyst devices once the fundamental processes for metal immobilisation onto graphene are investigated further, as currently they are not fully understood.

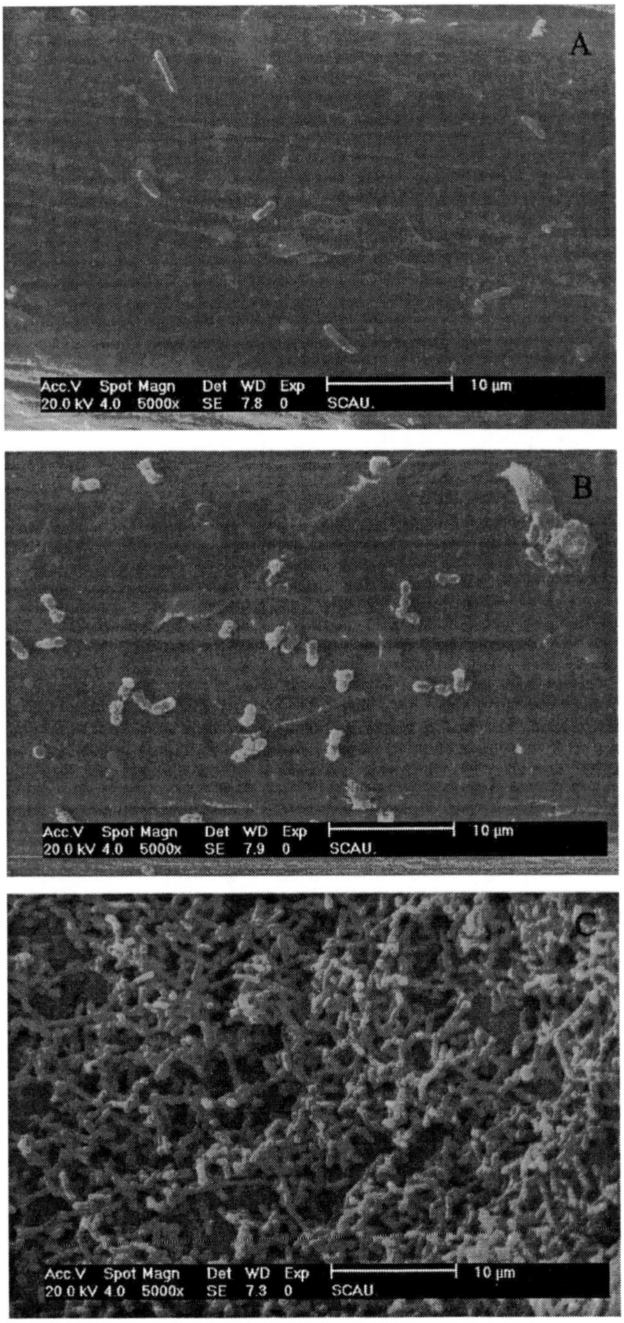

Figure 8. SEM micrographs of *E. coli* cells adhered on SSM (A), PMS (B), and GMS (C). Reproduced from Ref. [18, 101] with permission from Elsevier.

Microbial bio-fuel cells

Microbial fuel cells (MFCs) offer a great opportunity for cleaner, more sustainable energy whilst at the same time utilising waste products and meeting increasing energy demands. [99] MFCs generate electricity from the degradation of organic substances with the aid of the metabolic activity of microorganisms (bacteria). [99, 100] As with other low-temperature fuel cells the slow kinetics of the ORR at the cathode limits the performance and energy capabilities of the MFC, thus in order to improve the cathodic performance the development of catalysts is essential. [100] Recently Wu *et al.* [100] identified the need to investigate the catalytic activity of graphene towards the ORR in neutral media for its potential application in MFCs where it was demonstrated that the use of reduced graphene sheets (prepared *via* chemical reduction of graphite oxide and modified onto a glassy carbon electrode – using the electrolyte 3.5% NaCl) resulted in an obvious positive shift of the peak potential (lowering the overpotential) and additionally exhibited a larger current density; demonstrating excellent potential for MFC application. [100] Exciting work by Zhang *et al.* [101] has shown that by utilising a graphene modified anode material the performance of a MFC based on *Escherichia coli* can be greatly enhanced (towards glucose). The authors modified a stainless steel mesh (SSM) with reduced graphene oxide and investigated its performance against a bare SSM and a polytetrafluoroethylene (PTFE) modified SSM (PMS) where the graphene modified SSM (GMS) exhibited an enhanced electrochemical performance, and a MFC equipped with a GMS anode delivered a maximum power density of 2668 mW m^{-2}, which is 18 times larger than that obtained when using the bare SSM (142mW m^{-2}) and 17 times larger than when utilising the PMS anode (159 mW m^{-2}). [101] The enhanced performance of the GMS anode was attributed to graphene's high specific surface area and rough/flexible texture (as opposed to the SSM's and the PMS's relatively smooth surfaces) which promotes bacteria adhesion onto the anode surface and thus plays a key role in improving electron transfer and power output; Figure 8 depicts an SEM image showing the different attachment capabilities of each electrode material towards bacterial growth. [101] Although this work is promising, it is clear that graphene has not yet been fully explored in this area, but we expect many more examples and applications within this field to appear, building on the already vibrant examples available utilising graphite and CNTs, for instance in the treatment of domestic and industrial wastewater for the generation of energy. [9, 99]

Enzymatic bio-fuel cells

Enzymatic biofuel cells (EBFCs) possess the potential to be employed as an 'in-vivo' power source for implantable medical devices such as pacemakers. [102] The most remarkable feature of EBFCs is that they can utilise glucose or other abundant carbohydrates present in the human body as a fuel. However, EBFCs have major issues to be rectified including low power densities and poor stability. The low power density of the EBFC in comparison with conventional inorganic fuel cells is due to the location of the active site of the enzyme buried deep under the protein shell, effectively hindering electron transfer between the enzyme's active site and the electrode. [9] Previously CNTs have been employed to improve electron transfer to sites 'buried' deep within the enzyme and other approaches involve the covalent binding of the enzyme, however, a complex chemical treatment process of CNTs has to be performed in order to create active binding sites on the edge of the CNTs, a process that likely hinders mass production. [102] It is also worth mentioning that graphene can be synthesised to possesses a number of surface active functional moieties; such as carboxylic, ketonic, quinonic and C=C. Of these, the carboxylic and ketonic groups are reactive and can easily bind covalently with, for example, glucose oxidase (GOx) (used when glucose is the fuel), in addition to aiding in the shuttling of electrons. [102]

EBFCs are a relatively new area of graphene research and consequentially there is limited literature available, however, one of the few papers currently available is highly exciting. Liu *et al.* [102] have used graphene to fabricate the anode and cathode in the construction of a membraneless EBFC. The anode of the biofuel cell consisted of a gold electrode on which the authors co-immobilised graphene with GOx using a silica sol-gel matrix while the cathode was constructed in the same manner except they employed bilirubin oxidase (BOD) as the cathodic enzyme. [102] Voltammetric measurements were conducted to quantitatively evaluate the suitability and power output of employing a GNS as an electrode dopant and its performance was compared with a similar EBFC system constructed using SWCNTs. The maximum current density of the graphene based EBFC was found to be 156.6 ± 25 $\mu A\ cm^{-2}$, and for the SWCNT based EBFC 86.8 ± 13 $\mu A\ cm^{-2}$, and additionally the graphene based biofuel cell exhibited a maximum power density of 24.3 ± 4 μW at 0.38 V (load 15 kΩ), which is nearly two times greater than that of the SWCNTs EBFC (7.8 ± 1.1 μW at 0.25 V (load 15 kΩ)). [102] To evaluate the stability of the graphene based EBFC the system was stored in pH 7.4 phosphate buffer solution at 4 ^{o}C and tested daily with a 15 kΩ external load, where after the first 24 hours, it had lost 6.2% of its original power output. Later, the power output was found to decay slowly and became 50% of its original power output after 7 days; which is substantially longer than other EBFC devices and outperforms the SWCNT based EBFC. [102] The authors stated that the enhanced performance was based upon the larger surface area of graphene in comparison to SWCNTs, as well as graphene's greater sp^2 character than other materials within its field (responsible for shuttling the electrons and assisting in the better performance of the EBFC) in addition to the larger number of dislocations and electro-active functional groups present in graphene. [102] Thus it is evident that graphene based electrodes are better suited for applications in EBFCs than CNT based alternatives. Although there is currently very little literature available on this topic, due to graphene's excellent conductivity, ballistic electron mobilities at room temperature, large surface area, and other unique properties (as discussed above), it is expected to be an optimal replacement and starting platform for further research in energy related fields, where high performance EBFCs are expected soon. [9]

CONCLUSION

The performance of graphene based electrode materials has been explored with regards to their application in many electrochemical sensing and energy related devices. The investigation of graphene within the context of electrochemistry thus far has led to the development of devices with great potential for utilisation and exploitation in the future, where it is likely that further progression in the field may develop from the ability to manipulate the orientation of graphene upon an electrode surface, the emergence of graphene based hybrid materials, or the utilisation of nano-architectonics. Note however, for such applications to be adopted commercially there are still many issues to overcome (such as scalability, reproducibility and continuity of data outputs, in addition to the reproducibility and characterisation of graphene itself). Thus, where graphene is utilised in electrochemical applications it is suggested that:

Adequate characterisation of the graphene is performed, and that the electrochemical response of graphene is compared to that of its nearest counterparts, typically graphite or CNTs, under identical analytical conditions. [2, 9, 39] This is of high importance when ensuring that inaccurate claims of the performance capability of graphene do not dominate literature, where in fact an alternative derivative such as graphite is the true origin of the observed experimental data.

Appropriate control experiments incorporating the role of surfactants upon the electrochemical process, in addition to the role of other variables present such as oxygenated species or various impurities, also need to be considered. [2, 18-20, 39]

Graphene has had a significant and diverse impact within electrochemistry. Although some literature reports suggest that caution must be taken when utilising this unique nanomaterial, the majority depict innovative and advantageous outcomes where graphene is far-superior over existing electrode materials; consequently future prospects for graphene appear vibrant, where the greatly enhanced storage and renewable energy production devices, and the rapid, inexpensive, sensitive, and portable detection apparatus of the future are expected.

REFERENCES

[1] Novoselov, K. S.; Geim, A. K.; Morozov, S. V.; Jiang, D.; Zhang, Y.; Dubonos, S. V.; Grigorieva, I. V.; Firsov, A. A., Electric Field Effect in Atomically Thin Carbon Films. *Science* 2004, 306, 666-669.

[2] Brownson, D. A. C.; Banks, C. E., Graphene electrochemistry: an overview of potential applications. *Analyst* 2010, 135, 2768-2778.

[3] Pumera, M., Electrochemistry of graphene: new horizons for sensing and energy storage. *Chem. Rec.* 2009, 9, 211-223.

[4] Geim, A. K.; Novoselov, K. S., The rise of graphene. *Nat. Mater.* 2007, 6, 183-191.

[5] McCreery, R. L., Advanced Carbon Electrode Materials for Molecular Electrochemistry. *Chem. Rev.* 2008, 108, 2646-2687.

[6] Wallace, G. G.; Chen, J.; Li, D.; Moulton, S. E.; Razal, J. M., Nanostructured carbon electrodes. *J. Mater. Chem.* 2010, 20, 3553-3562.

[7] Metters, J. P.; Kadara, R. O.; Banks, C. E., New directions in screen printed electroanalytical sensors: an overview of recent developments. *Analyst* 2011, 136, 1067-1076.

[8] Compton, R. G.; Banks, C. E., *Understanding Voltammetry*. 1 ed.; World Scientific: Singapore, 2007.

[9] Brownson, D. A. C.; Kampouris, D. K.; Banks, C. E., An overview of graphene in energy production and storage applications. *J. Power Sources* 2011, 196, 4873-4885.

[10] Soldano, C.; Mahmood, A.; Dujardin, E., Production, properties and potential of graphene. *Carbon* 2010, 48, 2127-2150.

[11] Pan, D.; Wang, S.; Zhao, B.; Wu, M.; Zhang, H.; Wang, Y.; Jiao, Z., Li Storage Properties of Disordered Graphene Nanosheets. *Chem. Mater.* 2009, 21, 3136-3142.
[12] Guo, H.-L.; Wang, X.-F.; Qian, Q.-Y.; Wang, F.-B.; Xia, X.-H., A Green Approach to the Synthesis of Graphene Nanosheets. *ACS Nano* 2009, 3, 2653-2659.
[13] Hummers, W. S.; Offeman, R. E., Preparation of Graphitic Oxide. *J. Am. Chem. Soc.* 1958, 80, 1339.
[14] Gilje, S.; Han, S.; Wang, M.; Wang, K. L.; Kaner, R. B., A Chemical Route to Graphene for Device Applications. *Nano Lett.* 2007, 7, 3394-3398.
[15] Reina, A.; Jia, X. T.; Ho, J.; Nezich, D.; Son, H. B.; Bulovic, V., Large Area, Few-Layer Graphene Films on Arbitrary Substrates by Chemical Vapor Deposition. *Nano Lett.* 2009, 9, 30-35.
[16] Li, X.; Zhu, Y.; Cai, W.; Borysiak, M.; Han, B.; Chen, D.; Piner, R. D.; Colombo, L.; Ruoff, R. S., Transfer of Large-Area Graphene Films for High-Performance Transparent Conductive Electrodes. *Nano Lett.* 2009, 9, 4359-4363.
[17] Zhu, C.; Guo, S.; Fang, Y.; Dong, S., Reducing Sugar: New Functional Molecules for the Green Synthesis of Graphene Nanosheets. *ACS Nano* 2010, 4, 2429-2437.
[18] Brownson, D. A. C.; Banks, C. E., Graphene electrochemistry: Surfactants inherent to graphene inhibit metal analysis. *Electrochem. Commun.* 2011, 13, 111-113.
[19] Brownson, D. A. C.; Metters, J. P.; Kampouris, D. K.; Banks, C. E., Graphene Electrochemistry: Surfactants Inherent to Graphene Can Dramatically Effect Electrochemical Processes. *Electroanalysis* 2011, 23, 894-899.
[20] Brownson, D. A. C.; Banks, C. E., Fabricating graphene supercapacitors: highlighting the impact of surfactants and moieties. *Chem. Commun.* 2012, DOI: 10.1039/c1cc11276g.
[21] Wu, W.; Liu, Z.; Jauregui, L. A.; Yu, Q.; Pillai, R.; Cao, H.; Bao, J.; Chen, Y. P.; Pei, S.-S., Wafer-scale synthesis of graphene by chemical vapor deposition and its application in hydrogen sensing. *Sens. Actuators, B,* 2010, 150, 296-300.
[22] Kang, X.; Wang, J.; Wu, H.; Liu, J.; Aksay, I. A.; Lin, Y., A graphene-based electrochemical sensor for sensitive detection of paracetamol. *Talanta, 2010, 81, 754* 2010, 81, 754-759.
[23] Alwarappan, S.; Erdem, A.; Liu, C.; Li, C.-Z., Probing the Electrochemical Properties of Graphene Nanosheets for Biosensing Applications. *J. Phys. Chem. C* 2009, 113, 8853-8857.
[24] Dai, J.; Wang, Q.; Li, W.; Wei, Z.; Xu, G., Properties of well aligned SWNT modified poly (methyl methacrylate) nanocomposites. *Mater. Lett.* 2007, 61, 27-29.
[25] Geim, A. K.; MacDonald, A. H., Graphene: Exploring carbon flatland. *Phys. Today* 2007, 60, 35-41.
[26] McCreery, R. L., *Electroanalytical Chemistry*. Marcel Dekker: New York, 1990.
[27] Ji, X.; Banks, C. E.; Crossley, A.; Compton, R. G., Oxygenated Edge Plane Sites Slow the Electron Transfer of the Ferro-/Ferricyanide Redox Couple at Graphite Electrodes. *Chem. Phys. Chem.* 2006, 7, 1337-1344.
[28] Wildgoose, G. G.; Banks, C. E.; Leventis, H. C.; Compton, R. G., Chemically modified carbon nanotubes for use in electroanalysis. *Microchim. Acta* 2006, 152, 187-214.

[29] Wang, Y.; Wan, Y.; Zhang, D., Reduced graphene sheets modified glassy carbon electrode for electrocatalytic oxidation of hydrazine in alkaline media. *Electrochem. Commun.* 2010, 12, 187-190.

[30] Banks, C. E.; Crossley, A.; Salter, C.; Wilkins, S. J.; Compton, R. G., Carbon Nanotubes Contain Metal Impurities Which Are Responsible for the "Electrocatalysis" Seen at Some Nanotube-Modified Electrodes. *Angew. Chem. Int. Ed.* 2006, 45, 2533-2537.

[31] Ji, X.; Banks, C. E.; Xi, W.; Wilkins, S. J.; Compton, R. G., Edge Plane Sites on Highly Ordered Pyrolytic Graphite as Templates for Making Palladium Nanowires via Electrochemical Decoration. *J. Phys. Chem. B* 2006, 110, 22306-22309.

[32] Ji, X.; Kadara, R. O.; Krussma, J.; Chen, Q.; Banks, C. E., Understanding the Physicoelectrochemical Properties of Carbon Nanotubes: Current State of the Art. *Electroanalysis* 2010, 22, 7-19.

[33] Wang, J.; Yang, S.; Guo, D.; Yu, P.; Li, D.; Ye, J.; Mao, L., Comparative studies on electrochemical activity of graphene nanosheets and carbon nanotubes. *Electrochem. Commun.* 2009, 11, 1892-1895.

[34] Lin, W.-J.; Liao, C.-S.; Jhang, J.-H.; Tsai, Y.-C., Graphene modified basal and edge plane pyrolytic graphite electrodes for electrocatalytic oxidation of hydrogen peroxide and β-nicotinamide adenine dinucleotide. *Electrochem. Commun.* 2009, 11, 2153-2156.

[35] Keeley, G. P.; O'Neill, A.; Holzinger, M.; Cosnier, S.; Coleman, J. N.; Duesberg, G. S., DMF-exfoliated graphene for electrochemical NADH detection. *Phys. Chem. Chem. Phys.* 2011, In Press, DOI: 10.1039/C1CP20060G.

[36] Keeley, G. P.; O'Neill, A.; McEvoy, N.; Peltekis, N.; Coleman, J. N.; Duesberg, G. S., Electrochemical ascorbic acid sensor based on DMF-exfoliated graphene. *J. Mater. Chem.* 2010, 20, 7864-7869.

[37] Schedin, F.; Geim, A. K.; Morozov, S. V.; Hill, E. W.; Blake, P.; Katsnelson, M. I.; Novoselov, K. S., Detection of individual gas molecules adsorbed on graphene. *Nat. Mater.* 2007, 6, 652-655.

[38] Kampouris, D. K.; Banks, C. E., Exploring the physicoelectrochemical properties of graphene. *Chem. Commun.* 2010, 46, 8986-8988.

[39] Brownson, D. A. C.; Banks, C. E., Graphene Electrochemistry: Fabricating Amperometric Biosensors. *Analyst* 2011, 136, 2084-2089.

[40] Goh, M. S.; Pumera, M., Single-, Few-, and Multilayer Graphene Not Exhibiting Significant Advantages over Graphite Microparticles in Electroanalysis. *Anal. Chem.* 2010, 82, 8367-8370.

[41] Li, J.; Guo, S.; Zhai, Y.; Wang, E., High-sensitivity determination of lead and cadmium based on the Nafion-graphene composite film. *Anal. Chim. Acta* 2009, 649, 196-201.

[42] Kefala, G.; Economou, A.; Voulgaropoulos, A., A study of Nafion-coated bismuth-film electrodes for the determination of trace metals by anodic stripping voltammetry. *Analyst* 2004, 129, 1082-1090.

[43] Xu, H.; Zeng, L. P.; Xing, S. J.; Xian, Y. Z.; Shi, G. Y.; Jin, L. T., Ultrasensitive Voltammetric Detection of Trace Lead(II) and Cadmium(II) Using MWCNTs-Nafion/Bismuth Composite Electrodes. *Electroanalysis* 2008, 20, 2655-2662.
[44] Li, J.; Guo, S.; Zhai, Y.; Wang, E., Nafion–graphene nanocomposite film as enhanced sensing platform for ultrasensitive determination of cadmium. *Electrochem. Commun.* 2009, 11, 1085-1088.
[45] Rangel, N. L.; Seminario, J. M., Single molecule detection using graphene electrodes. *J. Phys. B: At. Mol. Opt. Phys.* 2010, 43, 115101.
[46] Lu, G. H.; Ocola, L. E.; Chen, J. H., Reduced graphene oxide for room-temperature gas sensors. *Nanotechnology* 2009, 40, 445502.
[47] Fowler, J. D.; Allen, M. J.; Tung, V. C.; Yang, Y.; Kaner, R. B.; Weiler, B. H., Practical Chemical Sensors from Chemically Derived Graphene. *ACS Nano* 2009, 3, 301-306.
[48] Guo, C. X.; Lu, Z. S.; Lei, Y.; Li, C. M., Ionic liquid–graphene composite for ultratrace explosive trinitrotoluene detection. *Electrochem. Commun.* 2010, 12, 1237-1240.
[49] Goh, M. S.; Pumera, M., Graphene-based electrochemical sensor for detection of 2,4,6-trinitrotoluene (TNT) in seawater: the comparison of single-, few-, and multilayer graphene nanoribbons and graphite microparticles. *Anal. Bioanal. Chem.* 2011, 399, 127-131.
[50] Kang, X.; Wang, J.; Wu, H.; Aksay, I. A.; Liu, J.; Lin, Y., Glucose Oxidase–graphene–chitosan modified electrode for direct electrochemistry and glucose sensing. *Biosens. Bioelectron.* 2009, 25, 901-905.
[51] Liu, Y.; Yu, D.; Zeng, C.; Miao, Z.; Dai, L., Biocompatible Graphene Oxide-Based Glucose Biosensors. *Langmuir* 2010, 26, 6158-6160.
[52] Kim, Y.-R.; Bong, S.; Kang, Y.-J.; Yang, Y.; Mahajan, R. K.; Kim, J. S.; Kim, H., Electrochemical detection of dopamine in the presence of ascorbic acid using graphene modified electrodes. *Biosens. Bioelectron.* 2010, 25, 2366-2369.
[53] Yin, H.; Zhou, Y.; Ma, Q.; Ai, S.; Ju, P.; Zhu, L.; Lu, L., Electrochemical oxidation behavior of guanine and adenine on graphene–Nafion composite film modified glassy carbon electrode and the simultaneous determination. *Process Biochem.* 2010, 45, 1707-1712.
[54] Du, D.; Zou, Z.; Shin, Y.; Wang, J.; Wu, H.; Engelhard, M. H.; Liu, J.; Aksay, I. A.; Lin, Y., Sensitive Immunosensor for Cancer Biomarker Based on Dual Signal Amplification Strategy of Graphene Sheets and Multienzyme Functionalized Carbon Nanospheres. *Anal. Chem.* 2010, 82, 2989-2995.
[55] Zhong, Z.; Wu, W.; Wang, D.; Wang, D.; Shan, J.; Qing, Y.; Zhang, Z., Nanogold-enwrapped graphene nanocomposites as trace labels for sensitivity enhancement of electrochemical immunosensors in clinical immunoassays: Carcinoembryonic antigen as a model. *Biosens. Bioelectron.* 2010, 25, 2379-2383.
[56] Wang, H.; Hao, Q.; Yang, X.; Lu, L.; Wang, X., Graphene oxide doped polyaniline for supercapacitors. *Electrochem. Commun.* 2009, 11, 1158-1161.
[57] Zhang, Y.; Li, H.; Pan, L.; Lu, T.; Sun, Z., Capacitive behavior of graphene–ZnO composite film for supercapacitors. *J. Electroanal. Chem.* 2009, 634, 68-71.

[58] Qu, L.; Liu, Y.; Baek, J.-B.; Dai, L., Nitrogen-Doped Graphene as Efficient Metal-Free Electrocatalyst for Oxygen Reduction in Fuel Cells. *ACS Nano* 2010, 4, 1321-1326.

[59] Zhang, K.; Zhang, L. L.; Zhao, X. S.; Wu, J., Graphene/Polyaniline Nanofiber Composites as Supercapacitor Electrodes. *Chem. Mater.* 2010, 22, 1392-1401.

[60] Paek, S.-M.; Yoo, E.; Honma, I., Enhanced Cyclic Performance and Lithium Storage Capacity of SnO2/Graphene Nanoporous Electrodes with Three-Dimensionally Delaminated Flexible Structure. *Nano Lett.* 2009, 9, 72-75.

[61] Dong, L.; Gari, R. R. S.; Li, Z.; Craig, M. M.; Hou, S., Graphene-supported platinum and platinum–ruthenium nanoparticles with high electrocatalytic activity for methanol and ethanol oxidation. *Carbon* 2010, 48, 781-787.

[62] Zhang, L. L.; Zhou, R.; Zhao, X. S., Graphene-based materials as supercapacitor electrodes. *J. Mater. Chem.* 2010, 20, 5983-5992.

[63] Yan, J.; Wei, T.; Fan, Z.; Qian, W.; Zhang, M.; Shen, X.; Wei, F., Preparation of graphene nanosheet/carbon nanotube/polyaniline composite as electrode material for supercapacitors. *J. Power Sources* 2010, 195, 3041-3045.

[64] Du, X.; Guo, P.; Song, H.; Chen, X., Graphene nanosheets as electrode material for electric double-layer capacitors. *Electrochim. Acta* 2010, 55, 4812-4819.

[65] Vivekchand, S. R. C.; Rout, C. S.; Subrahnabyam, K. S.; Govindaraj, A.; Rao, C. N. R., Graphene-based electrochemical supercapacitors. *J. Chem. Sci.* 2008, 120, 9-13.

[66] Wang, Y.; Shi, Z.; Huang, Y.; Ma, Y.; Wang, C.; Chen, M.; Chen, Y., Supercapacitor Devices Based on Graphene Materials. *J. Phys. Chem. C* 2009, 113, 13103-13107.

[67] Yan, J.; Wei, T.; Shao, B.; Fan, Z.; Qian, W.; Zhang, M.; Wei, F., Preparation of a graphene nanosheet/polyaniline composite with high specific capacitance. *Carbon* 2010, 48, 487-493.

[68] Lu, T.; Zhang, Y.; Li, H.; Pan, L.; Li, Y.; Sun, Z., Electrochemical behaviors of graphene–ZnO and graphene–SnO2 composite films for supercapacitors. *Electrochim. Acta* 2010, 55, 4170-4173.

[69] Stoller, M. D.; Park, S.; Zhu, Y.; An, J.; Ruoff, R. S., Graphene-Based Ultracapacitors. *Nano Lett.* 2008, 8, 3498-3502.

[70] Chen, Y.; Zhang, X.; Yu, P.; Ma, Y., Electrophoretic deposition of graphene nanosheets on nickel foams for electrochemical capacitors. *J. Power Sources* 2010, 195, 3031-3035.

[71] Kim, K.-S.; Kim, I.-J.; Park, S.-J., Influence of Ag doped graphene on electrochemical behaviors and specific capacitance of polypyrrole-based nanocomposites. *Synth. Met.* 2010, 160, 2355-2360.

[72] Shi, W.; Zhu, J.; Sim, D. H.; Tay, Y. Y.; Lu, Z.; Zhang, X.; Sharma, Y.; Srinivasan, M.; Zhang, H.; Hng, H. H.; Yan, Q., Achieving high specific charge capacitances in Fe3O4/reduced graphene oxide nanocomposites. *J. Mater. Chem.* 2011, 21, 3422-3427.

[73] Wang, H.; Casalongue, H. S.; Liang, Y.; Dai, H., Ni(OH)2 Nanoplates Grown on Graphene as Advanced Electrochemical Pseudocapacitor Materials. *J. Am. Chem. Soc.* 2010, 132, 7472-7477.
[74] Wu, Z.-S.; Wang, D.-W.; Ren, W.; Zhao, J.; Zhou, G.; Li, F.; Cheng, H.-M., Anchoring Hydrous RuO2 on Graphene Sheets for High-Performance Electrochemical Capacitors. *Adv. Funct. Mater.* 2010, 20, 3595-3602.
[75] Chen, S.; Zhu, J.; Wang, X., One-Step Synthesis of Graphene–Cobalt Hydroxide Nanocomposites and Their Electrochemical Properties. *J. Phys. Chem. C* 2010, 114, 11829-11834.
[76] Goh, M. S.; Pumera, M., Multilayer graphene nanoribbons exhibit larger capacitance than their few-layer and single-layer graphene counterparts. *Electrochem. Commun.* 2010, 12, 1375-1377.
[77] Miller, J. R.; Outlaw, R. A.; Holloway, B. C., Graphene double-layer capacitor with ac line-filtering performance. *Science* 2010, 329, 1637-1639.
[78] Kim, T.; Lim, S.; Kwon, K.; Hong, S.-H.; Qiao, W.; Rhee, C. K.; Yoon, S.-H.; Mochida, I., Electrochemical Capacitances of Well-Defined Carbon Surfaces. *Langmuir* 2006, 22, 9086-9088.
[79] Lian, P.; Zhu, X.; Liang, S.; Li, Z.; Yang, W.; Wang, H., Large reversible capacity of high quality graphene sheets as an anode material for lithium-ion batteries. *Electrochim. Acta* 2010, 55, 3909-3914.
[80] Guo, P.; Song, H.; Chen, X., Electrochemical performance of graphene nanosheets as anode material for lithium-ion batteries. *Electrochem. Commun.* 2009, 11, 1320-1324.
[81] Liang, M.; Zhi, L., Graphene-based electrode materials for rechargeable lithium batteries. *J. Mater. Chem.* 2009, 19, 5871-5878.
[82] Zhou, G.; Wang, D.-W.; Li, F.; Zhang, L.; Li, N.; Wu, Z.-S.; Wen, L.; Lu, G. Q.; Cheng, H.-M., Graphene-Wrapped Fe3O4 Anode Material with Improved Reversible Capacity and Cyclic Stability for Lithium Ion Batteries. *Chem. Mater.* 2010, 22, 5306-5313.
[83] Yoo, E. J.; Kim, J.; Hosono, E.; Zhou, H.-S.; Kudo, T.; Honma, I., Large Reversible Li Storage of Graphene Nanosheet Families for Use in Rechargeable Lithium Ion Batteries. *Nano Lett.* 2008, 8, 2277-2282.
[84] Uthaisar, C.; Barone, V., Edge Effects on the Characteristics of Li Diffusion in Graphene. *Nano Lett.* 2010, 10, 2838-2842.
[85] Takamura, T.; Endo, K.; Fu, L.; Wu, Y.; Lee, K. J.; Matsumoto, T., Identification of nano-sized holes by TEM in the graphene layer of graphite and the high rate discharge capability of Li-ion battery anodes. *Electrochim. Acta* 2007, 53, 1055-1061.
[86] Bhardwaj, T.; Antic, A.; Pavan, B.; Barone, V.; Fahlman, B. D., Enhanced Electrochemical Lithium Storage by Graphene Nanoribbons. *J. Am. Chem. Soc.* 2010, 132, 12556-12558.
[87] Yao, J.; Shen, X.; Wang, B.; Liu, H.; Wang, G., In situ chemical synthesis of SnO2-graphene nanocomposite as anode materials for lithium-ion batteries. *Electrochem. Commun.* 2009, 11, 1849-1852.

[88] Wang, X.; Zhou, X.; Yao, K.; Zhang, J.; Liu, Z., A SnO2/graphene composite as a high stability electrode for lithium ion batteries. *Carbon* 2011, 49, 133-139.

[89] Wang, H.; Cui, L.-F.; Yang, Y.; Casalongue, H. S.; Robinson, J. T.; Liang, Y.; Cui, Y.; Dai, H., Mn3O4–Graphene Hybrid as a High-Capacity Anode Material for Lithium Ion Batteries. *J. Am. Chem. Soc.* 2010, 132, 13978-13980.

[90] Ding, Y.; Jiang, Y.; Xu, F.; Yin, J.; Ren, H.; Zhuo, Q.; Long, Z.; Zhang, P., Preparation of nano-structured LiFePO4/graphene composites by co-precipitation method. *Electrochem. Commun.* 2010, 12, 10-13.

[91] Shen, L.; Yuan, C.; Luo, H.; Zhang, X.; Yang, S.; Lu, X., In situ synthesis of high-loading Li4Ti5O12–graphene hybrid nanostructures for high rate lithium ion batteries. *Nanoscale* 2011, 3, 572-574.

[92] Chen, D.; Tang, L.; Li, J., Graphene-based materials in electrochemistry. *Chem. Soc. Rev.* 2010, 39, 3157-3180.

[93] Jafri, R. I.; Rajalakshmi, N.; Ramaprabhu, S., Nitrogen doped graphene nanoplatelets as catalyst support for oxygenreduction reaction in proton exchange membranefuelcell. *J. Mater. Chem.* 2010, 20, 7114-7117.

[94] Zhang, S.; Shao, Y.; Li, X.; Nie, Z.; Wang, Y.; Liu, J.; Yin, G.; Lin, Y., Low-cost and durable catalyst support for fuel cells: Graphite submicronparticles. *J. Power Sources* 2010, 195, 457-460.

[95] Xin, Y.; Liu, J.-G.; Zhou, Y.; Liu, W.; Gao, J.; Xie, Y.; Yin, Y.; Zou, Z., Preparation and characterization of Pt supported on graphene with enhanced electrocatalytic activity in fuel cell. *J. Power Sources* 2011, 196, 1012-1018.

[96] Li, Y. M.; Tang, L. H.; Li, J. H., Preparation and electrochemical performance for methanol oxidation of pt/graphene nanocomposites. *Chem. Commun.* 2009, 11, 846-849.

[97] Bong, S.; Kim, Y.-R.; Kim, I.; Woo, S.; Uhm, S.; Lee, J.; Kim, H., Graphene supported electrocatalysts for methanol oxidation. *Electrochem. Commun.* 2010, 12, 129-131.

[98] Shang, N.; Papakonstantinou, P.; Wang, P.; Silva, S. R. P., Platinum integrated graphene for methanol fuel cells. *J. Phys. Chem. C* 2010, 114, 15837-15841.

[99] Pant, D.; Bogaert, G. V.; Diels, L.; Vanbroekhoven, K., A review of the substrates used in microbial fuel cells (MFCs) for sustainable energy production. *Bioresour. Technol.* 2010, 101, 1533-1543.

[100] Wu, J.; Wang, Y.; Zhang, D.; Hou, B., Studies on the electrochemical reduction of oxygen catalyzed by reduced graphene sheets in neutral media. *J. Power Sources* 2011, 196, 1141-1144.

[101] Zhang, Y.; Mo, G.; Li, X.; Zhang, W.; Zhang, J.; Ye, J.; Huang, X.; Yu, C., A graphene modified anode to improve the performance of microbial fuel cells. *J. Power Sources* 2011, In Press, doi:10.1016/j.jpowsour.2011.02.067.

[102] Liu, C.; Alwarappan, S.; Chen, Z.; Kong, X.; Li, C.-Z., Membraneless enzymatic biofuel cells based on graphene nanosheets. *Biosens. Bioelectron.* 2010, 25, 1829-1833.

[103] Yin, H.; Ma, Q.; Zhou, Y.; Ai, S.; Zhu, L., Electrochemical behavior and voltammetric determination of 4-aminophenol based on graphene–chitosan

composite film modified glassy carbon electrode. *Electrochim. Acta* 2010, 55, 7102-7108.

[104] Sun, J.-Y.; Huang, K.-J.; Wei, S.-Y.; Wu, Z.-W.; Ren, F.-P., A graphene-based electrochemical sensor for sensitive determination of caffeine. *Colloids Surf., B* 2011, 84, 421-426.

[105] Parvin, M. H., Graphene paste electrode for detection of chlorpromazine. *Electrochem. Commun.* 2011, 13, 366-369.

[106] Shan, C.; Yang, H.; Han, D.; Zhang, Q.; Ivaska, A.; Niu, L., Electrochemical determination of NADH and ethanol based on ionic liquid-functionalized graphene. *Biosens. Bioelectron.* 2010, 25, 1504-1508.

[107] Yin, H.; Zhou, Y.; Ma, Q.; Ai, S.; Chen, Q.; Zhu, L., Electrocatalytic oxidation behavior of guanosine at graphene, chitosan and Fe3O4 nanoparticles modified glassy carbon electrode and its determination. *Talanta* 2010, 82, 1193-1199.

[108] Zhou, K.; Zhu, Y.; Yang, X.; Luo, J.; Li, C.; Luan, S., A novel hydrogen peroxide biosensor based on Au–graphene–HRP–chitosan biocomposites. *Electrochim. Acta* 2010, 55, 3055-3060.

[109] Wu, J.-F.; Xu, M.-Q.; Zhao, G.-C., Graphene-based modified electrode for the direct electron transfer of Cytochrome c and biosensing. *Electrochem. Commun.* 2010, 12, 175-177.

[110] Seger, B.; Kamat, P. V., Electrocatalytically Active Graphene-Platinum Nanocomposites. Role of 2-D Carbon Support in PEM Fuel Cells. *J. Phys. Chem. C* 2009, 113, 7990-7995.

In: Graphene: Properties, Synthesis and Applications
Editor: Zhiping Xu
ISBN: 978-1-61470-949-7

Chapter 8

DEVELOPING GRAPHENE LIKE ELECTRODES FOR CAPACITIVE DEIONISATION

Linda Zou*

SA Water Centre for Water Management and Reuse
University of South Australia, Adelaide, Australia

Abstract

Capacitive deionisation (CDI) is a promising alternative technology in desalination. It targets to remove the salt ions which are only a small percentage of the feed solution, as compared to most other technologies that aim to shift water which accounts of 90% of the feed solution. As a result, CDI requires less energy to operate and the electrodes are easily regenerated. Porous carbon materials are the most important component in the CDI processes, as they are used as the electrodes that play a significant role in the efficiency of the desalting process. The ideal electrode materials for CDI should be both highly conductive and of high surface area and suitable pore structures. The currently available carbon electrodes limit the desalination efficiency of CDI due to their low conductivity and non-ideal pore structure and pattern. Previous research has demonstrated that the efficiency of CDI strongly depends upon the surface properties of the carbon electrodes, such as surface area and pore microstructure. Many kinds of carbon materials have been investigated as CDI electrodes such as carbon aerogel, carbon cloth, carbon nanotubes and mesoporous carbons. Recently, graphene-like nano-flakes (GNFs) with relatively high specific surface area have been prepared and used as electrodes for capacitive deionization. The GNFs were synthesized by a modified Hummers' method using hydrazine for reduction. The GNFs prepared by this process had the specific surface area as high as 222.01 m^2/g. It was found that the ratio of nitric acid and sulfuric acid plays a vital role in determining the specific surface area of GNFs. They were characterized by atomic force microscopy, N_2 adsorption at 77 K and electrochemical workstation. Its electrosorption performance was much better than commercial activated carbon (AC), suggesting a great potential in capacitive deionisation application. Further, the

* E-mail address: linda.zou@unisa.edu.au

electrosorptive performance of GNFs electrodes at various operational conditions, include different electrical potentials, flow rates and ionic strengths were measured and the electrosorption isotherm and kinetics were investigated. It is found that the specific electrosorptive capacity of the GNFs was 23.18 μmol/g for sodium ions (Na^+) when the initial concentration was at 25 mg/l, which was higher than that of previously reported data using graphene and AC under the same experimental condition. In addition, the equilibrium electrosorption capacity was determined as 73.47 μmol/g at 2.0 V by fitting data through the Langmuir isotherm, and the rate constant was found to be 1.01 min^{-1} by fitting data through pseudo-first-order adsorption. The results suggested that GNFs synthesized by a modified chemical oxidation and reduction method can be used as effective electrode materials in CDI process for brackish water desalination. It was found that highly charged multivalent ions with smaller hydrated radius would be strongly attracted to the electrodes. For cations with same charge, the one with smaller hydrated radius would be more effectively removed.

INTRODUCTION

CDI is a promising alternative technology in desalination [1-3], which is particularly suitable for small-scale inland brackish water desalination due to its lower energy demand and maintenance requirements. CDI targets the removal of the salt ions, which are only a small percentage of the feed solution, unlike most other technologies that aim to shift water, which accounts for 90% of the feed solution. As a result, CDI operates on low energy and the electrodes are easily regenerated (Wade, 1993). Porous cabon electrodes play a critical role in determining CDI performance. The ideal electrode materials for CDI should be highly conductive, with a high surface area and suitable pore size distribution. The current research problems faced by CDI is the available carbon electrodes limit the desalination efficiency of CDI due to their low conductivity and non-ideal pore structure that is difficult for ions to access. Previous research has shown that the efficiency of CDI strongly depends on the surface properties of the carbon electrodes, including their surface area and pore microstructure [4-6]. Many kinds of carbon materials have been investigated as CDI electrodes, including carbon aerogel, activated carbon, carbon cloth, carbon nanotubes and mesoporous carbons [7-12]. Tailor-designed specific characteristics of porous carbon materials will lead to improved electrosorption and desorption capacity in desalination applications.

Graphene is considered 'the thinnest material in our universe' [13]. Its exceptional electrical conductivity and two-dimensional flat structure have stimulated much research to optimise its properties and open pathways for various applications. Researchers have reported several breakthroughs in fundamental aspects and some promising practical applications [14-18]. The recent rapid improvement in chemical methodologies for synthesising graphene has made large-scale preparation feasible. Further, its potential applications in various fields have been explored [19, 20]. Graphene has a large theoretical specific area of 2600 m^2/g, which is twice that of the finely divided activated carbon [21]. It presents an excellent tensile modulus up to 35 GPa and a superior room temperature electric conductivity of 7200 sm^{-1}. These intriguing mechanical and electrical properties enable graphene to be used as a freestanding electrode for energy storage electrodes as well as in electrosorption. Chemically modified graphene has been studied in the context of many applications, due to its excellent electrical, mechanical and thermal properties [22]. Chemical modification of graphene oxide, which is generated from graphite powder as start material, has been a promising route to achieve mass production of graphene platelets [23-27]. Hummers and Offeman [28] developed a powerful oxidation method involving reacting graphite with a mixture of potassium permanganate ($KMnO_4$) and concentrated sulfuric acid (H_2SO_4) to achieve an efficient level of oxidation.

In this chapter, GNFs have been synthesized by a modified Hummers' method using hydrazine for reduction. The working conditions of the capacitive desalting tests, such as flow rate and bias potential, are firstly discussed. The electrosorptive behavior of GNFs, as well as commercial activated carbon (AC), was investigated under the same experimental conditions for comparison. In addition, further investigations on the electrosorptive isotherm and kinetics of GNFs electrodes as well as ion selective electrosorptive experiments were carried out and the obtained data are validated.

MATERIALS AND METHODS

Synthesis of Graphene Nanoflakes (GNFs)

Graphite oxide (GO) powders were synthesized by a modified Hummers method [29, 30]. 2.5 g of Graphite powder (< 20 micro, Sigma-Aldrich, Inc) was added into an 80°C solution of a concentrated nitric acid and sulphuric acid mixture with volume ratio of 1 : 1 (120 ml each), the suspension was mixed for 4.5 h. After that, the mixture was then cooled down to the room temperature and diluted with 0.5 L of deionized (DI) water and left overnight. Later, the mixture was filtered and dried in ambient condition for 24 hours. The reaction vessel contained 120 ml of concentrated H_2SO_4 and pre-treated graphite powder was immersed in an ice bath, and potassium permanganate was added slowly. Successively, the mixture was stirred for 2 h, then, after the dilution with DI water, 20 ml 30% H_2O_2 was added to the mixture. The colour of mixture changed into brilliant yellow along with bubbling. Finally, the mixture was filtered and washed with HCl aqueous solution (1:10 in volume) to remove residual metal ions, and washed with DI water until pH equaled 7. Exfoliation was carried out by adding hydrazine and mixing for 24 h at a temperature of 80°C. Successively, the as-prepared graphene suspension was washed several times and filtered carefully, followed by air dry at 60°C, the GNFs obtained were ready to use.

In order to investigate the impact of acid pre-treatment on the specific surface area of graphene, a series of different volume ration of nitric acid and sulfuric acid as 0 : 1, 1 : 1 and 2 : 1 were used, the samples obtained were names as GNFs-1, GNFs-2 and GNFs-3 respectively.

Assemble of Graphene Electrodes

The electrodes used for CDI experiements were fabricated by the following method: use 72% of the as-synthesized GNFs as actual electrode materials, 20% of graphite as conductive material, and 8% of PTFE powder as binder. The mixture of the above materials were then pressed into the dimension of 70 mm wide × 140 mm long × 0.3 mm thick, and had a flow-through hole with a diameter of 4 mm. It is necessary to finely grid the mixture of GNFs, graphite and PTFE for several hours, followed by adding 10-20 ml ethanol into the mixture to make it moist and it was then pressed on the graphite sheet. It is improtnat to secure the adhesion between the electrode piece and the graphite supporting paper. Finally, the electrodes were assembled into capacitive deionization device for testing, as shown in Figure 1.

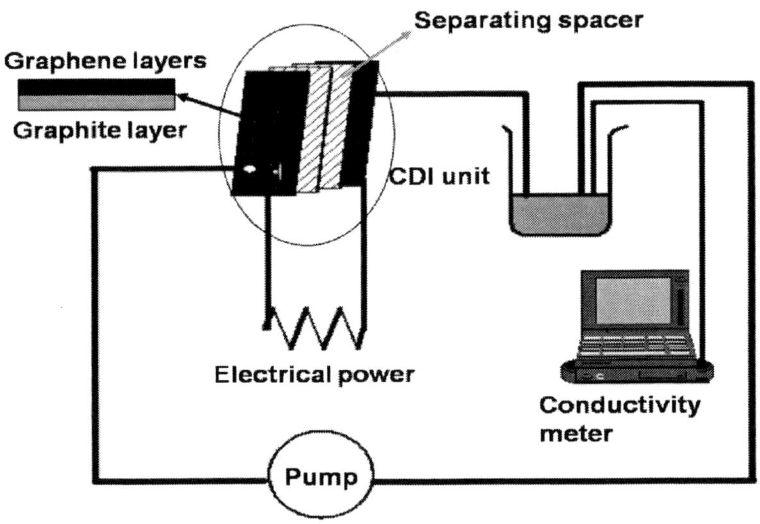

Figure 1. Schematic CDI set up. The reprint permission is from reference [31].

Characterization of GNFs

The structure of the electrode material was characterized by scanning electron microscopy (SEM, XL30) and transmission electron microscopy (TEM, CM200) imaging. The thickness of as-prepared GNFs sample was determined by atomic force microscopy (AFM, SPA-400) analysis. The N_2 adsorption-desorption isotherms of the carbon materials were performed at -196°C by Belsorp Mini (BEL JAPAN, INC, calculation method in SI). Electrochemical measurements of GNFs were examined by the potentiostat analyser (Autolab PGSTAT128N) in a typical three-electrode cell at ambient temperature. GNFs electrodes, glass carbon rod, and saturated KCl electrode were used as the working electrode, the counter electrode, and the reference electrode, respectively. The distance between the working electrode and counter electrode is about 2 cm. The potential sweep Cyclic Voltammetry (CV) measurement was carried out in the 1 mol/l NaCl aqueous solutions. The specific capacitance (F/g) in each scan rate (mv/s) has been estimated from the voltametric charge surrounded by the CV curves according to $= (q_a + |q_c|)/m_c\Delta v_d$, where q_a, q_c, m_c and Δv_d are the anodic and cathodic voltametric charges on positive and negative sweeps, the weight of GNFs loaded in the electrode and the potential range of CV, respectively.

Desalination Performance of GNF Electrodes

To investigate the electrosorptive capacity of GNFs and AC electrodes, batch-mode experiments were conducted in a bench scale electrosorptive unit cell as shown in Figure 1. In each experiment, the solution was continuously circulated by a peristaltic pump into the unit cell and the effluent returned to the feed tank. To determine the optimum flow rate and working voltage for GNFs-based CDI, the solution volume was maintained at 50 ml and the solution temperature was kept at 298 K, respectively. Meanwhile, the applied voltage was maintained at 2.0 V and no electrolysis of water occurred. The flow rate was varied in the range from 15 to 45 ml/min. In other experiments, the flow rate was kept at a constant value but the applied voltage was increased in steps of 0.2 V from 1.0 V to 2.0 V to find the optimum working voltage. The above-mentioned experiments were performed in a NaCl solution, which had an initial conductivity of 55 μS/cm. The relationship between conductivity and concentration was obtained according to a calibration table made prior to the experiment. The concentration variation of NaCl solution was continuously monitored and measured at the outlet of the unit cell using conductivity meter. In our experiment, the electrosorptive capacity was defined as follows:

$$Electrosorptive\ capacity\ (\mu mol/l) = \frac{(C-C_0)V}{58.44 \times M}. \quad (1)$$

Where C and C_0 (mg/l) represent the final and initial concentration, respectively, and V is the volume of the container (ml), and M is the mass of GNFs or AC.

Results and Discussions

Morphology and Specific Surface Area of GNFs

Figure 2(a) shows the TEM image of GNFs [31, 32]. It can be seen that the GNFs sheets shown a transparent character due to its thin layered structure with some wrinkles and folds on the surface and edge. An AFM image of a typical GNFs film with several-nanometers thickness deposited on the silicon nitride-on-silicon substrate is shown in Figure 2(b) and 2(c). This film is made up of overlapped graphite oxide nano-flakes so that a given region might consist of one layer (about 0.86 nm thick) or multiple layers (1.62 nm thick, not shown in Figure 2), the overlapped regions typically have three or more nano-flake layers. By observing the layers' structure from the TEM, SEM and AFM image, it is found that the GNFs used in this experiment were mainly made up of multiple graphene layers.

In addition, the investigation on the effects of specific surface area on the electrosorption capacities of GNF electrodes is the focus in the experiments. As discussed elsewhere [19], the acid treatment in the oxidation step, especially the amount of nitric acid, plays an important role in the final graphene product, Thus, the ratio of nitric acid and sulfuric acid was increased from 0 : 1 to 2 : 1, aimed to achieve the higher BET surface area results. It is found that the measured pore size associated with GNFs by BJH method was mainly contributed by the interlayer gaps between the GNFs. If the original graphite was not adequately exfoliated during the acid pre-treatment, it increased the possibility of serious aggregation and therefore resulted in the decreased specific surface area. In addition, no peak was found in pore size distribution suggested such gaps between GNFs varied widely in size. Conversely, once the raw graphite was treated by an excessive acid in oxidation process, it could also induce the aggregation due to the recombination between graphite oxide particles and thus lead to the decrease of the specific surface area.

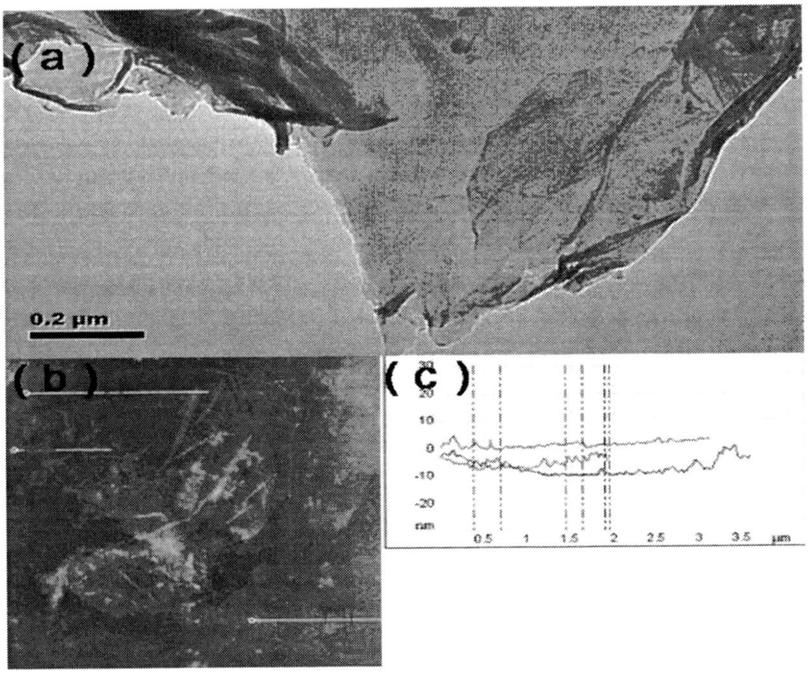

Figure 2. (a) TEM image of GNFs-2. The folded edge exhibits a relative height of 1.22 nm obtained from (b) and (c), which depict the AFM image and cross sectional analysis of GNFs-2, respectively (the scale bar is 10μm). The reprint permission is from reference [31].

Figure 3(a) shows the typical N_2 adsorption-desorption isotherm of AC and GNFs samples at 77 K, respectively. The pore size distribution of the AC and GNFs samples are shown in Figure 3(b). The inset presents the BET surface area and pore volume of the AC and GNFs samples. The BET specific surface area of AC is 989.54 m^2/g while the maximum BET specific surface area of GNFs is 222.01 m^2/g, corresponding to the ratio of nitric acid and sulfuric acid of 1 : 1. As shown in Figure 3(b), the BET surface area of GNFs-1 is 71.76 m^2/g and the total pore volume is 0.16 cm^3/g, indicating that the severe aggregation in GNFs-1 sample. Once the nitric acid was added into the mixture of graphite and sulphuric acid at the ratio of nitric acid and sulphuric acid 1 : 1, the BET surface area for GNFs-2 reached the higher value of 222.01 m^2/g and the total pore volume increased to 51.01 cm^3/g as shown in Figure 3(b).

If the ratio of nitric acid and sulphuric acid was chance from 1 : 1 to 2 : 1, the BET surface area of GNFs-3 was decreased to 44.7 m^2/g with a lower total pore volume of 10.27 cm^3/g. This indicated that the excessive nitric acid could dilute the concentration of sulfuric acid and thereby reduce its strength. The obtained surface areas were only less than 20% of theoretical value and it suggested that some graphitized carbon were remained from the incomplete exfoliation of graphite precursor. As having a high specific surface area is very important to electrosorptive capacity, the GNFs-2 with the highest specific surface area among the GNFs samples was employed to fabricate the CDI electrodes.

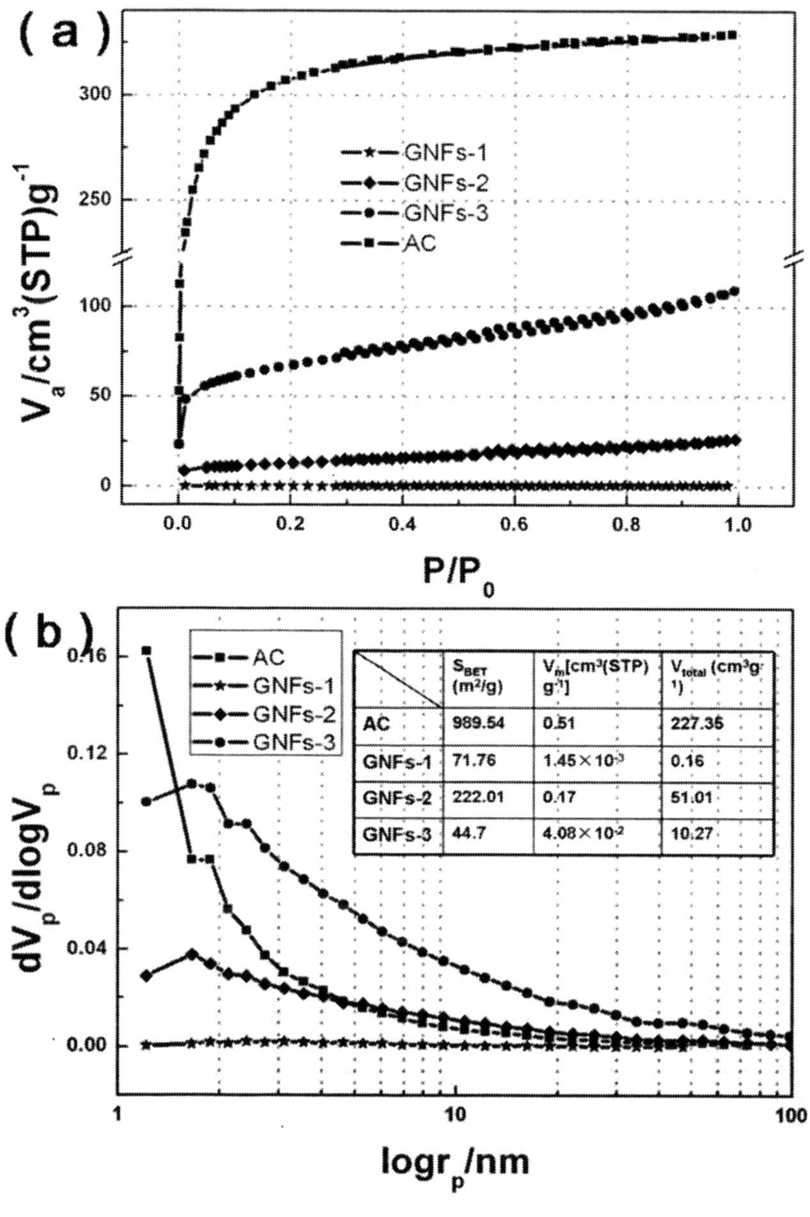

Figure 3. (a) N_2 adsorption-desorption isotherm of AC and GNFs sample at 77 K. (b) Pore size distribution of AC and different GNFs samples, inset presents the SBET and pore volume of corresponding sample. The reprint permission is from reference [31].

Effect of Working Conditions on Electrsorptive Performance of GNFs

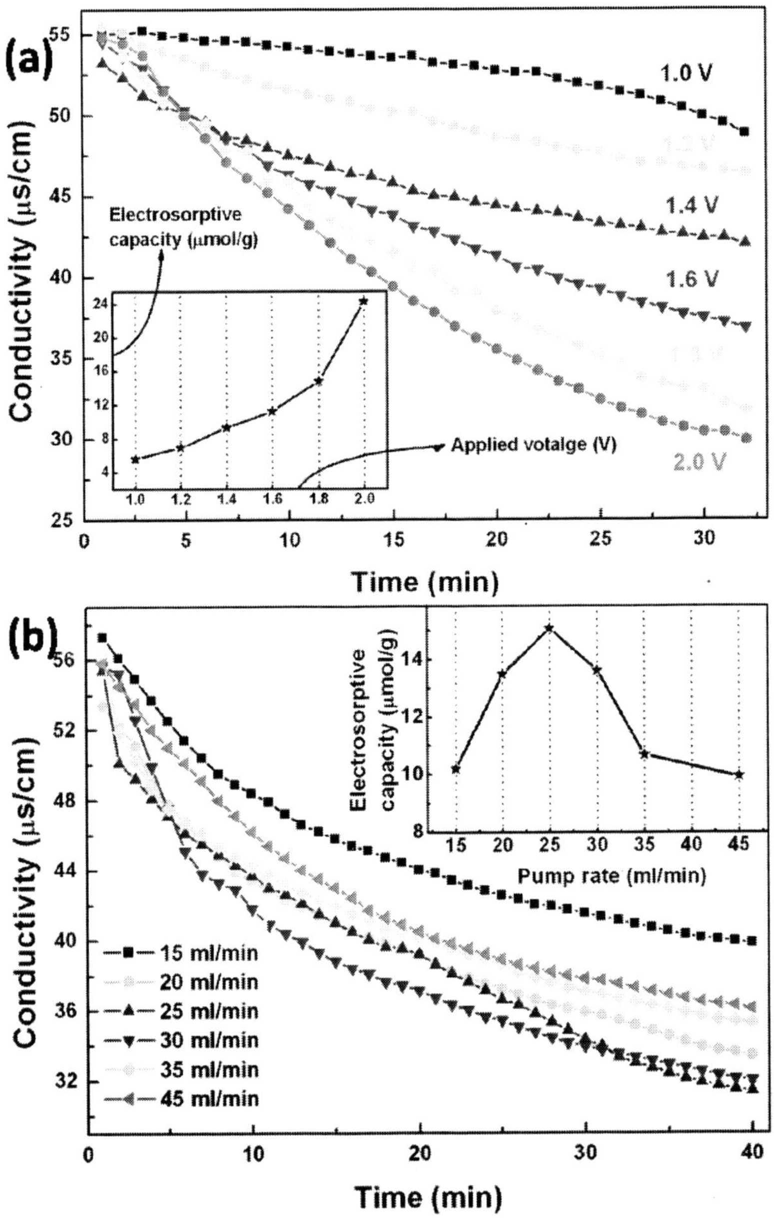

Figure 4. The electrosorption of Na^+ onto GNFs electrode at (a) different bias potentials, (b) different flow rates, respectively. The reprint permission is from reference [31].

Flow rate and bias potential are the two key factors that affect the electro-adsorbing of Na^+ on the surface of GNFs. As expected, higher voltage leads to higher electrosorption because of stronger Coulombic interaction. Figure 4(a) shows the electrosorptive performance of GNFs at different bias potentials, and the inset represents the electrosorptive capacity as a function of voltage. With the voltage increased from 0.8 V to 2.0 V, the electrosorptive capacity is gradually increased from 4.44 μmol/g to 23.18 μmol/g, which is nearly a five-fold increase. When the applied voltage is at 2.0 V, no visible gas bubbles were observed, indicating that no water electrolysis was taking place. When taking both electrosorptive capacity and energy consumption into account, the optimum working voltage for GNFs based CDI was 2.0 V. Figure 4(b) depicts the variation of conductivity in solution with flow rate, and the inset demonstrates to the correlation between the electrosorptive capacity and flow rate. It is clear from the figures that at a lower flow rate, e.g. 15 ml/min, GNFs has a lower electrosorptive capacity of 9.03 μmol/g. When the flow rate was increased from 15 ml/min to 25 ml/min, the electrosorptive capacity reached the maximum value of 13.9 μmol/g. However, the electrosorptive capacity of GNFs electrode was reduced from 13.9 μmol/g to 8.49 μmol/g when the flow rate was further increased. This was due to the equilibrium between electrostatic force and the driving force in the flow rate of 25 ml/min.

Effect of Initial Concentration on Electrsorptive Performance of GNFs

The electrosorption isotherm and kinetic models of GNFs to remove Na^+ was obtained at different initial Na^+ concentrations at a constant temperature of 298 K. The initial concentrations of the NaCl solutions were 25, 50, 100, 250, 400 and 500 mg/l, respectively, as shown in Figure 5. Langmuir Equation (2) and Freundlich Equation (3) isotherms were used to validate the experimental data for electrosorption of Na+ onto GNFs, respectively.

$$q = \frac{q_m K_L C}{1+K_L C}, \tag{2}$$

$$q = K_F C^{1/n}. \tag{3}$$

Where C is the equilibrium concentration (mg/L), q is the amount of adsorbed Na^+ (in micromoles per gram of GNFs), q_m is the maximum adsorption capacity corresponding to complete monolayer coverage. Table 1 shows the determined parameters and regression coefficients r^2, K_L and K_F of Langmuir and Freundlich isotherms for CDI, respectively.

It was found that the Langmuir isotherm correlated better with the experimental data according to the r^2. This phenomenon suggested that the monolayer adsorption was dominant during the electrosorption process. The parameter q_m in the Langmuir isotherm model is considered as the maximum adsorption capacity, as mentioned above. Thus the equilibrium electrosorption capacity, at polarization of 2 V and the flow rate of 25 ml/min, was 73.47 μmol/g, which is higher than previously reported data [16].

Table 1. Valance and hydrated radius of each target ions [33]

Ions	Valence	Hydrated radius (Å)
Sodium	+1	3.58
Calcium	+2	4.12
Magnesium	+2	4.28
Ferric	+3	4.57

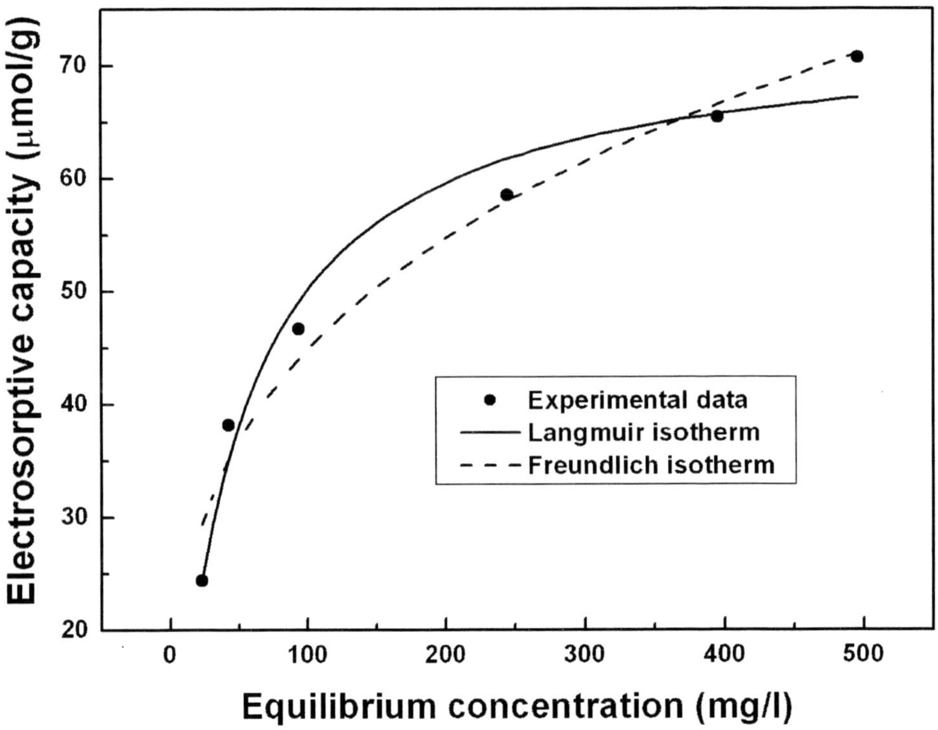

Figure 5. The electrosorptive isotherm of Na^+ onto GNFs electrode at 2.0 V. The reprint permission is from reference [31].

Electrosorption Kinetics of GNFs

Adsorption kinetics, indicating the adsorption rate, is an important characteristic of adsorbents. The adsorption rate constants of Na+ were determined by the Lagergren equation, which is often called the pseudo-first-order adsorption kinetics Equation (4):

$$\log(q_e - q) = log q_e - \frac{kt}{2.303}. \quad (4)$$

Where *k* is the adsorption rate constant (min-1), and qe and q the amount of Na^+ adsorbed at equilibrium (μmol/g) and time t (min), respectively.

The adsorption kinetics was fitted with the Lagergren law by nonlinear regression using the method of least squares. The result is shown in Figure 6 and the inset displays the fitting parameters, including rate constant k and regression coefficient r^2. It can be seen that the r^2 is close to 1, indicating that the experimental data fit to the pseudo-first-order adsorption kinetics equation very well. Additionally, the rate constant from the figure is of 1.01 min^{-1}, which is higher than that of reference data [16]. That means the GNFs with high specific surface area better facilitate the ion diffusion.

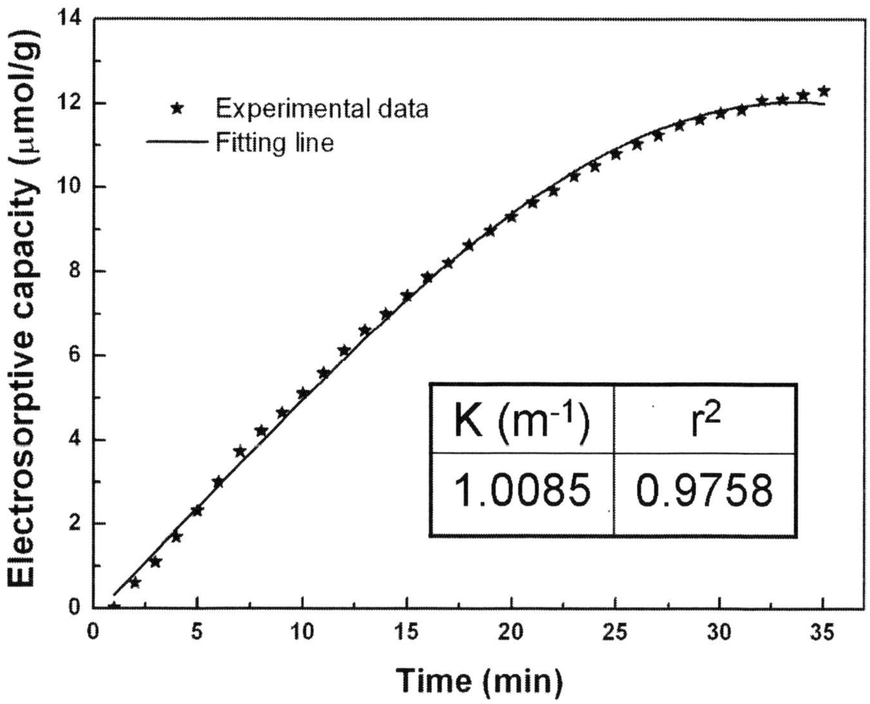

Figure 6. The electrosorptive kinetics of Na^+ onto GNFs electrode at 2.0 V. The reprint permission is from reference [31].

Electrochemical Properties of GNFs

As the working principle of capacitive deionization is similar to supercapacitor, Cyclic Voltammery (CV) is often used as an important tool to evaluate the potentials of the materials used in supercapacitor as well as capacitive deionization applications. The electrochemical properties such as specific capacitance can be obtained from the CV resutls. The Cyclic voltammograms of graphene obtained at a sweep rate from 1 mv/s to 100 mv/s in 1 mol/L NaCl solutions by using GNFs and Activated carbon (AC) electrode for compare are illustrated in Figure 7 (a) and (b), respectively. Figure 7 (c) depicts the CV curve of GNFs and AC in sweep rate of 1 mv/s while (d) shows the comparative results of GNFs and AC in various sweep rates. It can be observed, with the decrease of sweep rate, the specific capacitance of GNFs has dramatically increased from 8 F/g to 134 F/g, indicating that the novel GNF material presents an accessible surface structure which allows a rapid diffusion of ions. From Figure 7 (c), the CV curves for the GNFs nearly retained a rectangular shape over a wide range of sweep rates. Furthermore, at each sweep rate, GNFs showed higher specific capacitance than activated carbon (AC). The comparative results of specific capacitance of GNFs and AC in Figure 7 (d) confirmed that GNFs had superior surface structure as well as better conductivity. Furthermore, GNFs provide an interlayered flat sheet surface that is easily accessed by ions. Although the BET value is not as high as activated carbon, it is more favorable for the electrosoprtion of ions.

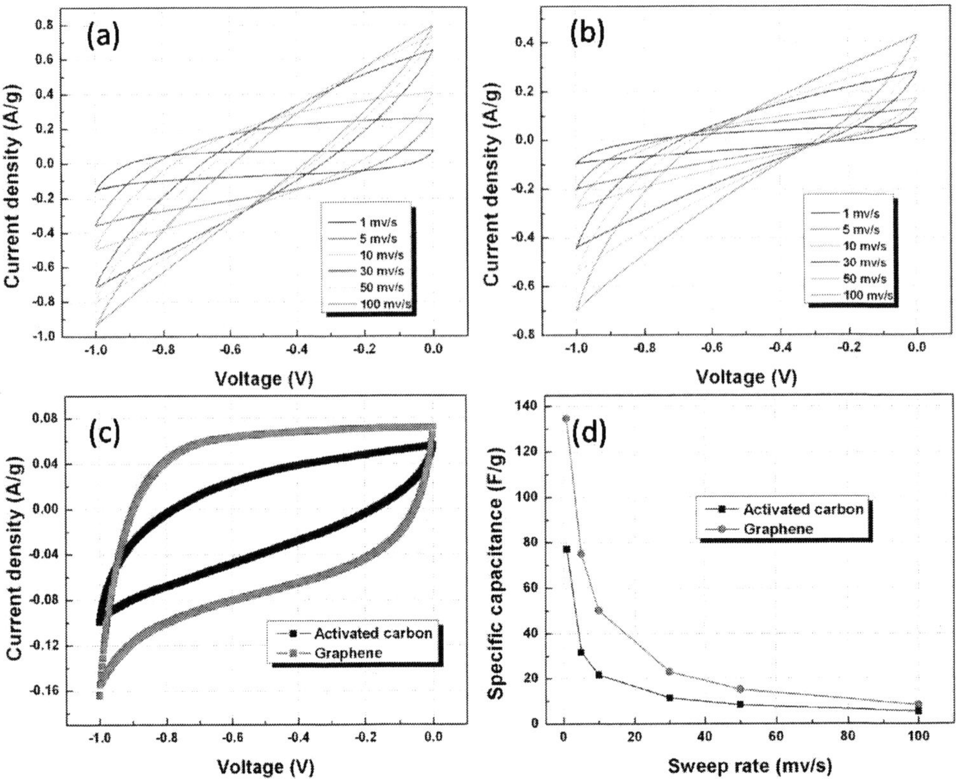

Figure 7. Cyclic voltammograms for the (a) GNFs and (b) AC, respectively. (c) Typical CV curve of GNFs and AC in the sweep rate of 1 mv/s (d) comparative results of specific capacitance for GNFs and AC electrodes. The reprint permission is from reference [31].

Comparison of Electrosorptive Capacities of GNFs and AC

Figure 8(a) shows the comparative results of the electrosorptive capacity as well as BET surface area between the GNFs and AC based CDI experiments. These results were obtained at the applied voltage of 2.0 V and the same experimental conditions. Although having the larger surface area (989.54 m^2/g) than GNFs (222.01 m^2/g), AC only has an electrosorptive capacity of 13.73 μmol/g, which is much lower than that of GNFs (whose electrosorptive capacity is 23.18 μmol/g). This can be attributed to the fact that GNFs have an interlayer flat sheet structure that is more accessible for ions, while AC has a large fraction of inaccessible small micropores. As a result, the effective surface area of GNFs is higher than that of AC. Figure 8(b) illustrates the principle of CDI and the mechanism of how ions are electrosorbed onto AC and GNFs electrodes, respectively. TEM and SEM images of AC and CNFs are used to confirm the hypothesis. The TEM image of GNFs indicates that GNFs are aggregated together, showing a semi-transparent flower shape interlayer pattern. It also shows that GNFs are homogenous flakes with micro-size that are beneficial to ions accessing and are adsorbed on the surface of the flakes. In contrast, the structure of AC on the TEM image shows that it presents a beehive-type pore structure so that the ions cannot gain access to the inner pores and therefore a high electrosorptive capacity is difficult to achieve.

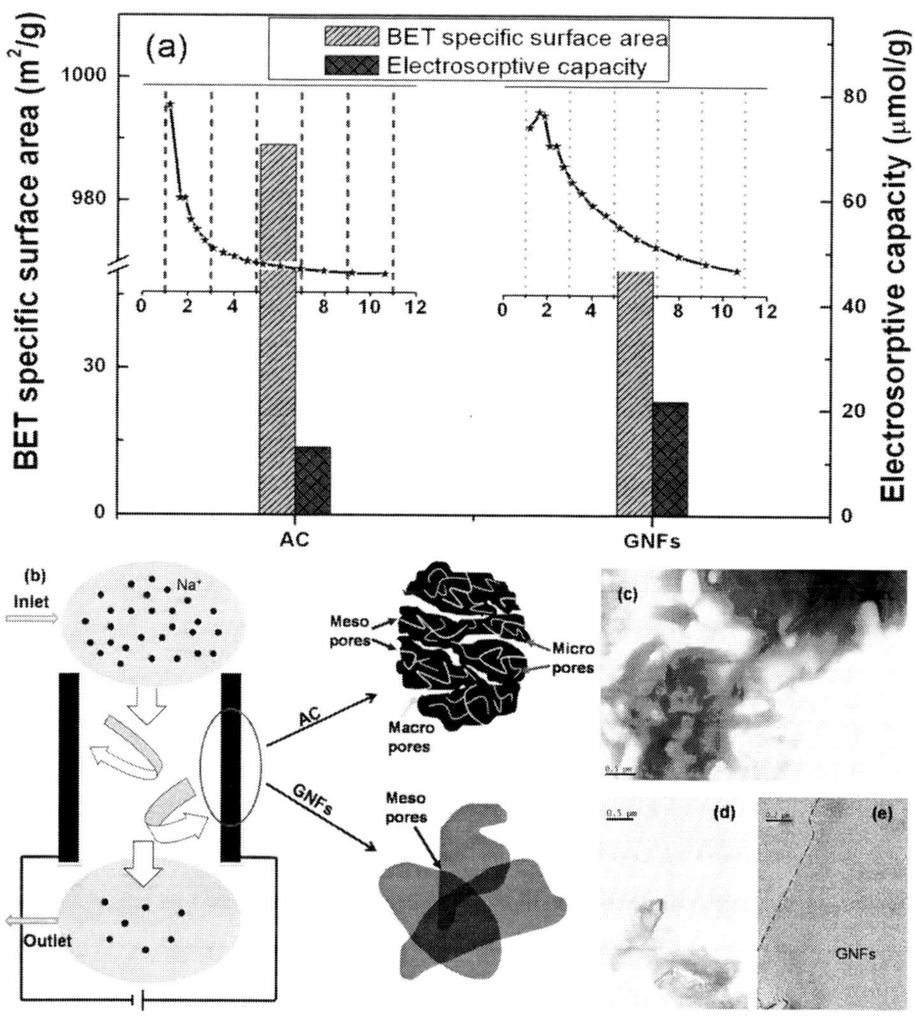

Figure 8. (a) Comparison of electrosorptive performance by employing GNFs and AC at the same experimental condition, the pictures at top-left and top-right depict the pore size distribution of AC and GNFs below 10 nm, respectively. (b) Mechanism of CDI employing AC and GNFs electrode. TEM observation images of AC (c) and GNFs in low (d) and high (e) magnification, respectively. The reprint permission is from reference [31].

In addition, it is believed that the conductivity of carbon materials also plays a vital role in the electrosorptive process. Having higher bulk conductivity is equivalent to having a higher applied voltage between the two electrodes. It is reported that the conductivity of graphene prepared via chemical approaches is normally above 200 s/m [20]. It is much higher than that of commercial AC whose electrical conductivity is normally between the orders of 10^{-6} and 10^{-3} sm^{-1}. Another possible reason for the high conductivity for GNFs is the presence of conductive graphitized chunks in the GNFs, which was caused by the unreacted graphite precursor from the graphene preparation process. When considering both effective specific surface area and electrical conductivity, it is believed that the GNFs with high specific surface area have the potential as an excellent candidate electrode material for the CDI.

Removal of Various Cations

To investigate the electrosorption performance of the fabricated GNFs in different ion solutions, batch electrosorptive experiments were further conducted at the above mentioned working conditions in NaCl, $CaCl_2$, $MgCl_2$ and $FeCl_3$ solutions, respectively. It is known generally for electrosorption process, the larger ions, i.e., ions with higher valance but smaller hydrated radius, can be more effectively removed because of stronger electrostatic force. Table 1 listed the valance and hydrated radius of each target ion. The preferential ion electrosorption for alkali and alkaline-earth metals are estimated as follows: $Fe^{3+} > Ca^{2+} > Mg^{2+} > Na^{+}$.

Figure 9 showed the results of removal of various ions at flow rate of 25 ml/min and applied voltage of 2.0 V. The electrosorptive amount of $FeCl_3$, $CaCl_2$, $MgCl_2$ and NaCl are 0.62, 0.55, 0.52 and 0.45 mg/g, respectively, following the above mentioned preference order. It means that when the hydrated cations are electrically adsorbed on a negatively charged carbon electrode, highly charged multivalent ions with smaller hydrated radius would be strongly attracted. For cations with same charge, the one with smaller hydrated radius would be more effectively removed.

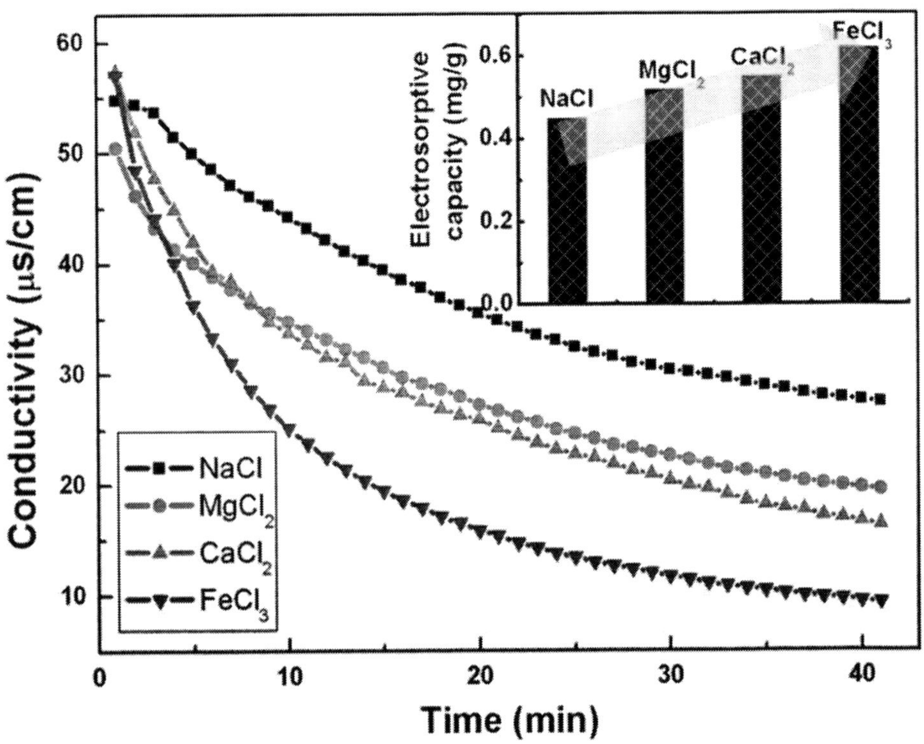

Figure 9. Effects of ion species on removal of conductivity. The applied electrical voltage and initial concentration was 2.0 V and 50 μs/cm and the feed solution were NaCl, $CaCl_2$, $MgCl_2$ and $FeCl_3$, respectively. The reprint permission is from reference [32].

CONCLUSION

In this chapter, GNFs with a relatively high specific surface area of 222.01 m^2/g were employed to fabricate electrodes for Na^+ and Cl^- removal by CDI. The electrosorption experimental results indicated that the GNFs had a better CDI performance than that of AC. It was found that the optimum working conditions of flow rate and bias potential are 25 ml/min and 2.0 V, respectively. Based on the above-mentioned conditions, the specific electrosorptive amount of the GNFs is 23.18 μmol/g for sodium ions. The electrosorptive performance of GNFs is in agreement with the Langmuir adsorption isotherm and pseudo-first-order adsorption kinetics. The equilibrium electrosorption capacity and rate constant at 2.0 V were 73.47μmol/g and 1.01 min^{-1} respectively. In addition, in comparison to AC and previously reported graphene under the same experimental condition, the electrosorptive performance of GNFs showed a promising potential for applications in brackish water desalination and drinking water purification. Further investigations on the ion selective electrosorption experiments were carried out. It was found that highly charged multivalent ions with smaller hydrated radius would be strongly attracted to the electrodes. For cations with same charge, the one with smaller hydrated radius would be more effectively removed.

ACKNOWLEDGMENTS

The author acknowledge the financial support from Australian Research Council Linkage Grant (LP0883282) and National Centre of Excellence in Desalination Australia.

REFERENCES

[1] Welgemoed, T.J. and Schutte, C.F. (2005). Capacitive Deionization TechnologyTM: An alternative desalination solution. *Desalination, 183*(1-3), 327-340.
[2] Farmer, J.C., Tran, T.D., Richardson, J.H., Fix, D.V., May, S.C. and Thomson, S.L. (1997a). The application of carbon aerogel electrodes to desalination and wast treatment. *Annual Meeting of the American Institute of Chemical Engineers*, Los Angeles, CA, November 16-21.
[3] Oren, Y. (2008). Capacitive deionization (CDI) for desalination and water treatment – past, present and future (a review). *Desalination, 228*, 10-29.
[4] Li, L., Zou, L. Song, H. and Morris, G. (2009). Ordered mesoporous carbons synthesized by a modified sol-gel process for electrosorptive removal of sodium chloride. *Carbon, 47*, 775-781.
[5] Zou, L., Li, L., Song, H. and Morris, G. (2008a). Using mesoporous carbon electrodes for brackish water desalination. *Water Research, 42*, 2340-2348.
[6] Zou, L., Morris, G. and Qi, D. (2008b). Using activated carbon electrode in electrosorptive deionisation of brackish water. *Desalination, 225*, 329-340.
[7] Ryoo, M., Kim, J. and Seo, G. (2003). Role of titania incorporated on activated carbon cloth for capacitive deionization of NaCl solution. *Journal of Colloid Interface Science, 264*, 414-419.

[8] Zhang, D., Shi, L., Fang, J., Dai, K. and Li, X. (2006a). Preparation and desalination performance of mutiwall carbon nanotubes. *Materials Chemistry and Physics, 97*, 415-419.

[9] Zhang, D., Shi, L., Fang, J., Dai, K. and Liu, J. (2006b). Influence of carbonization of hot-pressed carbon nanotubes on removal of NaCl from saltwater solution. *Materials Chemistry and Physics, 96*, 140-144.

[10] Dai, K., Shi, L., Zhang, D. and Fang, J. (2006). NaCl adsorption in multi-walled carbon nanotube/active carbon combination electrode. *Chemical Engineering Science, 61*, 428-433.

[11] Xu, P., Drewes, J.E., Heil, D. and Wang, G. (2008). Treatment of brackish produced water using carbon aerogel-based capacitive deionization technology. *Water Research, 42*, 2605-2617.

[12] Li., H., Gao, Y., Pan, L., Zhang, Y., Chen, Y. and Sun, Z. (2008). Electrosorptive desalination by carbon nanotubes and nanofibres electrodes and ion-exchange membranes. *Water Research, 42*, 4923-4928.

[13] Geim, A.K. and MacDonald, A.H. (2007). Graphene: Exploring the carbon flatland. *Physics Today, 60*, 35-41.

[14] Geim, A.K. and Novoselov, K.S. (2007). The rise of graphene. *Nature Materials, 6*(30), 183-191.

[15] Novoselov, K.S., Geim, A.K., Morozov, S.V., Jiang, D., Zhang, Y.S., Dubonos, V., Grigorieva, I.V. and Firsov, A.A. (2004). Electric field effect in atomically thin carbon films. *Science, 306*(5696), 666-669.

[16] Gilje, S., Han, S., Wang, M.S., Wang, K.L. and Kaner, R.B. (2007). A chemical route to graphene for device applications. *Nano Letters, 7*(11), 3394-3398.

[17] Lui, C.H., Liu, L., Mak, K.F., Flynn, G.W. and Heinz, T.F. (2008). Ultraflat graphene. *Nature, 462*(11), 339-341.

[18] Fan, X.B., Peng, W.C., Li, Y., Li, X.Y., Wang, S.L., Zhang, G.L. and Zhang, F.B. (2008). Deoxygenation of exfoliated graphite oxide under alkaline conditions: A green route to graphene preparation. *Advanced Materials, 20*(23), 4490-4493.

[19] Choucair, M., Thordarson, P. and Stride, J.A. (2009). Gram-scale production of graphene based on solvothermal synthesis and sonication. *Nature Nanotechnology, 4*(12), 30-33.

[20] Hao, R., Qian, W., Zhang, L.H. and Hou, Y.L. (2008). Aqueous dispersions of TCNQ-stabilized graphene sheets. *Chemical Communications, 48*, 6576-6578.

[21] Stoller, M.D., Park, S.J., Zhu, Y.W., An, J.B. and Ruoff, R.S. (2008). Graphene-Based Ultracapacitors. *Nano Letters, 8*(10), 3498-3502.

[22] Dreyer, D.R., Park, S., Bielawski, C.W. and Ruoff, R.S. (2010). The chemistry of graphene oxide. *Chemical Society Reviews, 39*(1), 228-240.

[23] Shioyama, H. (2001). Cleavage of graphite to graphene. *Journal of Materials Science Letters, 20*, 499-500.

[24] Wang, J.J., Zhu, M.Y., Outlaw, R.A., Zhao, X., Manos, D.M. and Holloway, B.C. (2004). Free-standing subnanometer graphite sheets. *Applied Physics Letters, 85*, 1265-1267.

[25] Stankovich, S., Piner, R.D., Chen, X., Wu, N., Nguyen, S.T. and Ruoff, R.S. (2006). Stable aqueous dispersions of graphitic nanoplatelets via the reduction of exfoliated

graphite oxide in the presence of poly(sodium 4-styrenesulfonate). *Journal of Materials Chemistry, 16*, 155-158.

[26] Eda, G., Fanchini, G. and Chhowalla, M. (2008). Large-area ultrathin films of reduced graphene ocide as a transparent and flexible electronic material. *Nature Nanotechnology, 3*, 270-274.

[27] Hernandez, Y., Nicolosi, V., Lotya, M., Blighe, F., Sun, Z., De, S., McGovern, I.T, Holland, B., Byrne, M., Gun'Ko, Y., Boland, J.J, Niraj, P., Duesberg, G., Krishanmurthy, S., Goodhue, R., Hutchison, J., Scardaci, V., Ferrari, A. and Coleman, J.N. (2008). High-yield production of graphene by liquid-phase exfoliation of graphite. *Nature Nanotechnology, 3*, 563-568.

[28] Hummers, W.S. and Offeman, R. E. (1958) *Journal of American Society, 80*, 1339.

[29] Li, H.B., Lu, T., Pan, L.K., Zhang, Y.P., Sun, Z. (2009). Electrosorption of behaviour of graphene in NaCl solutions. *J. Mater. Chem., 19*, 6773-6779.

[30] William, S., Hummers, J., Offeman, R.E. (1958). Preparation of graphitic oxide. *J. Am. Chem. Soc., 80*, 1339-1341.

[31] Li, H., Zou, L., Pan, L. and Sun, Z. (2010a). Novel graphene-like electrodes for capacitive deionization. *Environmental Science & Technology, 44*, 8692-9687.

[32] Li, H., Zou, L., Pan, L. and Sun, Z. (2010b). Using graphene nano-flakes as electrodes to remove ferric ions by capacitive deionisation. *Separation and Purification Technology, 75*, 8-14.

[33] Nightingale, E. R. (1959). Phenomenological theory of ion solvation: effective radii of hydrated ions. *J. Phys. Chem. 63*, 1381-1387.

In: Graphene: Properties, Synthesis and Applications
Editor: Zhiping Xu
ISBN: 978-1-61470-949-7

Chapter 9

METAL NANOSTRUCTURES/GRAPHENE NANOHYBRIDS: ELECTROCHEMICAL PREPARATION, CHARACTERIZATION, AND ELECTROCATALYTIC CHARACTERISTICS TOWARD OXYGEN REDUCTION AND SMALL ORGANIC MOLECULES OXIDATION

Yaojuan Hu and Chenxin Cai*
Jiangsu Key Laboratory of New Power Batteries
Jiangsu Key Laboratory of Biofunctional Materials
Laboratory of Electrochemistry
College of Chemistry and Materials Science
Nanjing Normal University, Nanjing, P. R. China

Abstract

Graphene, a recently discovered carbon nanomaterial with carbon atoms tightly packed into a two dimensional honeycomb lattice, possesses many novel and unique physical and chemical properties because of its unusual monolayer atomic structure. It has found potential applications in many fields such as nanocomposites, batteries, supercapacitors, nanoelectronic devices, and sensors, etc. This chapter describes the application of graphene in the field of electrocatalyst preparation, and presents the fabrication, characterization and electrocatalytic characteristics of the metal–graphene nanocomposites. The electrochemically deposition of Au and Pt–Au nanostructures on the surface of graphene sheets with the formation of the Au–graphene and Pt–Au–graphene nanocomposites, respectively, and characterization of these nanocomposites with scanning electron microscopy (SEM), energy–dispersive spectroscopy (EDS), X–ray powder diffraction (XRD), and electrochemical methods are depicted. The dependence of the morphology, size, and composition of the Au and Pt–Au nanostructures on the

* E-mail address: cxcai@njnu.edu.cn, caichenxiin@njnu.edu.cn

experimental parameters, such as electrodeposition time, the concentration of the metal precursors, and the molar ratio of these precursors is also described. The electrocatalytic activities of the Au–graphene and Pt–Au–graphene nanocomposites toward oxygen reduction and some small organic molecules (glucose, methanol) oxidation were investigated by cyclic voltammetry. For comparison, the Au and Pt–Au nanostructures were also electrodeposited on the surface of the conventional carbon materials, such as glassy carbon and Vulcan XC–72 carbon, and the electrocatalytic activities of these nanocatalysts were also studied. These studies presented in this chapter can provide a general route for fabrication of graphene–based metal nanocatalysts, which could be potential utility to fuel cells and the related fields.

1. Introduction

Carbon is the fourth most abundant element on earth. The diversity of bonding nature in addition to the ease of synthesis make carbon the most important element for biological and physiological activities as well as for material applications. Carbon, in the ground state, has four valence electrons, two in the *2s* subshell and two in the *2p* subshell. When forming bonds with other carbon atoms, it will promote one of its *2s* electrons into its empty *2p* orbital and then form bonds with other carbon atoms via *sp* hybrid orbitals. Depending on the number of *p* orbitals (1 to 3) mixing with the *s* orbital, there are three kinds of *sp* hybrid orbitals, i.e., *sp*, sp^2, and sp^3 hybrid orbitals. Carbon atoms with sp^2 and sp^3 hybrid orbitals are able to form three and four bonds with neighboring carbon atoms, respectively, which form the bases of graphene and diamond. The properties of a material are determined not only by the nature of its chemical bonds but also its dimensionality and shape. Over the last 20 years, new members of the carbon nanostructure family arose, and more are coming. A new form of carbon, fullerence C_{60}, was discovered in 1985 by Smalley, Kroto and Curl [1], and led to the Nobel Prize in chemistry in 1996. Besides graphite, diamond, and C_{60}, quasi one–dimensional carbon nanotube is another form of carbon first discovered by Iijima in 1991 by arc-discharge method [2]. The discovery of C_{60} and carbon nanotube attracted the attention of researchers worldwide.

In 2004, Geim and co–workers at Manchester University first isolated single–layer graphene from graphite by mechanical exfoliation technique (Figure 1) [3]. Graphene, which consists of a single sheet of conjugated sp^2 carbon atoms, and has opened up a new era of two–dimensional (2D) fundamental science and potential technology [4–9]. The discovery of graphene led to an explosion of interest, in part because 2D crystals were thought to be thermodynamically unstable at finite temperatures. [10, 11] Quasi–2D films grown by molecular beam epitaxy (MBE) are stabilized by a supporting substrate, which often plays a significant role in growth and has an appreciable influence on electrical properties [12]. In contrast, the mechanical exfoliation technique used by the Manchester group isolated the 2D crystals from 3D graphite. Resulting single– and few–layer flakes were pinned to the substrate by only van der Walls forces and could be made free–standing by etching away the substrate [13–16]. This minimized any induced effects and allowed scientists to probe graphene's intrinsic properties. As the basic structure of all graphitic forms, graphene is a building block for carbon materials of all other dimensionalities, it can be wrapped to form 0D fullerenes, rolled to form 1D nanotubes, and stacked to form 3D graphite (Figure 2) [17].

Long–range π-conjugation in graphene yields many extraordinary properties, such as high values of its Young's modulus (~1,100 GPa) [18], fracture strength (125 GPa) [18], thermal conductivity (~5,000 $Wm^{-1}K^{-1}$) [19], mobility of charge carriers (200,000 $cm^2V^{-1}S^{-1}$) [14] and specific surface area (2630 m^2g^{-1}) [20], plus fascinating transport phenomena such as the quantum Hall effect [21]. Because of these novel properties, graphene has found potential applications in many fields such as nanocomposites [22–24], batteries [25, 26], supercapacitors [27], nanoelectronics [28] and sensors [29–31], etc.

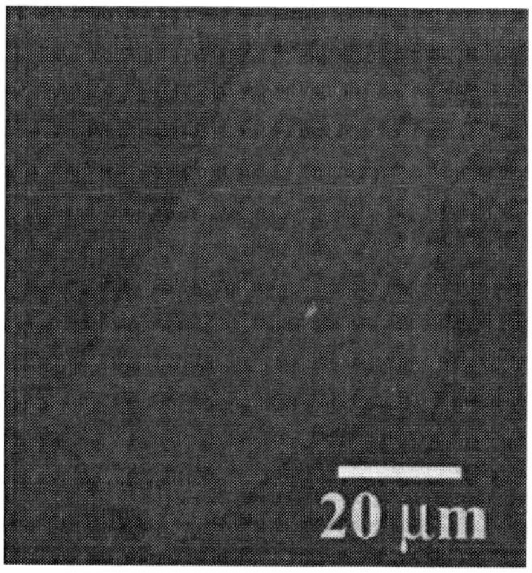

Figure 1. Single layer graphene first observed by Geim and co-workers at Manchester University [3].

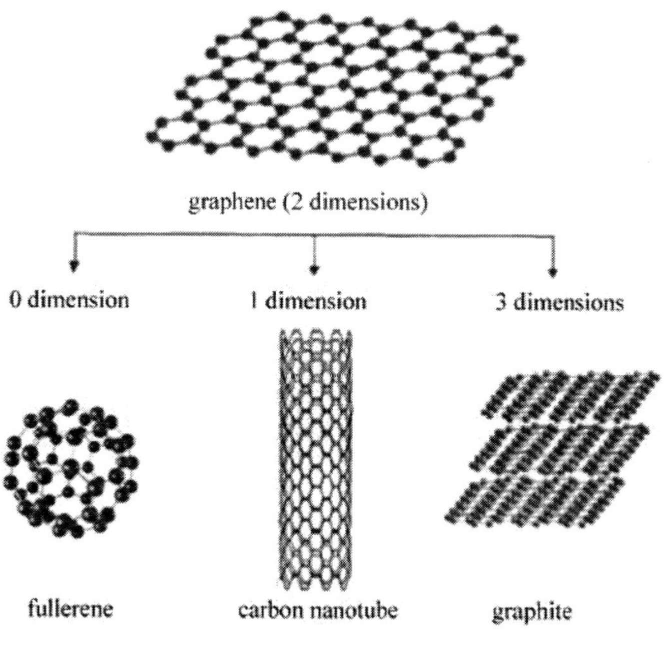

Figure 2. Schematic diagrams of graphene and its derivatives [17]

Carbon materials have been widely utilized in both analytical and industrial electrochemistry, where in many areas they have outperformed the traditional noble metals. This diversity and success stems largely from carbons structural polymorphism, chemical stability, low cost, wide potential windows, relatively inert electrochemistry, rich surface chemistry and electrocatalytic activities for a variety of redox reactions [32]. Carbon–based nanomaterials including carbon nanotubes [33–38], carbon nanofibers [39], highly ordered mesoporous carbons [40, 41], etc., have drawn much more attention because of their unusual electronic properties and ability to improve catalytic properties. Recently, graphene–based hybrid materials have become a hot research topic in material science because the hybridization can effectively enhance the functionality of materials, and can provide a new way to develop catalytic, magnetic and optoelectronic materials. Noble metal nanoparticles (NPs) are of great interest because of their unusual properties and have been widely used as electrocatalysts in fuel cells [42–45], and biosensing [46].

To date, several graphene–based noble metal nanocomposites have been prepared and used in the fields of biosensing [47–51], fuel cells [52–59], and other related fields. For example, Niu and coworkers [47–49] fabricated a series of graphene-based noble metal nanocomposites, for example they prepared chemically converted graphene (CCG)/3,4,9,10–perylene tetracarboxylic acid (PTCA)/Au–ionic liquid (Au–IL) composites (CCG/PTCA/Au–IL), and the CCG/PTCA/Au–IL composites exhibited good electrocatalytic behavior toward oxygen reduction [47]. They also fabricated graphene/AuNPs/chitosan composites film, and the composites exhibited good electrocatalytical activity toward H_2O_2 and O_2. Moreover, with glucose oxidase (GOD) as a model, the graphene/AuNPs/GOD/chitosan compostite–modified electrode shows prominent electrochemical response to glucose [48]. Furthermore, they synthesized Pt NPs/polyelectrolyte–functionalized ionic liquid (PFIL)/graphene sheets nanocomposites, and the obtained Pt/PFIL/GS nanocomposite–modified electrode was fabricated to simultaneously determine ascorbic acid (AA) and dopamine (DA) by cyclic voltammetry [49]. Shi and coworkers used gold nanoparticles (GNPs)–functionalized graphene (PEG) composite (GNP–PEG) modified electrode to detect uric acid (UA) [50]. Wang and coworkers demonstrated the use of cationic polyelectrolyte poly (diallyldimethyl ammonium chloride) (PDDA) functionalized graphene nanosheets (GNs) as the building block in the self–assembly of GNs/Au–NPs heterostructure, and they employ the GN/Au–NPs as the electrochemical enhanced material for H_2O_2 sensing [51], and they also demonstrated a wet–chemical approach for the synthesis of high–quality 3D Pt–on–Pd bimetallic nanodendrites supported on graphene sheets, and used as a nanoelectrocatalyst for methanol oxidation reaction (MOR) [52]. Kim et al. [53] and Hou et al. [54] synthesized graphene–supported Pt–Ru catalyst, which showed good electrochemical activity toward MOR. Li et al., [55] Honma et al., [56] and Ajayan et al. [57] deposited Pt NPs on surface of graphene, and used for MOR. Lin and coworkers [58, 59] loaded Pt NPs on functionalized graphene sheets (FGSs), and used the Pt–FGSs for oxygen reduction reaction (ORR).

This chapter describes main results of the recent studies in our group in use of graphene for fabrication of electrocatalysts. Specifically, we will report the fabrication, characterization and electrocatalytic characteristics of the metal–graphene nanocomposites. We mainly depict the fabrication of Au–graphene and Pt–Au–graphene nanocomposites by electrodeposition method, characterization of these nanocomposites with scanning electron microscopy (SEM), energy–dispersive spectroscopy (EDS), X–ray powder diffraction (XRD), and electrochemical methods, and the electrocatalytic activities of the Au–graphene and Pt–Au–

graphene nanocomposites toward oxygen reduction and some small organic molecules (glucose, methanol) oxidation. The work presented in this chapter would provide a useful method for preparing graphene–based noble metal nanocomposites and electrocatalysts, which could be potential useful in the fields of electrocatalysis, fuel cells, and other related fields.

2. Synthesis and Characterization of the Graphene Sheets

Graphene was synthesized by a route involving the steps of graphite oxidation, exfoliation, and chemical reduction. Graphite oxide (GO) was prepared by a modified Hummers method, [60] starting from graphite powder. Graphite was oxidized by concentrated H_2SO_4, $K_2S_2O_8$, and P_2O_5 to produce preoxidized graphite, which was then subjected to reoxidization by comcentrated H_2SO_4 and $KMnO_4$. Exfoliation was carried out by sonicating GO dispersion (0.05 wt.%) under ambient temperature for ca. 30 min. The resulted homogeneous yellow–brown dispersion was then subjected to 30 min of centrifugation at 3000 rpm to remove any unexfoliated graphite oxide (usually present in a very small amount). 3.5 μL of hydrazine solution and 40 μL of ammonia solution were added into the obtained GO dispersion (5 mL). After being vigorously shaken or stirred for a few minutes, the vial was put in a bath (95℃) for 1 h, graphene sheets were obtained by filtration of the product with a nylon membrane (0.22 μm) and drying in vacuum.

TEM image shows large graphene sheets with a length up to 4–5 μm and a width of ca. 3–4 μm are synthesized (Figure 3). The graphene sheets are rippled and resemble crumpled silk veil waves. AFM image indicates that the thickness of the prepared graphene is ca. 0.7 nm (Figure 4), which matches well with the reported apparent thickness of graphene sheets [61,62], suggesting the single–sheet nature of graphene is obtained in this work. The AFM image reveals that no sheets either thicker or thinner than 0.7 nm are observed, demonstrating that complete exfoliation of graphene oxide down to an individual graphene sheet is indeed achieved under our experimental conditions.

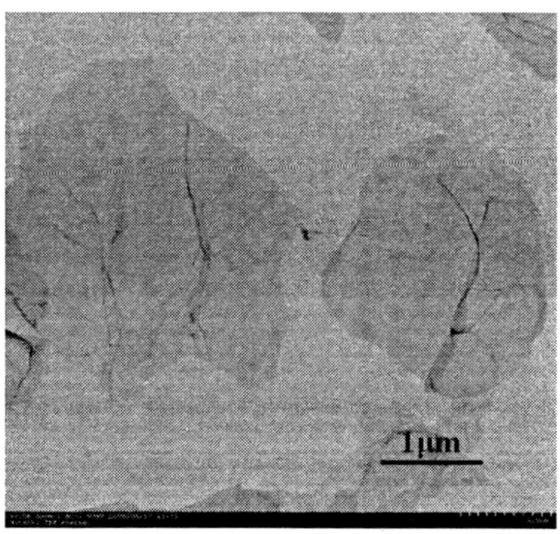

Figure 3. Typical TEM image of graphene sheets.

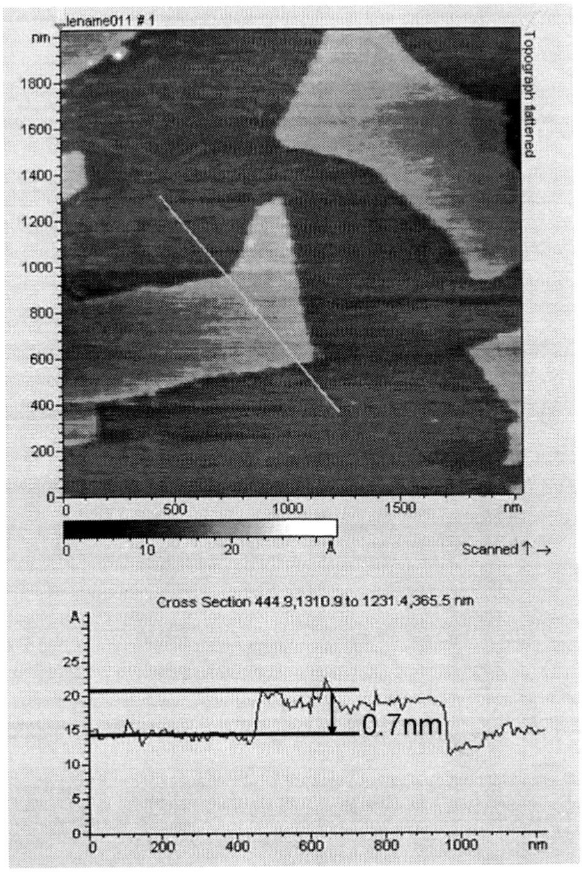

Figure 4. Typical tapping mode AFM image of graphene sheets.

After reduction with hydrazine, graphene oxide has been well deoxygened and the graphene is formed as verified by IR (Figure 5) and Raman spectra (Figure 6). Graphene and the GO together with the original graphite powder were also characterized by FT–IR spectroscopy. The IR spectrum of graphite does not show any observable peak (Figure 5a). In an IR spectrum of GO (Figure 5b), the characteristic stretching vibration modes of the O–H groups (~3440 cm^{-1}), C = O groups (~1730 cm^{-1}), the deformation peak of O–H groups (~1410 cm^{-1}), the stretching vibration peak of C–OH (1225 cm^{-1}), and C – O – C (1070 cm^{-1}) are obvious, demonstrating the successful oxidation of graphite. These results indicate that there are some oxygen–containing functional groups such as hydroxyl groups, carboxyl groups, and epoxy groups presenting on the GO surface. The peak at 1630 cm^{-1} is assigned to the vibrations of the adsorbed water molecules and the skeletal vibrations of unoxidized graphite domains [63]. After reduction with hydrazine, the residual oxygen functionalities are still present on the graphene surface; however, the IR intensities of those oxygen–containing groups decrease significantly (Figure 5c), indicating that graphene oxide has been well deoxygened and the graphene is successfully prepared.

Raman spectra of GO and graphene were recorded for obtaining further information on the structure of the prepared graphene. As showed in Figure 6, the Raman spectrum of GO exhibits two bands, the D band at ca.1350 cm^{-1} and the G band at ca.1590 cm^{-1} (Figure 6a).

The D band represents the symmetry A_{1g} mode of the crystalline graphite, and the G band represents the E_{2g} mode of sp^2 carbon atoms [64]. The intensity of G band is higher than that of D band. After reduction with hydrazine, the Raman spectrum of the graphene also contains two bands with the positions similar to those of GO (Figure 6b). However, the intensity of G band is weaker than that of D band. The increase of D/G intensity ratio compared to that for GO indicates that the GO has been well reduced and the graphene is formed [64, 65]. This conclusion is in good agreement with that obtained from IR spectra.

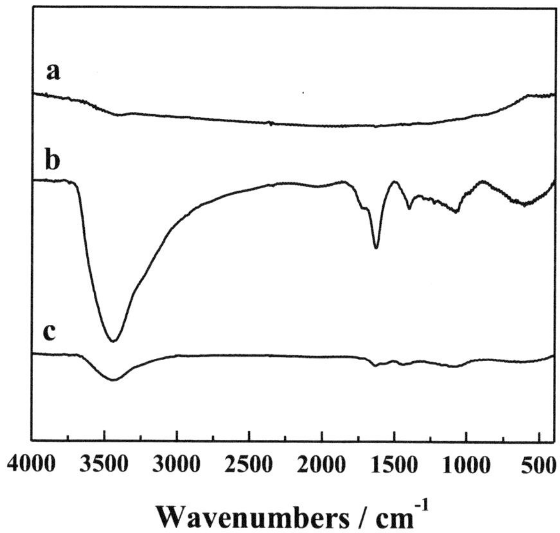

Figure 5. FT–IR spectra of the graphite (a), graphene oxide (b), and graphene (c).

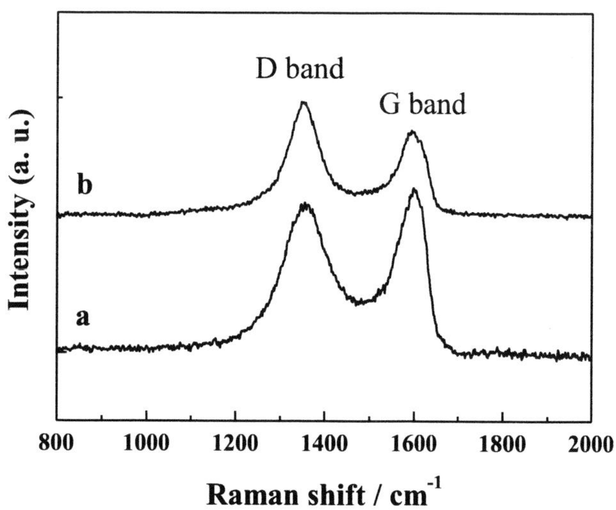

Figure 6. Raman spectra of the graphene oxide (a), and graphene (b).

3. FABRICATION AND CHARACVTERIZATION OF THE GRAPHENE/GC ELECTRODE

Prior to use, the GC electrode (3 mm in diameter, CH Instruments) was polished sequentially with metallographic abrasive paper (No. 6) and slurries of 0.3 and 0.05 μm alumina to create a mirror finish, and then sonicated with absolute ethanol and double distilled water for about 1 min, respectively. It was rinsed thoroughly with double distilled water and dried under ambient temperature. The experimental parameters were optimized to obtain the best voltammetric response. Typically, the as–prepared graphene sheets (1 mg) were first dispersed into double distilled water (1 mL) to form a black homogeneous suspension (1 mg mL^{-1}). Then, the suspension (6 μL) was cast onto the surface of the pretreated GC electrode with a microsyringe, and the solvent was allowed to be evaporated at ambient temperature before use. Graphene can easily form a thin film on the GC electrode surface through π–π electronic interactions between graphene and GC (denoted as graphene/GC electrode). SEM image of graphene on the GC electrode surface reveals the typical crumpled and wrinkled graphene sheet structure (Figure 7).

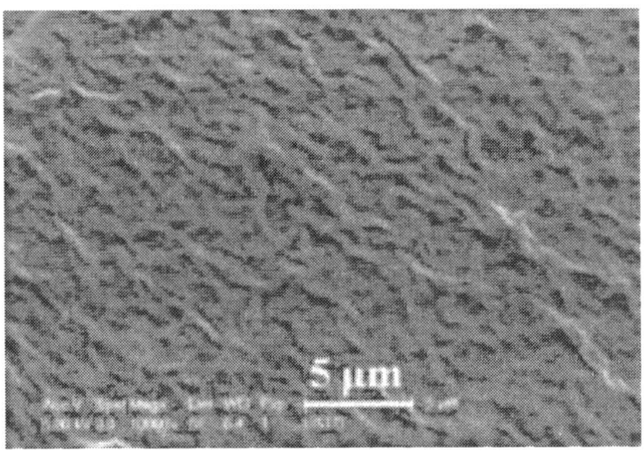

Figure 7. Typical SEM images of graphene film on GC electrode surface.

The cyclic voltammograms of graphene/GC and GC electrode in 0.1 M PBS (pH 7.0) are shown in Figure 8. No any electrochemical reactions can be observed from the cyclic voltammograms of GC electrode (curve a) and graphene/GC electrode (curve b). However, the background currents of Figure 8b is much larger than those of Figure 7a, which might be attributed to the reason that the apparent surface area of graphene/GC electrode is larger than that of a bare GC electrode.

To further characterize the electrochemical characteristics of the prepared graphene/GC electrode, the electrochemical performance of $K_3Fe(CN)_6$ at the electrode was carried out. Figure 9 is the cyclic voltammograms of 4 mM $K_3Fe(CN)_6$ in 0.1 M KCl solution at a bare GC electrode (curve a) and a graphene/GC electrode (curve b) at a scan rate of 100 mV/s. It can be seen that the anodic and cathodic peak potentials of $K_3Fe(CN)_6$ at a graphene/GC electrode slightly shift toward the negative and positive direction, respectively, compared with those at a bare GC electrode. The peak separation (ΔE_p) obtained for $Fe(CN)_6^{4-/3-}$ at graphene/GC electrode is 87 mV, which is smaller than that of GC electrode, suggesting the faster electron transfer kinetics of the $Fe(CN)_6^{4-/3-}$ redox probe at the graphene/GC electrode.

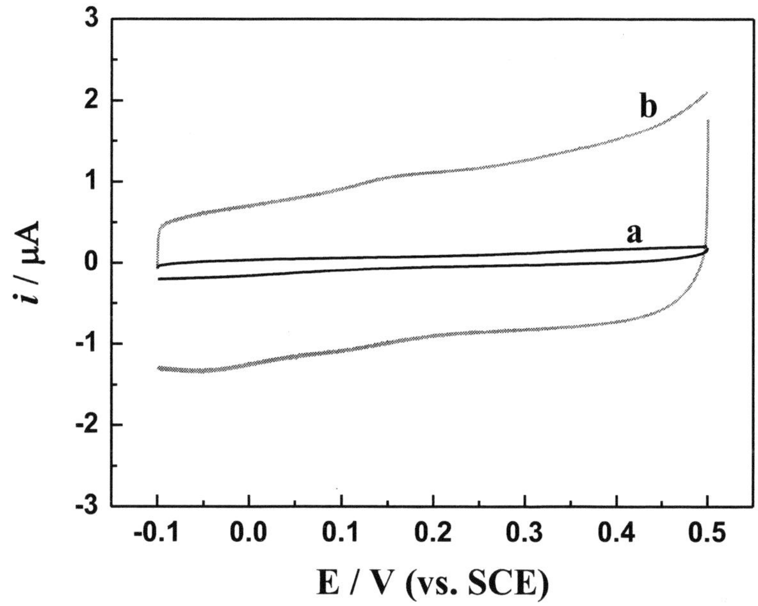

Figure 8. The cyclic voltammograms of GC (a) and graphene/bare GC (b) electrode in 0.1 M PBS (pH 7.0) at a scan rate of 10 mV s^{-1}.

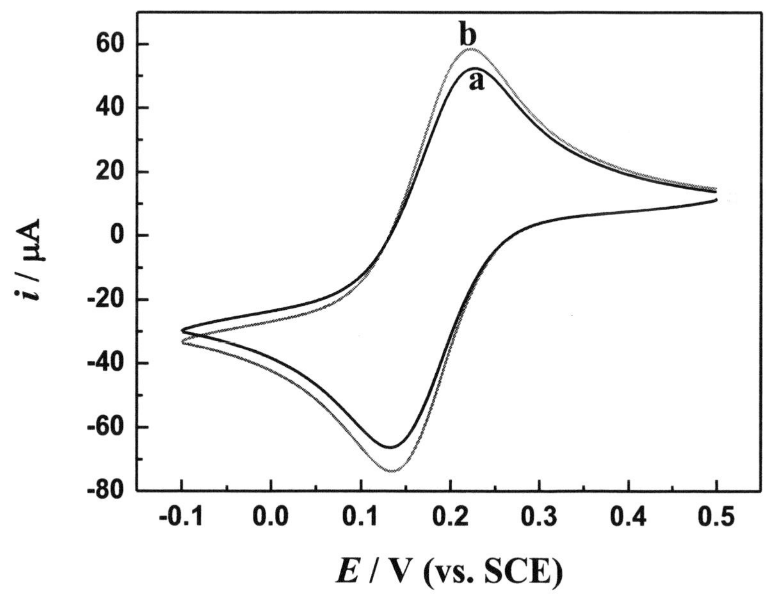

Figure 9. Cyclic voltammograms of 4 mM of $K_3Fe(CN)_6$ in 0.1 M KCl solution on a bare GC electrode (curve a) and a graphene/GC electrode (curve b) at a scan rate of 100 mV s^{-1}.

4. Graphene–Gold Nanostructure Composites Fabricated by Electrodeposition and their Electrocatalytic Activity toward the Oxygen Reduction and Glucose Oxidation

Because of their unique physical and chemical properties, gold NPs, as one of the most important metal nanomaterials, have been widely applied in optical, electronic, magnetic, catalytic, and biomedical fields [66–75]. Electrodeposition method is one of the mostly useful approaches to prepare gold nanostructures [76]. By this means, gold NPs can be facilely synthesized on the surface of conducting surfaces, and the sizes and shapes of the prepared nanoparticles can be easily controlled by altering the conditions of electrochemical deposition.

The Au NPs were electrodeposited on the graphene/GC electrode from aqueous solutions of 0.5 M H_2SO_4 containing 1 mM $HAuCl_4$ as precursor at a constant potential of –0.5 V (versus SCE) for 10 min forming the Au–graphene/GC electrode. Before deposition, the solution was deaerated by purging high–purity nitrogen, and the nitrogen environment was then kept over the solution to prevent the solution from oxygen. For study the effects of concentration of the precursor and the deposition time on the morphology of the deposited Au NPs, a variety of concentration of $HAuCl_4$ (from 0.1 mM up to 10 mM) and deposition time (from 1 min up to 30 min) were used. After thoroughly rinsed with double distilled water, the Au–graphene/GC electrode was ready to be used for studying the electrocatalytic characteristics toward the reduction of oxygen and the oxidation of glucose.

To study the effects of substrate on the morphology of the deposited Au NPs, the Au/GC electrode was also fabricated by employing the similar procedures as above, and its electrocatalytic characteristics toward the oxygen reduction and glucose oxidation were recorded and compared with those of the Au–graphene/GC electrode. Figure 10 A and B show the SEM images of the graphene/GC (A) and the bare GC electrodes (B) before (panel A1, and panel B1) and after (panel A2, A3 and panel B2, B3) electrodeposition of Au NPs. The surface of graphene film is quite uniform before electrodeposition of Au NPs. After electrodeposition of Au NPs, the morphology of the electrode changes significantly. There are many Au NPs wrapped with graphene sheets on the surface of the electrode (panel A2, Figure 10). The Au NPs integrate uniformly with the graphene forming the Au–graphene nanocomposites. Each Au NP, which has a spherical shape and a diameter of ca. (1.3 ± 0.2) μm, is built up by numerous small primary Au nanoparticles with average diameter of ca. 50 nm (panel A3, Figure 10). It appears that these small primary nanoparticles are interconnected with one another to form larger Au NPs architectures with recognizable boundaries or voids between the component subunits. These results suggest that the Au NP does not grow from a single seed. During the formation process, the seed nucleation is slow and new nuclei can be continuously formed. The deposition of Au nanostructures proceeds by a progressive nucleation growth mode [77–79]. These processes lead to the formation of nanostructures consisting of many small primary nanoparticles as shown in Figure 10.

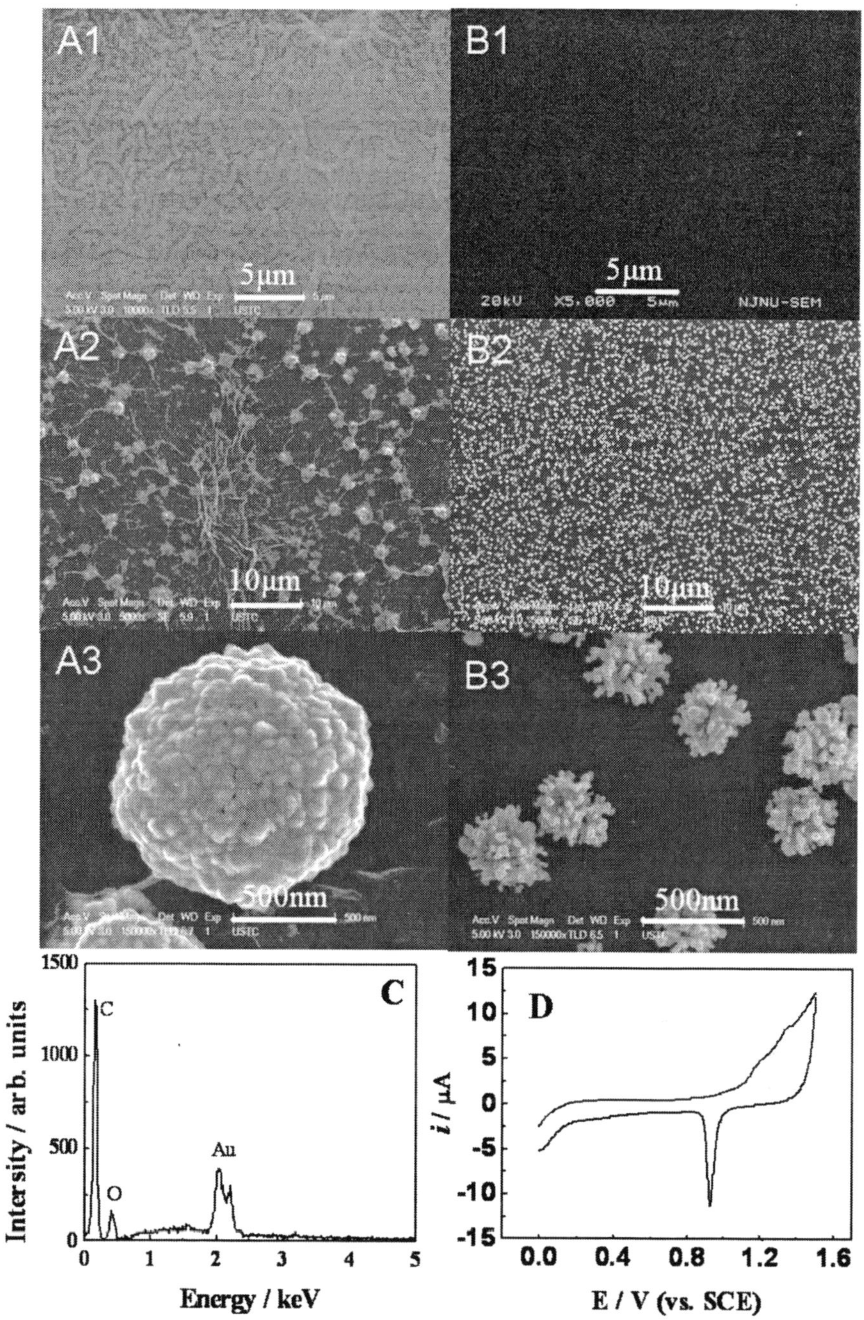

Figure 10. Typical SEM images of the graphene film on the GC electrode surface (panel A1), the bare GC electrode surface (panel B1), the Au NPs electrochemically deposited on the surface of graphene/GC electrode (panel A2, A3), and on bare GC electrode surface (panel B2, B3) under a constant potential of –0.5 V for 10 min in 0.5 M H_2SO_4 solution containing 1 mM $HAuCl_4$. Panel C shows the EDS of the Au–graphene nanocomposites, and panel D is the cyclic voltammogram of the Au–graphene/GC electrode in N_2–saturated 0.5 M H_2SO_4 at a scan rate of 10 mV s^{-1}.

The growth of Au NPs on the surface of graphene was further studied by analyzing the experimental current ~ time ($i \sim t$) transients based on theoretical transients for instantaneous and progressive nucleation. The dependence of current transients for the instantaneous and progressive nucleation (for three dimensional (3D) multiple nucleation) can be expressed in terms of non-dimensional quantities by Eqs (1) and (2), respectively [77, 78].

$$\left(\frac{i}{i_m}\right)^2 = \frac{1.9542}{(t/t_m)}\left\{1-\exp\left[-1.2564(t/t_m)\right]\right\}^2, \quad (1)$$

$$\left(\frac{i}{i_m}\right)^2 = \frac{1.2254}{(t/t_m)}\left\{1-\exp\left[-2.3367(t/t_m)\right]\right\}^2, \quad (2)$$

where i is the current at time t after the potential step, i_m is the maximum value of the current, and t_m is the time at which i_m is observed. The plots of $(i/i_m)^2$ against t/t_m for the experimental and theoretical transients are shown in Figure 11. Except for short time (where double layer charging is occurring), the transients follow quite closely the theoretical characteristic for the progressive nucleation process. At a longer time, the experimental currents decrease more slowly than that predicated by theoretical mode (Figure 11a and 11c). This phenomenon can be explained in terms of enhanced diffusion and rapid replacement of the discharging ion through hemispherical diffusion to growth centers [79].

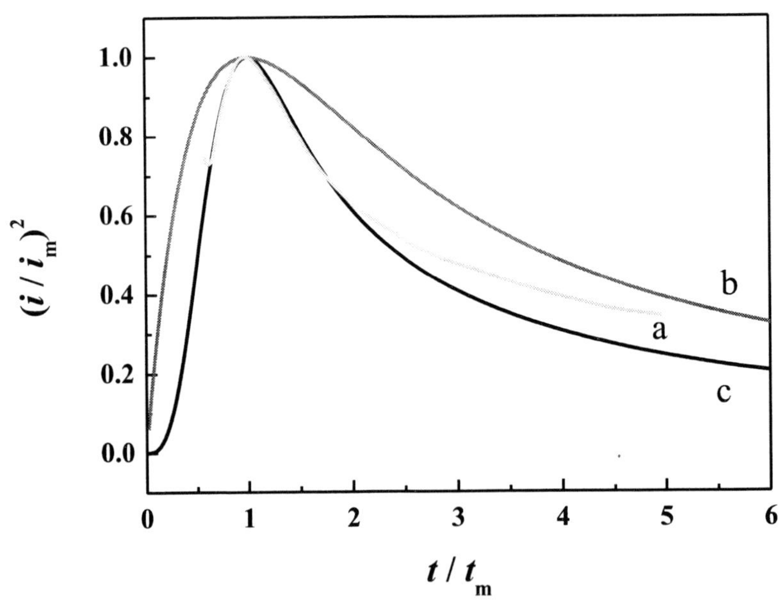

Figure 11. The experimental $(i/i_m)^2$ as a function of t/t_m for the Au nanostructures deposition at the graphene/GC electrode in 0.5 M H_2SO_4 solution containing 1 mM $HAuCl_4$ at –0.5 V (a). Curves (b) and (c) are theoretical curves corresponding to instantaneous (b) and progressive nucleation processes (c), respectively, with the 3D growth limited by diffusion.

The Au–graphene nanocomposites were also characterized with the use of EDS and cyclic voltammetry (CV). The EDS results in panel C of Figure 10 reveal that the presence of carbon, oxygen, and Au, and no other element is detectable, demonstrating that the NPs integrated with the graphene are Au nanoparticles. Further characterization of the Au NPs was achieved by recording the CV response of the Au–graphene/GC electrode in N_2–saturated 0.5 M H_2SO_4 solution in the potential range of 0 to 1.5 V at a scan rate of 10 mV s^{-1}. As shown in panel D of Figure 10, the electrode exhibits the characteristic feature of the redox reaction of Au with two oxidation peaks at ca. +1.1 and +1.3 V (versus SCE) corresponding to the formation of Au oxide, and a reduction peak at ca.+0.9 V corresponding to the reduction of the Au oxide [80]. These results further verify the formation of Au–graphene nanocomposites.

For comparison, the Au NPs were also deposited on the surface of bare GC electrode. The Au NPs formed on bare GC electrode surface have an average size of ca. 250 nm (panel B2, Figure 10), much smaller than those electrodeposited on the graphene/GC electrode surface. However, the density of Au NPs deposited on the bare GC electrode is much higher than that on the graphene/GC electrode surface. Moreover, the morphology is also significantly different from that of obtained at graphene/GC electrode (panel A3 and panel B3 in Figure 10). The image in panel B3 (Figure 10) clearly indicates that the obtained Au NPs at bare GC electrode surface are all uniform 3D dendritic structures, which are built up by tens of primary nanoparticles with an average dimension of ca. 20 nm. These results suggest that the shape, size, and the density of the electrochemically deposited Au NPs on the electrode surface are greatly affected by the grown substrate (for example bare GC or graphene). The reasons of the different nature and morphology of Au NPs grew on the bare and graphene modified GC electrode are unclear at present and need to be further studied. These differences may originate from the different surface nature of these electrodes.

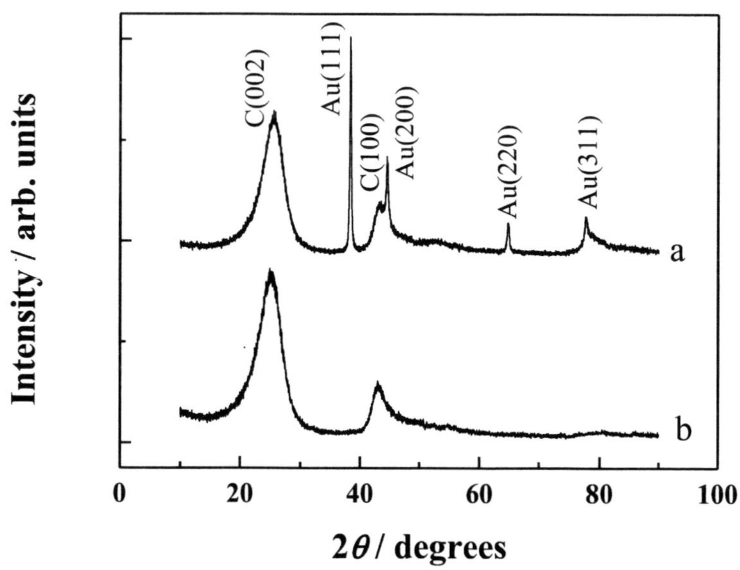

Figure 12. Typical XRD patterns of Au–graphene nanocomposites (a) and graphene (b).

The crystalline nature of the Au nanostructures in Au–graphene composites is confirmed by XRD studies, as shown in Figure 12. The patterns of Au–graphene nanocomposite (Figure 12a) show all of the major peaks of Au at 38.1 (111), 44.4 (200), 64.6 (220), and 77.6 ° (311). The peaks at ca.24.9 and 43.1° in Figure 12a come from the pattern of carbon (002), and (100) [81], respectively, as shown in Figure 12b for the XRD patterns of graphene. The XRD patterns of Au nanostructures are in good agreement with those of the reference patterns for the fcc (face centered–cubic) Au (JCPDS 04–0784). The sharp diffraction peak of the (111) with high intensity indicates that the deposited gold nanostructure has a tendency to grow with the surfaces dominated by the lowest energy (111) facet.

The electrochemical reduction of oxygen is importance in view of its practical applications in chemical processing, energy conversion, and chemical sensing. It has been well–established that noble metal nanostructures such as Pd, Pt, Au, etc., and their alloys have good catalytic activity toward the reduction of oxygen [50, 51, 82–84]. Au NPs supported on various carbon substrate materials such as carbon black [85], highly oriented pyrolytic graphite (HOPG) [86], and carbon nanotube (CNT) [87] have been prepared, and used for catalyzing the reduction of oxygen. Herein, the electrocatalytic activity of the Au–graphene nanostructures toward the reduction of oxygen in H_2SO_4 solution (0.5 M) is studied by cyclic voltammetry.

The electrochemical reduction of oxygen at the bare GC (panel A, Figure 13), graphene/GC (panel B, Figure 13), Au/GC (panel C, Figure 13), and Au–graphene/GC electrode (panel D, Figure 13) was studied, and their electrocatalytic characteristics were compared. Panel A in Figure 13 shows the cyclic voltammograms of the bare GC electrode in the air–saturated (curve b) and N_2–saturated PBS (curve a). No redox peaks can be observed in the cyclic voltammogram of the bare GC electrode (curve a), suggesting that bare GC electrode cannot undergo the redox reaction in the potential range of interest. The cyclic voltammograms presented in panel A also clearly show that the presence of O_2 in solution almost does not affect the voltammetric features of the bare GC electrode so much, only a very weak reduction current can be observed at the potential negative than −100 mV, demonstrating that the electrocatalytic activity of the bare GC electrode toward the reduction of O_2 is too low to be detected in the potential range of interest. The cyclic voltammetric characteristics of the graphene/GC electrode (panel B in Figure 13) is almost the same as that of the bare GC electrode except a slight increase in the background current because of the significant increase in the apparent surface area of the electrode after modifying graphene. These results indicate that graphene also does not have obvious electrocatalytic activity toward the reduction of O_2 in the experimental potential ranges. The cyclic voltammograms presented in panel C show the voltammetric features of the Au/GC electrode. In the absence of O_2, there isn't any redox peaks (curve a). However, the cyclic voltammogram (curve b) of the Au/electrode in air–saturated solution is significantly different from that of in N_2–saturated solution. There is an obvious reduction peak at ca.–50 mV, suggesting that the Au NPs deposited on the bare GC electrode exhibits electrocatalytic activity toward the reduction of O_2. The cyclic voltammogram of the Au–graphene/GC electrode in the air–saturated solution is characterized with a large cathodic peak at ca. –30 mV (panel D). These characteristics are the typical features of electrocatalytic reactions. The reduction of O_2 starts at ca. 220 mV, and the observed reduction current increases quite rapidly during negative potential scanning to reach the current of more than 17.5 μA at ca. –30 mV. The onset potential, which has a ca.100 mV positive shift compared with that obtained at the Au/GC electrode (ca. 120 mV), is 250 mV more positive than the reduction peak actually appears. This demonstrates that the positive potential for reduction of O_2 obtained at the Au–graphene/GC electrode is due to the electrocatalytic nature of Au–graphene nanocomposites. In addition, the starting potential for the reduction of oxygen at the Au–graphene/GC electrode is similar to that obtained at the electrode of Au NPs chemically deposited on multi–walled carbon nanotubes (ca. 200 mV) [87]. However, the onset potential observed at the Au–graphene/GC electrode is 120, and 220 mV more positive than those obtained for oxygen electroreduction at the electrode of carbon black supported Au NPs (ca. 100 mV) [85], and at Au NPs electrodeposited on highly oriented pyrolytic graphite (HOPG) electrode (ca. 0 mV) [86], respectively. The catalytic currents (17.5 μA) are more than 4 times of those obtained at the Au/GC electrode (4.0 μA). These results might be attributed to the synergistic effect of graphene and Au NPs. The high electrocatalytic current of the Au-graphene/GC electrode may also be due to its larger effective surface area in comparison with the Au/GC electrode because the capacitive current of the Au–graphene/GC electrode (curve a in panel D) is much higher than that of the Au/GC electrode (curve a in panel C).

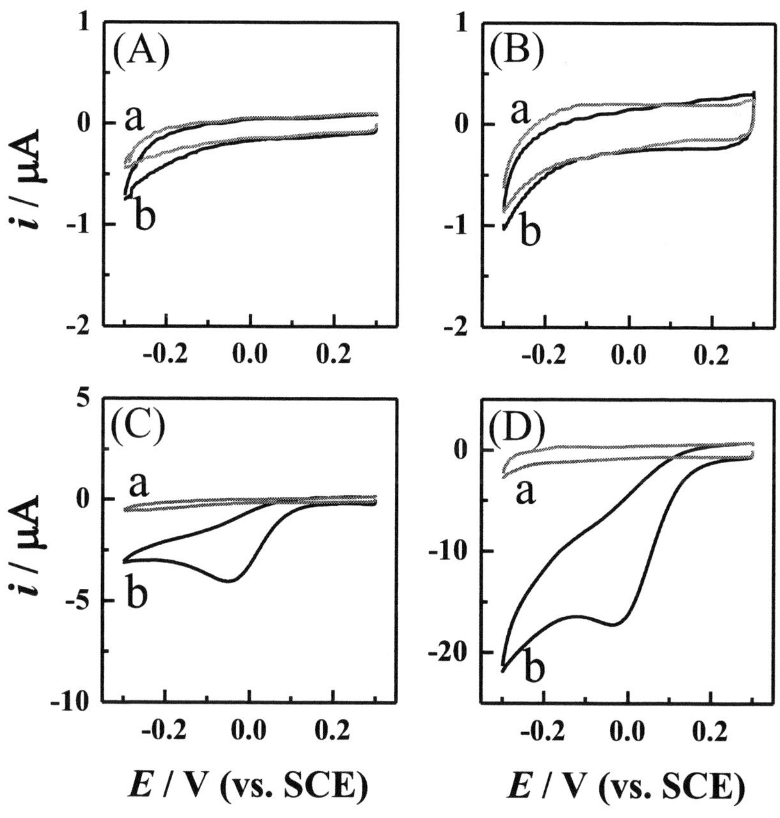

Figure 13. Cyclic voltammograms of the bare GC (panel A), graphene/GC (panel B), Au/GC (panel C), and Au–graphene/GC (panel D) in (a) N_2–saturated, and (b) air-saturated H_2SO_4 solution (0.5 M). Scan rate is 10 mV s^{-1}.

Glucose detection is of great interest from several points of view ranging from medical applications of blood glucose sensing to ecological approaches, such as in wastewater treatment, in food and textile industries, in environmental monitoring [88, 89], and so forth. It has been well-addressed that Au nanostructures such as nanoporous Au [76], chitosan-gold nanoparticles film [89], and so forth, show high electrocatalytic activity for oxidation of glucose. In this case, we investigated the electrocatalytic activity of the prepared Au–graphene nanocomposites toward the oxidation of glucose under physiological condition.

The electrocatalytic oxidation of glucose at the GC (panel A, Figure 14), graphene/GC (panel B, Figure 14), Au/GC (panel C, Figure 14), and Au–graphene/GC electrode (panel D, Figure 14) was studied in N_2–saturated PBS (pH 7.4), and their electrocatalytic characteristics were compared. The bare GC electrode cannot catalyze the oxidation of glucose in the potential ranges because the size and shape of the cyclic voltammogram of the electrode in glucose-contained solution (curve b, panel A in Figure 14) are almost the same as those in glucose–free solution (curve a, panel A in Figure 14). The results depicted in panel B (Figure 14) demonstrate that the graphene also does not have obvious electrocatalytic activity toward the oxidation of glucose.

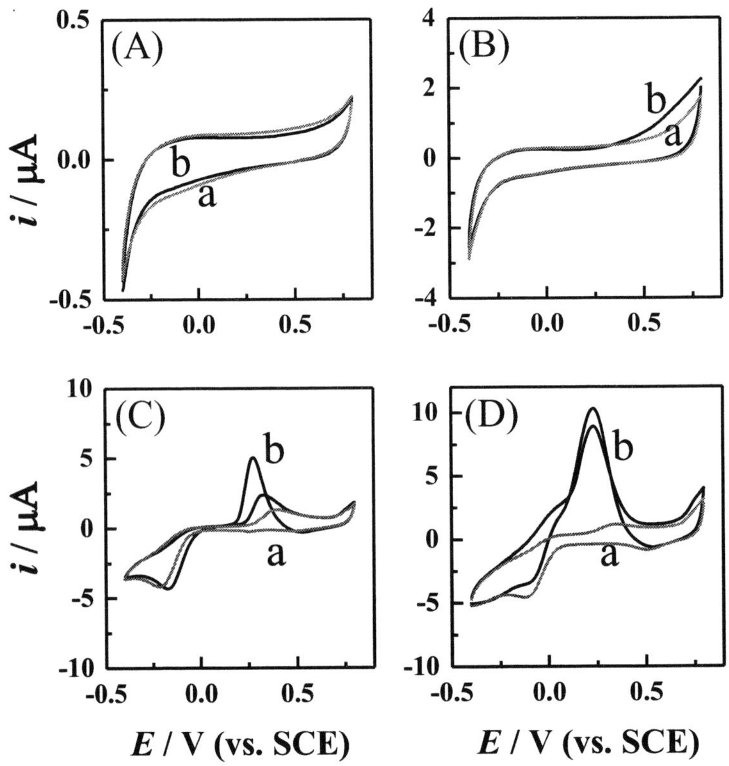

Figure 14. Cyclic voltammograms of the bare GC (panel A), the graphene/GC (panel B), the Au/GC (panel C), and the Au–graphene/GC (panel D) electrode in PBS (0.1 M, pH 7.4) with (curve b) and without (curve a) 5 mM glucose. Scan rate is 10 mV s^{-1}.

The cyclic voltammograms presented in panel D (Figure 14) show the voltammetric features of the Au–graphene/GC electrode in PBS (pH 7.4) with (curve b) and without (curve a) glucose (5 mM). In the absence of glucose, the cyclic voltammogram of the electrode exhibits the characteristic features of gold oxide formation (ca. 0.36 V) in the positive potential scan and the reduction of the oxide (ca. –0.12 V) in the negative potential scan [89]. Upon addition of glucose (5 mM), the voltammetric characteristics of the electrode change significantly and show a complicated electrochemical behavior. The cyclic voltammogram in the positive potential scan shows one anodic current peak at ca. 0.22 V, which corresponds to the direct oxidation of glucose. The decrease in current after the peak should be due to the formation of gold oxide, which competes for adsorption sites with glucose and in turn inhibits the electrooxidation of glucose as well. In the negative potential scan, the surface gold oxide was reduced and enough surface active sites will be available for the direct oxidation of glucose. Therefore, there is an anodic current peak appearing at ca. 0.22 V again. When the potential is moved to more negative values, a cathodic current peak corresponding to the reduction of gold oxide appears at ca. –0.10 V.

There are three types of glucose, two hemiacetal–types (α– and β–glucose) and an aldehyde–type (γ–glucose) of glucose. The two hemiacetal–types of glucose are converted to each other via aldehyde–type glucose. The electrochemical oxidation of glucose involves the dehydrogenation at C1 carbon [90]. β–glucose is the most reactive species among the possible anomeric forms. The poor reactivity of α–glucose is supposedly attributed to the geometric orientation of the hydrogen atom bound to the anomeric carbon. The product of electrochemical oxidation of α– and β–glucose is glucono–δ–lactone, which is hydrolyzed into gluconic acid. The electrochemical oxidation of γ–glucose produces gluconic acid directly. Therefore, regardless of whether the glucono–δ–lactone in involved as an intermediate or not, gluconic acid is the final stable product of two electron oxidation of glucose.

The electrocatalytic characteristics of the Au/GC electrode (panel C of Figure 14) are similar to those of the Au–graphene/GC electrode. However, the electrocatalyitc current is less than 1/2 of that obtained the Au–graphene/GC electrode, indicating that the electrocatalytic oxidation of glucose at the surface of the Au–graphene catalyst is effective and the electrocatalytic activity of Au–graphene is much higher than that of Au nanostructures alone.

The electrocatalytic activity of Au–graphene catalyst can be considered to result from the combination of two factors: one is the catalytic activity of the Au NPs, and the other one is the electrical network formed through their direct integrating with the graphene. The high catalytic characteristics of Au NPs certainly result in their good electrocatalytic power. The high density and well–distributed Au NPs on the surface of graphene would induce more active sites for the catalytic redox reaction and bring about an efficient electrical network through their direct integrating with graphene, which would greatly increase the electrocatalytic activity. Recently, it has been reported that the oxygen–containing functional groups of single–walled carbon nanotubes (SWNTs) such as the hydroxyl and carboxyl groups etc. can improve the electrocatalytic activity of the Pt–SWNT [91] and Pd–SWNTs catalyst [92], and these groups can play the role of Ru as secondary catalyst in the case of the Pt–Ru catalyst [91]. Therefore, the small amount of the oxygen–containing groups presented on the surface of graphene (please see the IR results depicted in Figure 3) is also believed to be responsible for better electrocatalytic activity of the catalyst, and they play a major role to remove those accumulated carbonaceous species like CO formed in the oxidation of glucose and contribute to the low catalyst poisoning in the Au–graphene catalyst, resulting in the high electrocatalytic activity of the catalyst.

It has been reported that the amount and the effective surface area of the Au NPs prepared by electrodeposition were determined by several parameters including the electrodeposition time, concentration of the precursor ($AuCl_4^-$ ions) [93], and so forth. Therefore, the effects of these parameters on the electrocatalytic activity of the prepared Au NPs toward the oxidation of the glucose were studied.

The effects of deposition time on the catalytic characteristics of the deposited Au NPs are presented in Figure 15. The Au–graphene/GC was fabricated by electrodepositing of Au NPs from a 0.5 M H_2SO_4 solution containing 1 mM $HAuCl_4$ under the applied voltage of –0.5 V with deposition time from 1 min up to 30 min. The catalytic current for the oxidation of 5 mM glucose increases with the increase of the deposition time up to 10 min, and reaches the maximum value at deposition time of 10 min (curve a–c). However, further increasing the deposition time leads to the decrease in the catalytic current (curve d).

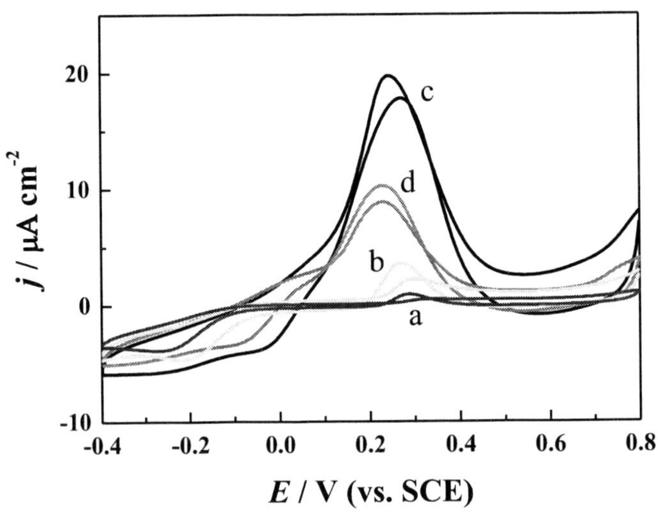

Figure 15. Cyclic voltammograms of the Au–graphene/GC electrode in 0.1 M PBS (pH 7.4) containing 5 mM glucose. The electrode was fabricated by electrodepositing of Au NPs from a 0.5 M H_2SO_4 solution containing 1 mM $HAuCl_4$ under the applied voltage of –0.5V with deposition time of (a) 1, (b) 5, (c) 10, and (d) 30 min. Scan rate is 10 mV s^{-1}.

The morphology of the Au–graphene electrode and the Au NPs obtained at the different deposition time were characterized by SEM (Figure 16). The density of Au NPs deposited on the surface of graphene/GC electrode increases with the deposition time from 1 min up to 30 min (panel A–D, Figure 16). With 30 min deposition, the graphene/GC electrode surface is almost completely coated by the Au NPs as viewed from panel D in Figure 16. The size of Au NPs almost remains unchangeable at the deposition time from 1 min up to 10 min (panel A–C). However, the size of Au NPs increases greatly at the deposition time of 30 min (panel D), moreover, the Au NPs aggregates together as revealed by the inset in panel D. The aggregation of Au NPs leads to decrease in the catalytic efficiency. Therefore, the combining effects of both the density of Au NPs on the surface of electrode and the aggregation of Au NPs results to the catalytic currents as depicted in Figure 15. The density of Au NPs on the electrode surface may be a key factor affected the catalytic currents of glucose oxidation at the Au–graphene/GC electrode fabricated by deposition time below 10 min. The enhancement in the Au NPs density leads to the increase of the catalytic currents. Too high density induces the aggregation of Au NPs (for example, at the deposition time of 30 min), which results to the decrease of the catalytic current.

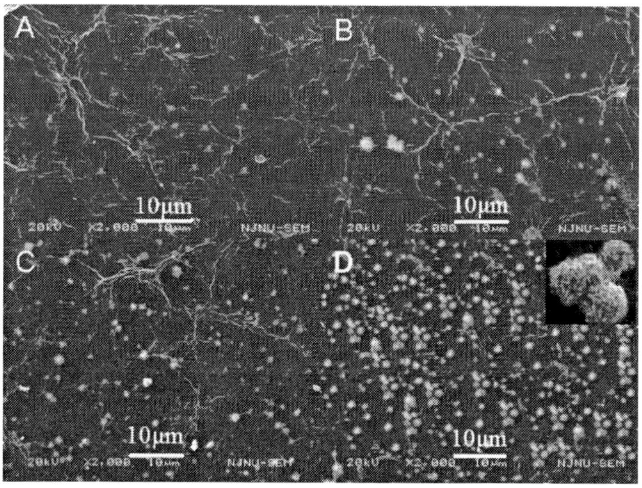

Figure 16. SEM images of the Au–graphene/GC electrode fabricated with electrodeposition of 1 (panel A), 5 (panel B), 10 (panel C), and 30 min (panel D) under the potential of –0.5 V in 0.5 M H_2SO_4 solution containing 1 mM $HAuCl_4$. The inset in panel D shows the images of Au NPs at a high magnification.

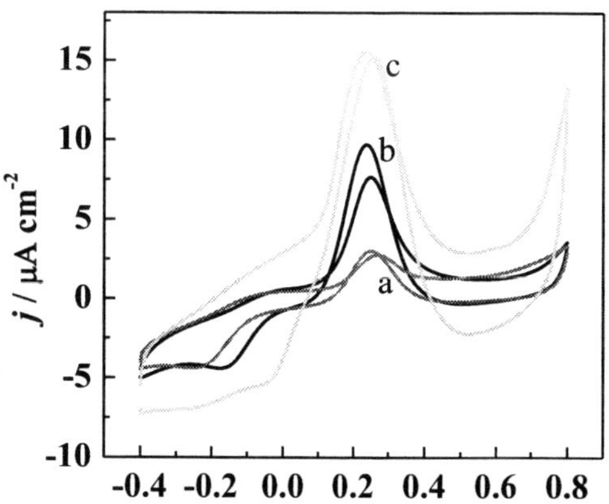

Figure 17. Cyclic voltammograms of the Au–graphene/GC electrode in 0.1 M PBS (pH 7.4) with 5 mM glucose. The electrode was fabricated by electrodepositing of Au NPs from a 0.5 M H_2SO_4 solution containing (a) 0.1, (b) 1, and (c) 10 mM $HAuCl_4$, respectively, under the applied voltage of –0.5 V with deposition time of 10 min. Scan rate is 10 mV s^{-1}.

The voltammetric results demonstrate that the catalytic currents of the oxidation of glucose at the Au–graphene/GC electrode are also significantly affected by the concentration of the $AuCl_4^-$ precursor, from which the Au NPs are deposited. As shown in Figure 17, the electrocatalytic current density of the glucose oxidation increases with the increase of the concentration of the precursor.

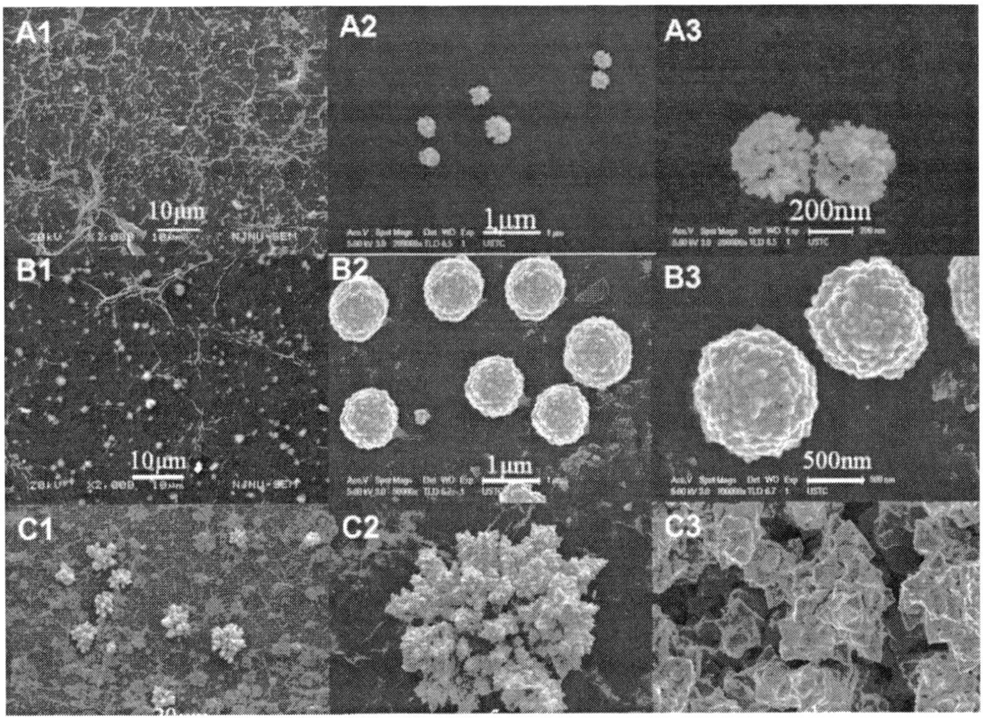

Figure 18. SEM images of Au NPs electrodeposited from 0.5 M H_2SO_4 solution containing 0.1 (panel A1–A3), 1 (panel B1–B3), and 10 mM $HAuCl_4$ (panel C1–C3) under a potential of –0.5 V for 10 min.

The results from SEM images indicate that the density of Au NPs is affected by the concentration of the precursor (panel A1–C1 in Figure 18), the density of Au NPs increases with increasing of the concentration of precursor. Moreover, the morphology and the size of the prepared Au NPs are also greatly affected by the concentration of the precursor (panel A2–C2, and panel A3–C3 in Figure 18). When the Au NPs are deposited from 0.1 mM $AuCl_4^-$ solution, the Au NPs are all uniform 3D dendritic structures with a diameter of about 250–300 nm. When the precursor concentration is increased to 1 mM, the obtained Au NPs have spherical shape with a diameter of ca. (1.3 ± 0.2) μm. The Au NPs are built up by numerous small primary Au nanoparticles with average diameter of ca.50 nm. With the concentration of the precursor is further increased to 10 mM, flowerlike Au NPs with the size of 15–20 μm are prepared, and the Au NPs are composed by numerous small flakes with the size of about 200 nm (panel C3 in Figure 18). The different density, size, and the morphologies of the obtained Au NPs prepared in various concentration of the precursor may be the main factors that determine the electrocatalytic activity of the Au–graphene/GC electrode toward the oxidation of glucose as depicted in Figure 17. The different effective surface area of the Au–graphene/GC electrode may also contribute the electrocatalytic activity for the glucose oxidation. The above results also suggest that the size and the morphology of the Au NPs deposited on the surface of graphene can be easily controlled by adjusting the concentration of $AuCl_4^-$ precursor.

5. Bimetallic Pt–Au Nanocatalystc Electrochemically Deposited On Graphene and their Electrocatalytic Characteristics Toward the Oxygen Reduction and Methanol

Due to their ease of handling, high energy density, and low operation temperatures [94, 95], direct methanol fuel cells (DMFCs) are considered as an attractive power source with potential applications in a variety of systems, ranging from automobiles to portable electronic devices [96, 97]. Generally, the basic operations of DMFCs include methanol oxidation reaction (MOR) and oxygen reduction reaction (ORR) on noble metal catalysts [98]. Pt catalysts are the most popular and effective electrocatalysts for both the anode and the cathode of DMFCs [99, 100]. However, the common Pt catalyst usually suffers from several disadvantages, such as high cost [100], easily being poisoned by CO–like intermediate species [101], and the kinetic limitation of the oxygen reduction [102], blocking the commercialization of DMFCs. Many efforts have been devoted to preparing various forms of Pt catalysts [103–105], such as mesoporous Pt [103], highly ordered Pt nanotube arrays [104], three–dimensional dendritic Pt nanostructures [105], etc., for reducing costs, providing higher poison tolerance, and increasing resistance to corrosion. Recently, bimetallic catalysts have attracted considerable attention because their properties often differ markedly from either of the constituent metals, with the mixtures frequently exhibiting enhanced catalytic stabilities, activities, or selectivities [106–108]. There are many reports of bimetallic Pt–M catalysts with increased electrocatalytic activity, where M is transition metals such as Cr, Mn, Fe, Co, Ni, Cu, Zn [109–118] or noble metals such as Ru, Pd, Ag and Au [119–134]. Au is a special metal in view of its inertness in bulk state and its high catalytic activity in nanoscale. This unique property of the nanoscaled Au has been found to show unprecedented electrocatalytic activity for CO oxidation [135–137]. Therefore, the Pt–Au nanocatalysts could be expected to provide synergistic catalytic effects in DMFCs. This bimetallic Pt–Au nanocatalyst could address the problem of poisoning of Pt catalysts by CO–like species, change the electronic band structure of Pt by modifying the strength of the surface adsorption [129], and significantly increase the electrochemical active surface area (ECSA) of Pt, which is an essential factor in improving the catalytic performance and utilization efficiency of Pt catalysts [124].

To further enhance the activity of catalysts and lower the use of noble metals, it is highly desirable to load catalysts on the surface of suitable supporting materials. Many forms of carbon materials have been investigated as catalyst supports for DMFCs, such as carbon black [138], carbon nanotubes (CNTs) [139], and carbon nanofibers (CNFs) [140] etc. These carbon materials not only significantly enhance the availability ECSA of electrocatalyst for electron transfer but also provide high mass transport of reactants to the electrocatalyst. Moreover, the presence of the oxygen–containing groups such as hydroxyl, carboxyl, etc. on the surface of carbon materials can remove those accumulated carbonaceous species like CO and contribute to the low catalyst poisoning, resulting in the high electrocatalytic activity of the catalyst [141]. It has been reported that the structures and properties of the carbon supports, such as surface functional groups [142], graphitizing structure [143], and surface area [144], have a significant effects on the activity and durability of the catalysts. Recently, gaphene has found potential applications as catalyst support in DMFCs because of its unique graphitized basal plane structure [145].

The Pt–Au nanostructures were electrodeposited on the graphene/GC electrode from aqueous solution of 0.5 M H_2SO_4 containing different Pt/Au molar ratios (the total concentration of $PtCl_6^{2-}$ to $AuCl_4^-$ was kept at 2 mM) as precursors at a constant potential of –0.4 V (versus SCE) for 10 min. Before deposition, the solution was deaerated by purging high–purity nitrogen, and the nitrogen environment was then kept over the solution to prevent oxygen from reaching the solution. After thoroughly rinsed with double distilled water, the Pt–Au–graphene/GC electrode with various molar ratios of Pt/Au was obtained and was ready to be used for studying the electrocatalytic characteristics for the MOR and ORR. For comparison, the Pt and Au nanostructures were electrodeposited on the graphene/GC electrode from H_2SO_4 (0.5 M) containing 2 mM $PtCl_6^{2-}$ and $AuCl_4^-$, respectively, as precursors at the same potential and the same time as that of the Pt–Au–graphene/GC electrode. The MOR and ORR were also studied at the Pt–graphene/GC and Au–graphene/GC electrode, respectively, the electrocatalytic characteristics were compared with those of the Pt–Au–graphene/GC electrode. To compare the electrocatalytic characteristics of Pt–graphene, Au–graphene, and Pt–Au–graphene with various Pt/Au molar ratios better, the electrocatalytic currents were normalized by the Pt loading (for the Au–graphene, the currents were normalized by the Au loading). The amount (mass) of Pt, Au, and Pt–Au nanostructures deposited on the graphene was estimated from the charge passing the electrochemical system under correction with blank system.

To study the role of graphene in the electrocatalytic activity of the Pt–Au–graphene nanocatalysts, the Pt–Au nanostructures with a Pt/Au molar ratios of 2:1 were also deposited on bare GC electrode and XC–72 carbon/GC electrode, respectively, using the similar procedures as above, and their electrocatalytic activity for the MOR were recorded and compared with that of the Pt–Au–graphene/GC electrode.

The typical SEM images of the Pt–Au nanostructures with different Pt/Au molar ratios deposited on the surface of the graphene/GC electrode were showed in Figure 19. The deposited Au nanostructures have a flowerlike structure with the size of 3–4 μm and are composed by numerous small particles with the size of about 150–200 nm (panels A1 to A3). When the reaction was carried out in the presence of $PtCl_6^{2-}$ (for example 0.5:1 for Pt/Au), the resulted Pt–Au nanostructures also have flowerlike structures as those for Au nanostructures with the size of about 9–10 μm (panel B2), much larger than that of Au. The high–magnification micrograph clearly shows that the Pt–Au nanostructures are also built up by numerous small particles with the average diameter of ca. 100 nm (panel B3). As the amount of Pt is increased in the Pt–Au nanostructures, the size of the Pt–Au nanostructures is becoming smaller. For example, the size of the Pt–Au nanostructures is about 8 (panel C2), 7 (panel D2), and 3.5 μm (panel E2) at the Pt/Au molar ratio of 1:1, 2:1, and 3:1, respectively. The morphology evolves from large flowerlike structures, to small flowerlike structures, to round particles (panels C to E). When the reaction was carried out in the presence of only $PtCl_6^{2-}$, the prepared Pt nanostructures have a spherical shape with an average size of 0.7 μm (panel F2), these spherical Pt nanostructures are built by nanoflakes (panel F3).

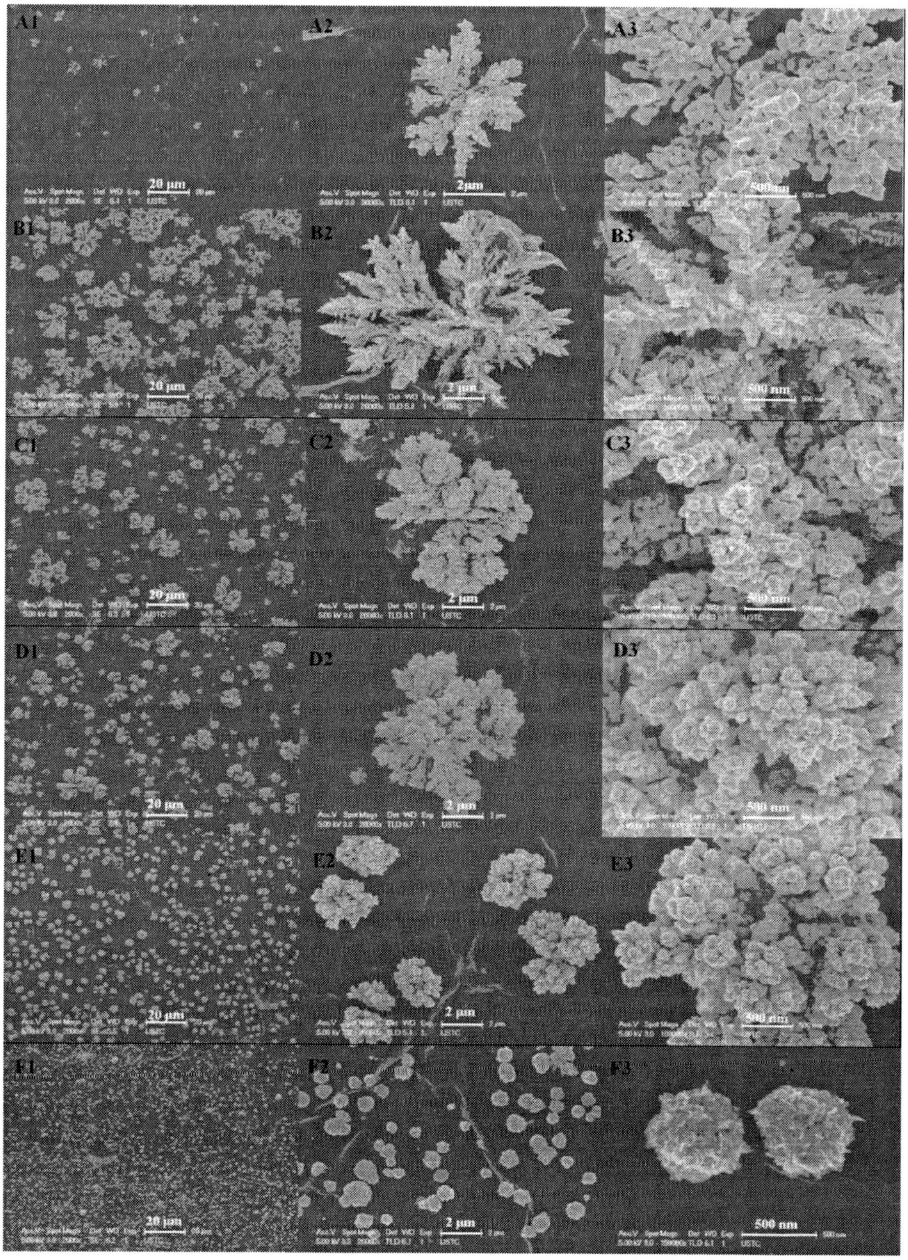

Figure 19. Typical SEM images of the Pt–Au nanostructures electrochemically deposited on the surface of graphene/GC electrode under a constant potential of –0.4 V in 0.5 M H_2SO_4 solution containing $AuCl_4^-$ (panel A), $PtCl_6^{2-}$ (panel F), and $PtCl_6^{2-}/AuCl_4^-$ molar ratio of 0.5:1 (panel B), 1:1 (panel C), 2:1 (panel D), and 3:1 (panel E), respectively.

The Pt–Au nanostructures were also characterized with the use of EDS. The typical EDS results in Figure 20 reveal that the presence of carbon, Au, and Pt, and no other element is detectable, demonstrating that the nanostructures deposited on the graphene/GC electrode surface are Pt–Au. In addition, the atomic ratio of Pt to Au in the bimetallic nanostructures is estimated based on the analysis of EDS. It demonstrates that the atomic compositions of the prepared Pt–Au nanostructures are in good agreement with the Pt/Au molar ratio of precursors in the solution for deposition (Table 1). The results suggest that the molar ratio of Pt to Au in the synthesized nanostructures can be easily controlled by adjusting the concentrations of those precursors involved in the reaction mixtures.

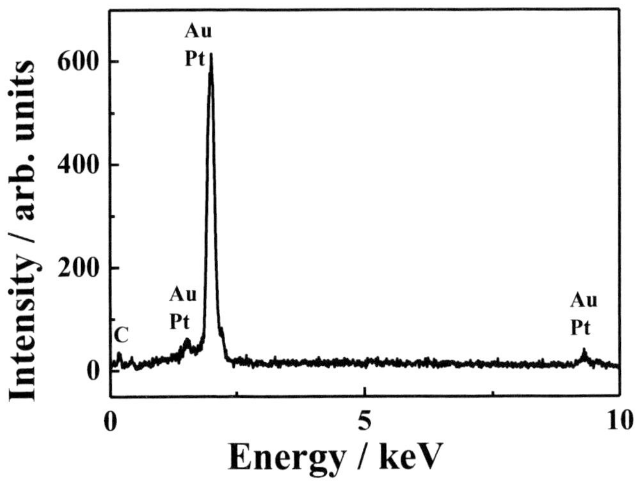

Figure 20. Typical energy–dispersive X–ray spectroscopy of the Pt–Au–graphene nanostructure with the Pt/Au molar ratio of 2:1.

Table 1. Comparison of Atomic Concentrations of Pt and Au in Pt–Au Nanostructures with those of in Deposition Solution

No.	molar ratio of Pt/Au in solution[a]	Pt atom (%)[b]	Au atom (%)[c]	ratio of Pt/Au
1	only $AuCl_4^-$ in solution	0	100	pure Au
2	0.5:1	31.54	68.46	0.46:1
3	1:1	51.63	48.37	1.07:1
4	2:1	65.97	34.03	1.94:1
5	3:1	75.29	24.71	3.05:1
6	only $PtCl_6^{2-}$ in solution	100	0	pure Pt

[a] The total concentration of $PtCl_6^{2-}$ and $AuCl_4^-$ in the deposition solution was kept at 2 mM.
[b] The Pt atomic concentration in prepared Pt–Au nanostructures estimated by EDS.
[c] The Au atomic concentration in prepared Pt–Au nanostructures estimated by EDS.

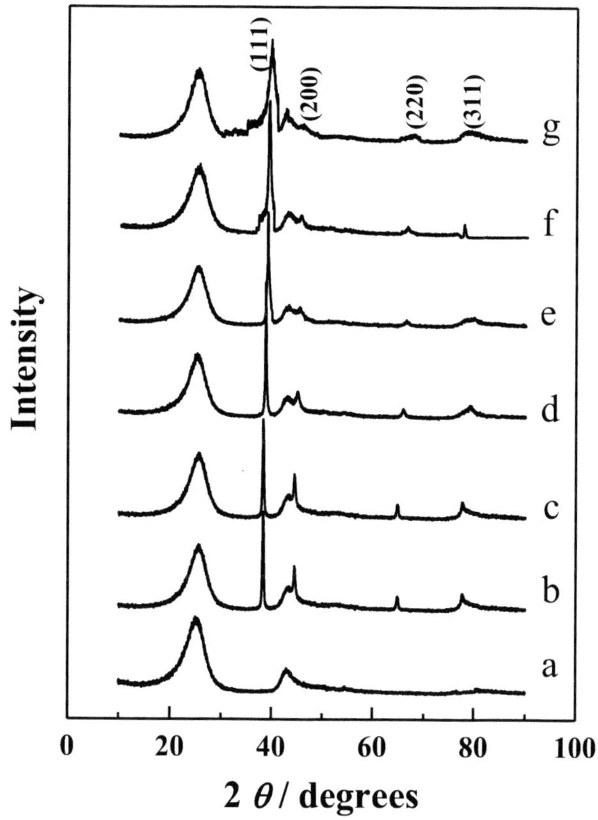

Figure 21. Typical XRD patterns of graphene (a), Au–graphene (b), Pt–graphene (g), and Pt–Au–graphene with Pt/Au molar ratio of 0.5:1 (c), 1:1 (d), 2:1 (e), and 3:1 (f), respectively.

The crystalline structures of Pt–Au nanostructures electrodeposited on graphene are confirmed by XRD measurements. The XRD patterns of Au (curve b, Figure 21) and Pt nanostructures (curve g, Figure 21) are in good agreement with those of the reference patterns for the fcc (face centered–cubic) Au (JCPDS 04–0784) and Pt (JCPDS 04–0802). The peak at ca.38.3, 44.5, 64.7, and 77.7 ° (curve b) can be indexed to Au (111), (200), (220), and (311) diffraction peaks, respectively, and the peaks at ca.39.8, 46.5, 67.9, and 79.1 ° (curve g) to Pt (111), (200), (220), and (311) diffraction peaks, respectively, confirming that the as–synthesized Pt and Au nanostructures have a highly crystalline face centered cubic phase. The peaks at ca.25.1 and 43.0 ° come from the pattern of carbon (002), and (100) [81], respectively, as shown in curve a (Figure 21) for the XRD patterns of graphene. The lattice parameters of Pt and Au nanostructures calculated based on the diffraction peaks are 4.065, and 3.919 Å, respectively, which is almost identical to the reported value of 4.078, and 3.923 Å for Au and Pt, respectively [146]. The well–resolved peaks depicted in curve c to f (Figure 21) can be indexed as (111), (200), (220), and (311) diffraction peaks, respectively, for Pt (or Au). Due to the similarity of the Pt and Au crystalline structures, the difference of the XRD peak positions for Pt and Au in Pt–Au nanostructures cannot be distinguished. In addition, with the increase of Pt/Au molar ratio, the peaks shifts to high degrees (curve c to f,

Figure 21), indicating the proportion of Pt increases progressively in the Pt–Au nanostructures.

The values of d–spacing, and lattice parameter for the bimetallic Pt–Au nanostructures at various molar ratios were also calculated and presented in Table 2. It can be found that the diffraction peak position, d–spacing, and lattice parameter of the nanostructures are in between pure Au and Pt (Table 2), and moreover, the lattice value is nonlinearly dependent on the composition of the nanostructures. These results demonstrate that the metallic constituents condense independently, not form alloy of Pt and Au [146], because the value of lattice parameter of the alloys should be linear dependence on their compositions as predicted by Vergard's law [147].

Table 2. The Values of d-spacing (*d*), and Lattice Parameter (*a*) of the Pt–Au Nanostructures at Various Pt/Au Molar Ratio

molar ration of Pt/Au	2θ for (111) (°)	d (Å)[a]	a (Å)[b]
pure Au	38.32	2.347	4.065
0.5:1	38.40	2.343	4.058
1:1	38.76	2.321	4.020
2:1	39.18	2.297	3.979
3:1	39.37	2.287	3.962
pure Pt	39.80	2.263	3.919

[a] The value of d was calculated using Bragg's equation: $2d\sin\theta = n\lambda$.

[b] The value of a was calculated using the equation of $d_{hkl} = \frac{a}{\sqrt{h^2 + k^2 + l^2}}$ for face centered-cubic crystal system.

The Pt oxide formation and reduction, as well as the characteristic hydrogen adsorption/desorption features of Pt–graphene/GC and Pt–Au–graphene/GC electrode were also studied as shown in panel A of Figure 22. The current associated with the hydrogen adsorption/desorption of the Pt–Au–graphene/GC electrode is larger than that of the Pt–graphene/GC electrode, suggesting that the Pt–Au nanostructures possess higher ECSA than that of pure Pt nanostructures. The values of ECSA for Pt–Au nanostructures can be estimated by integrating the charge in the adsorption/desorption region of hydrogen, assuming that hydrogen is adsorbed on Pt sites and not on Au particles [148, 149]. The calculated value of ECSA is ca.1.8 m^2 (g Pt)$^{-1}$ for pure Pt nanostructure, and 10.7, 35.1, 82.2, and 38.6 m^2 (g Pt)$^{-1}$ for Pt–Au nanostructures at the Pt/Au molar ratio of 0.5:1, 1:1, 2:1, and 3:1, respectively. These data indicate that the growth of Pt on the surface of Au nanostructures can enhance the value of ECSA of Pt, and the ECSA reaches the highest value when the Au nanostructures are covered by a thin layer of Pt atom at the Pt/Au molar ratio of 2:1. A thicker Pt layer growing on the Au nanostructures causes the decrease in ECSA. These features may greatly affect the electrocatalytic characteristics of Pt–Au nanostructures because the catalytic activities of catalysts are closely related to their ECSA.

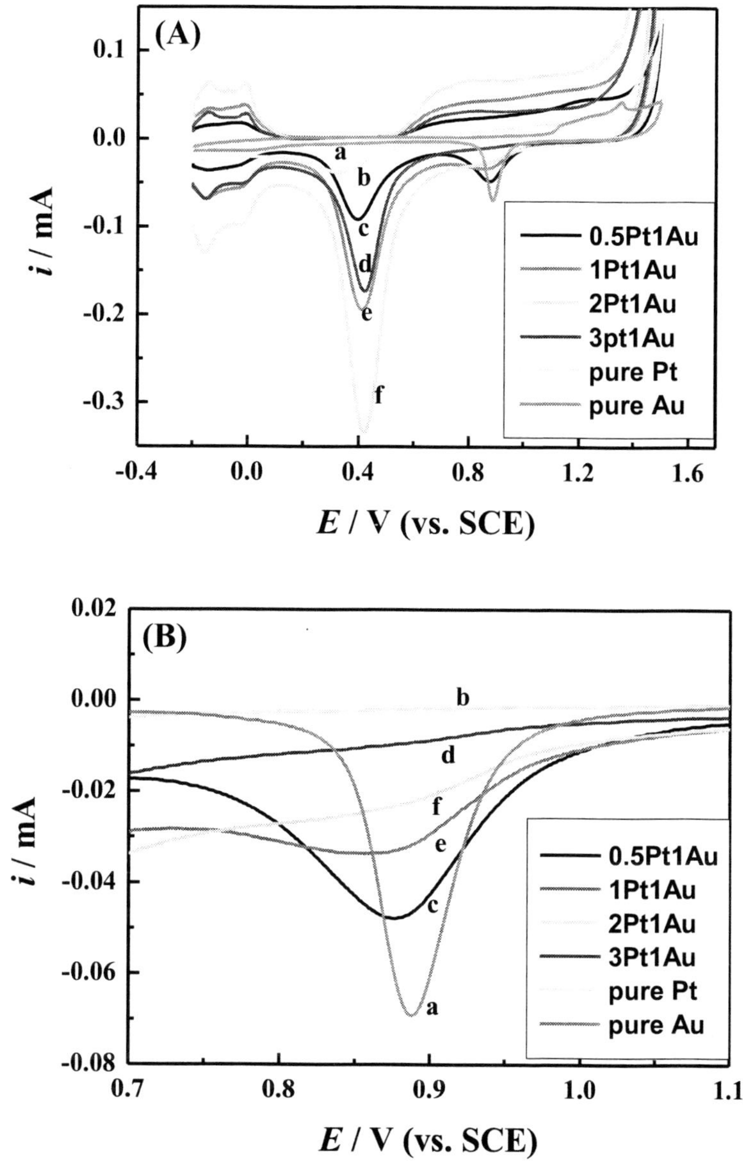

Figure 22. (panel A) Cyclic voltammograms of the Au–graphene/GC (a), Pt–graphene/GC (b), and Pt–Au–graphene/GC electrode with Pt/Au molar ratio of 0.5:1 (c), 3:1 (d), 1:1 (e), and 2:1 (f), respectively, in N_2-saturated 0.5 M H_2SO_4 solution. (panel B) Linear scanning voltammograms for the reduction of Au oxide in the potential range of 0.7 to 1.1 V. The scanning rate is 50 mV s^{-1}.

Figure 23 displays the polarization curves of the ORR catalyzed by the Pt–Au–graphene nanocatalysts with different Pt/Au molar ratios in an O_2–saturated 0.5 M H_2SO_4 solution at a scan rate of 10 mV s^{-1}. These curves were recorded at a RDE under a rotation rate of 1600 rpm. For comparison, the electrocatalytic features of ORR at the graphene/GC, Au–graphene/GC and Pt–graphene/GC electrode are also recorded and presented in Figure 23. No cathodic current is observed at the graphene/GC electrode, indicating that graphene cannot catalyze the reduction of O_2 in the potential range of the experiments (curve a). The onset potential for the reduction of O_2 starts at ca.0.45 and 0.78 V (versus RHE) on Au and Pt nanocatalysts (curves b, c, Figure 23), respectively, and the highest mass specific current is ca.1.7 and 6.6 mA, respectively, indicating that Pt–graphene nanocatalyst exhibits higher catalytic activity for the ORR than Au–graphene catalyst. Figure 23 also shows that the ORR at the Pt–Au–graphene nanocatalysts exhibits a diffusion controlled region at the potential negative than 0.6 V, and a mixed diffusion-kinetic controlled region in the potential range of 0.6 to 1.0 V (curves d–g, Figure 23). The onset potential of the ORR at Pt–Au–graphene nanocatalysts, compared with Pt–graphene catalyst, shifts toward positive potentials and the reduction current is enhanced significantly (curves d–g), implying that Au is a catalytically enhancing agent. The half–wave potentials of ORR catalyzed by Au–graphene, Pt–graphene, and Pt–Au–graphene catalysts with different Pt/Au molar ratios are estimated to be 0.33 (pure Au), 0.54 (pure Pt), 0.73 (0.5:1), 0.76 V (3:1), 0.77 (1:1), and 0.83 (2:1), respectively, increase in the order of pure Au < pure Pt < Pt–Au (Pt/Au = 0.5:1) < Pt–Au (Pt/Au = 3:1) ≈ Pt–Au (Pt/Au = 1:1) < Pt–Au (Pt/Au = 2:1). This means that the half–wave potential obtained at Pt–Au–graphene catalyst with Pt/Au molar ratio of 2:1 has a more than 290 mV positive shift compared with that obtained at the Pt–graphene catalyst (0.54 V). It is also 80 mV more positive than that obtained at PtAu/C catalyst (0.75 V, versus RHE) [126], indicating the higher catalytic activity of the Pt–Au–graphene catalyst (at Pt/Au molar ratio of 2:1) toward ORR.

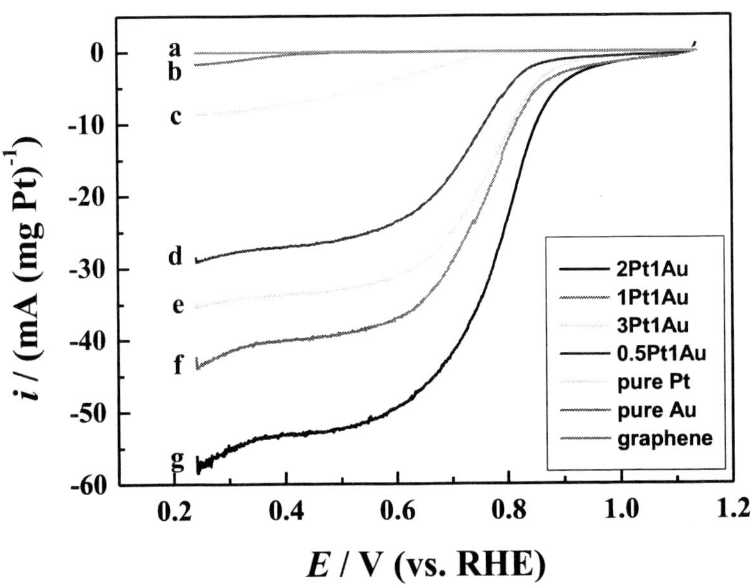

Figure 23. Polarization curves for the ORR catalyzed by graphene (a), Au–graphene (b), Pt–graphene (c), and Pt–Au–graphene nanostructures with Pt/Au molar ratios of 0.5:1 (d), 3:1(e), 1:1(f), and 2:1 (g) in oxygen-saturated 0.5 M H_2SO_4 at scan rate of 10 mV s^{-1}. Rotation rate: 1600 rpm. Curve a is obtained by normalizing the experimental current of the graphene/GC electrode with the mass of graphene coated on the GC surface.

The surface catalytic reaction of ORR on Pt–based catalysts significantly depends on the ECSA of catalyst [150]. As shown in Figure 22, the growth of Pt on the surface of Au nanostructures can significantly enhance the value of ECSA of Pt catalyst, and enhance the electrocatalytic activity as presented in Figure 23. When the Pt content is low, the value of ECSA increases with the increasing of the content of Pt/Au molar ratio, leading to the enhancement of the catalytic activity. The highest value of ECSA is reached when the Au nanostructures are completely covered by a thin layer of Pt atoms at the Pt/Au molar ratio of 2:1, resulting to the highest catalytic activity for ORR (Figure 23). However, thicker Pt layer grew on Au nanocatalyst leads to the decrease of the ECSA, and consequently, the electrocatalytic activities of the catalyst decrease as presented in Figure 23.

The different catalytic activity of the Pt–Au nanostructures can also be understood based on the synergistic role of the Au and Pt in ORR. It is reported that the O_2 reduction reaction involves both the breaking of the O–O bond and the formation of an O–H bond [151]. Based on simple thermodynamic principles, Fernandez et al [152] provided guidelines for the design of binary and multicomponent electrocatalytic materials for the reduction of oxygen, which combine one metal that easily breaks the O–O bond of O_2 forming M–OH adsorbed atomic oxygen with another metal that will facilitate the formation of O–H bond by breaking the M–OH bond. The Au nanostructures are hardly active as an ORR electrocatalyst because Au is not capable of providing adsorption sites for the formation of Au–OH. For pure Pt electrocatalysts, Pt–OH is easily formed on the surface of Pt, however, the Pt–OH cannot be broken easily. High concentration of Pt–OH species can be considered as poisoning species in the ORR since their presence reduces the number of active sites for the activation of oxygen via dissociative chemisorption, or splitting of the O–O bond [150]. Hence, the electrocatalytic activity of the pure Pt nanostructures is not high (Figures 23). The density functional theory (DFT) studies have shown that the catalytic activity of Pt–based catalysts is closely related the electronic interaction between the Pt and its substrate [151, 153]. When the Au is introduced into the Pt catalyst, Au can facilitate the breaking of Pt–OH and thus allows Pt to act efficiently in the electrocatalytic process. Therefore, in comparison with that on pure Pt catalyst, the concentration of Pt–OH on the surface of Pt–Au is significantly reduced. This characteristic could help to improve the kinetics and to enhance the ORR activity of the catalyst (Figure 23). However, the Pt/Au molar ratio has a significant effect on the ORR activity since the Pt is the active component for ORR. As shown in Figures 22, if the Pt content is low, the surface of Au cannot be totally covered by Pt nanostructures, and therefore the ORR activity is not very high. With the increase of Pt/Au molar ratio, more and more Pt nanostructures formed on the surface of Au nanostructures, resulting in the enhancement in the electrocatalytic activity (Figures 23). As shown in Figure 22, the Au is completely covered by Pt nanostructures when the Pt/Au molar ratio reaches 2:1, and the Pt–Au nanostructures with Pt/Au ratio of 2:1 exhibit the highest catalytic activity for ORR among the prepared electrocatalysts (Figures 23). However, the higher Pt content results in the increase of thickness of Pt, and the top layer of Pt cannot interact well with Au and therefore the oxygen reduction will be hindered (Figures 23). A similar phenomenon has also been observed in Au@Pt [123] and Pd@Pt nanocatalysts [120].

The electrocatalytic activity of the Pt–Au–graphene catalyst not only depends on the Pt/Au molar ratios, but also depends on the amount (mass) of the catalyst deposited on the surface of the GC electrode. Therefore, for a further understanding of the observed catalytic activities of the as–prepared catalysts for ORR, the specific mass activity at 0.80 V was calculated based on the polarization curves and amount of the Pt–Au nanostructures loading. The specific mass activities of Pt–Au–graphene catalysts are much higher than that of Pt–graphene and Au–graphene catalysts (Figure 24). The value for Pt–Au–graphene with Pt/Au ratio of 2:1 is ca.50 times higher than that of pure Pt, underlining the importance of Pt–Au–graphene electrocatalysts in reducing the amount of Pt utilization in the fuel cell.

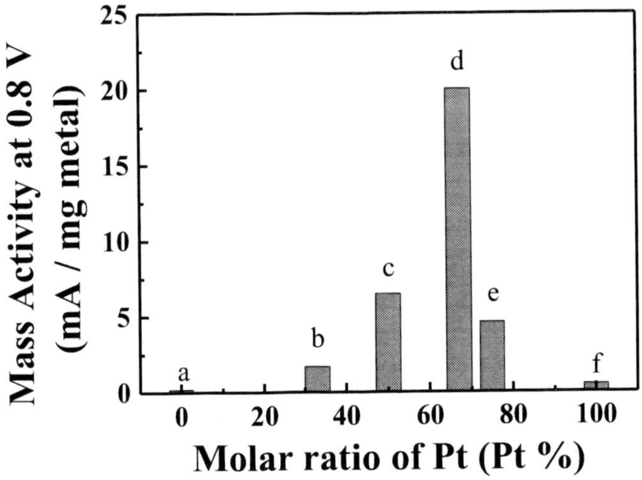

Figure 24. Comparison of mass activity for Au–graphene (a), Pt–graphene (f), and Pt–Au–graphene nanostructures with Pt/Au molar ratios of 0.5:1 (b), 1:1 (c), 2:1(d), and 3:1 (e) at potential of 0.8 V.

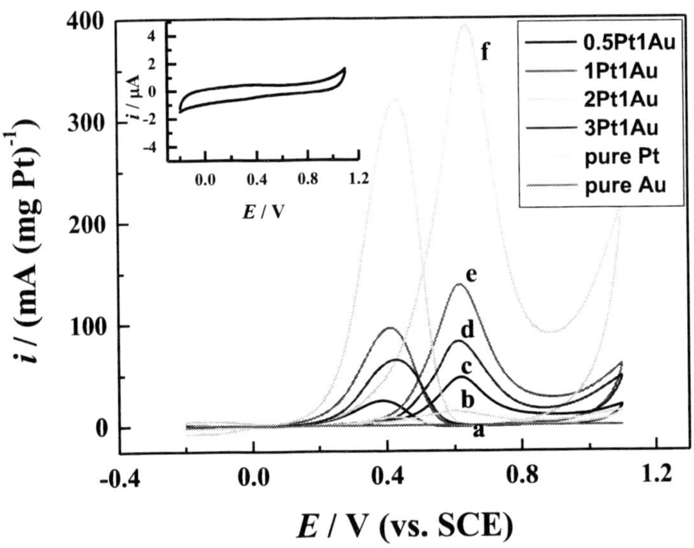

Figure 25. Cyclic voltammograms of oxidation of 0.5 M methanol at the Au–graphene/GC (a), Pt–graphene/GC (b), and Pt–Au–graphene/GC electrode with Pt/Au molar ratio of 0.5:1 (c), 3:1 (d), 1:1 (e), and 2:1 (f) in 0.5 M H_2SO_4 solution. The current was normalized with the mass of Pt in the catalyst. The inset shows the voltammetric response of the graphene/GC electrode in 0.5 M H_2SO_4 solution containing 0.5 M methanol. The scan rate was 50 mV s^{-1}.

Electrocatalytic oxidation of methanol in acidic solution (0.5 M H_2SO_4) was also used as a model reaction to study the catalytic activity of the prepared Pt–Au nanocatalysts deposited on the surface of graphene. The electrocatalytic currents were normalized by the Pt loading as depicted in Figure 25. No anodic current for the oxidation of methanol is observed at the graphene/GC (inset in Figure 25), or Au–graphene/GC electrode (curve a), indicating that the graphene, and Au are inert for MOR under the present conditions. However, for the Pt and Pt–Au catalysts, the well–defined anodic peaks are observed at about 0.64 V in the forward sweep and 0.45 V in the backward sweep (curves b–f, Figure 25). The Pt–Au catalyst shows much higher catalytic activity than pure Pt. The peak current is enhanced with the increase of Pt content in Pt–Au catalysts (curves c, e, and f), and reaches the highest value at a Pt/Au molar ratio of 2:1 (curve f). However, the further increasing the Pt/Au molar ratio (for example, 3:1) leads to the decrease in anodic current (curve d). These results are in good agreement with the values of ECSA of Pt–Au nanostructures as shown in Figure 22.

The specific mass peak current is ca.394 mA (mg Pt)$^{-1}$ for Pt–Au electrocatalyst with a Pt/Au molar ratio of 2:1 (for 0.5 M methanol), slightly higher than 376 and 323.6 mA (mg Pt)$^{-1}$ for the oxidation of 0.5 M methanol at Pd@Pt catalyst [120], and at graphene/Pt–on–Pd bimetallic nanodendrite hybrids catalyst [52], respectively. It is higher than 756.7 mA (mg Pt)$^{-1}$ for the oxidation of 1 M methanol at the Pd/Pt core-shell nanowire arrays electrode [154] by considering the difference of the methanol concentration used for measurements. However, this value is much higher than 51 and 199.6 mA (mg Pt)$^{-1}$ obtained at commercial carbon supported platinum (Pt/XC–72R, E–Tek, 60% Pt, 6 nm in diameter for Pt) [155], and Pt/graphene catalyst [55], respectively. The remarkable high catalytic activity of the Pt–Au catalyst with a Pt/Au ratio of 2:1 is directly related to its surface features. As shown in Figure 25, the oxidation methanol at the Pt–Au catalyst starts at potential of ca. 0.15 V, and the oxidation current increases with the potential scanning to more positive value until reaches the highest current at ca.0.64 V. During this process, CO, HCOOH, HCOH, etc. intermediates are formed, and adsorbed on the surface of the catalyst. These adsorbed intermediates will poison the catalyst, inhibit the oxidation of methanol [127]. It is reported that Au atoms surrounding the Pt atom play an important role in removing the intermediate CO–like species in MOR [129] because the nanoscaled Au is highly active for CO oxidation [135–137]. Therefore, during the negative scan, the CO and other adsorbed species can be removed from the nanocatalysts surface [124]. The oxidative removal of the adsorbed CO is greatly accelerated by the availability of a large number of active oxygen–containing species on the underlying Au nanostructures. However, too low Pt atom content decreases the probability of finding the neighboring Pt atoms for methanol chemisorptions and hinders methanol oxidation as shown in curves c and e (Figure 25). When the Pt/Au molar ratio is 3:1 or higher, there is no Au atoms surrounding the Pt at the surface of the Pt–Au nanocatalyst for removing CO–like species, leading to decrease of the catalytic activity of the catalyst as shown in curve d (Figure 25). Therefore, the highest catalytic activity will be reached at a suitable Pt/Au molar ratio, for example 2:1 for Pt–Au–graphene catalyst. This phenomenon has also been observed in other bimetallic catalysts [120, 123, 128].

It is reported that the structures and properties of the catalyst supports have a great effect on the activity and durability of the catalysts [142–144]. To clarify the role of graphene in the Pt–Au–graphene catalyst, the Pt–Au nanostructures (with Pt/Au molar ratio of 2:1) were also electrodeposited on the surface of the Vulcan XC–72 carbon (Cabot, Boston, MA) and bare GC electrode, respectively, and the electrocatalytic activity toward the oxidation of 0.5 M methanol was studied and compared that of Pt–Au–graphene catalyst. The cyclic voltammetric features (such as the position and shapes of the oxidation peaks) of the oxidation of methanol at the Pt–Au–XC–72 carbon/GC and Pt–Au/GC electrode are similar to those of the Pt–Au–graphene/GC electrode (Figure 26). However, the highest specific mass current obtained at the Pt–Au–graphene/GC electrode (ca.394 mA (mg Pt)$^{-1}$, curve a, Figure 26) is ca. 2, and 4 times of those obtained at the Pt–Au–XC–72 carbon/GC (ca. 210 mA (mg Pt)$^{-1}$, curve b, Figure 26) and Pt–Au/GC electrode (ca.105 mA (mg Pt)$^{-1}$, curve c, Figure 26), respectively, indicating that the electrocatalytic activity of Pt–Au–graphene is much higher than that of Pt–Au–XC–72 carbon and Pt–Au catalyst, suggesting that graphene has a significant role in enhancing the catalytic activity of the Pt–Au–graphene electrocatalyst.

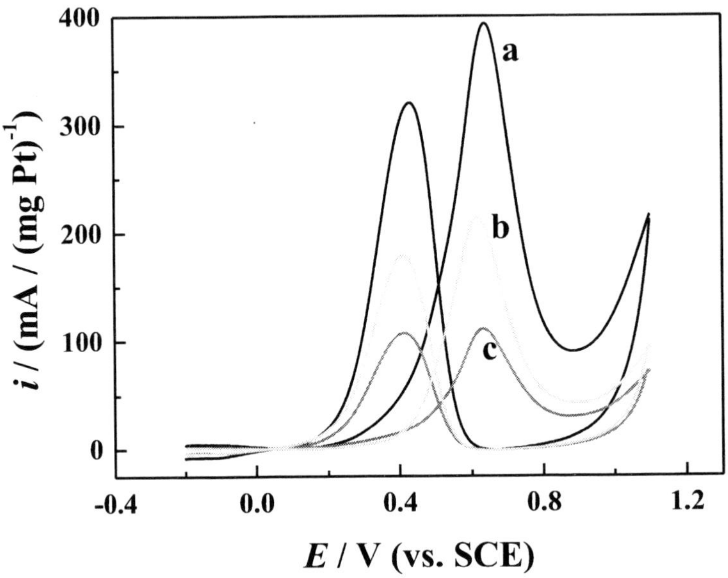

Figure 26. Cyclic voltammograms of oxidation of 0.5 M methanol at the Pt–Au–graphene/GC electrode (a), Pt–Au–XC–72 carbon/GC (b), and Pt–Au/GC electrode (b) (with Pt/Au molar ratio of 2:1) in 0.5 M H_2SO_4 aqueous solution at a scan rate of 50 mV s^{-1}. The mass of Pt–Au deposited on graphene, XC–72 carbon, and bare GC surface was controlled to be the same.

To explain the reason of the graphene enhancing the catalytic activity of the catalyst, the morphology of the Pt–Au deposited on the graphene, XC–72 carbon, and bare GC electrode surface is recorded. The typical morphology of Pt–Au deposited on the graphene characterized with large flower-like structures built by numerous of small nanoparticles (panel A and the inset, Figure 27), while Pt–Au deposited on the XC–72 carbon and bare GC surface has a morphology of small and irregular nanoparticles (panels B and C, Figure 27). The different morphology of the Pt–Au nanostructures results to their different ECSA, which can be estimated by integrating the charge in the adsorption/desorption region of hydrogen. The flower-like Pt–Au nanostructures possess higher ECSA value (82.2 m^2 (g Pt)$^{-1}$) than those irregular nanoparticles deposited on XC–72 carbon (47.6 m^2 (g Pt)$^{-1}$) and bare GC surface (34.7 m^2 (g Pt)$^{-1}$). This higher ECSA of the Pt–Au–graphene catalyst mainly contributes to its higher catalytic activity.

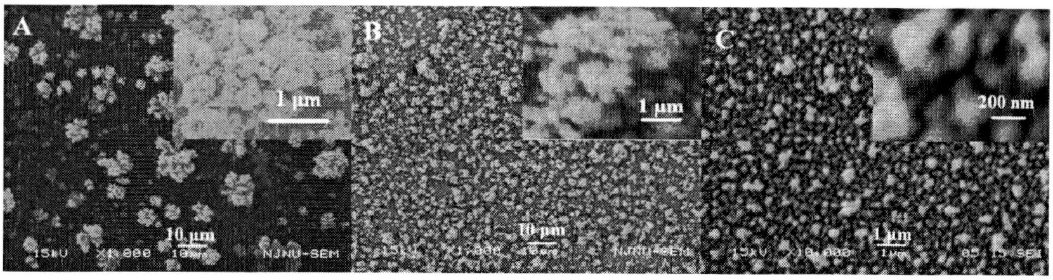

Figure 27. Typical SEM images of the Pt–Au nanostructures with Pt/Au molar ratio of 2:1 electrodeposited on the surface of graphene (panel A), XC–72 carbon (panel B), and bare GC (panel C) under a constant potential of –0.4 V in 0.5 M H_2SO_4 solution. The insets show the images at high magnification.

Graphene can also enhance the tolerance of the catalyst to the carbonaceous species poisoning. For the methanol oxidation, the ratio of the forward anodic peak current (i_f) to the backward anodic peak current (i_b) can be used for evaluating the catalyst tolerance to the intermediate carbonaceous species accumulated on electrode surface [156]. The higher i_f/i_b value indicates more effective in removing the poisoning species on the catalyst surface, which means methanol can be oxidized to carbon dioxide much more efficiently. The value of i_f/i_b for Pt–Au–graphene catalyst is ca.1.25, which is higher than ca. 1.16, 1.03 for Pt–Au–XC–72 carbon, and Pt–Au catalyst, respectively, suggesting that Pt–Au–graphene surface has less carbonaceous species accumulation and therefore has a high ability to tolerate CO poisoning. Recently, it has been reported that the oxygen-containing functional groups of single-walled carbon nanotubes (SWNTs) such as the hydroxyl and carboxyl groups etc. can improve the electrocatalytic activity of the Pt–SWNT [91] and Pd–SWNTs catalyst [92], and these groups can play the role of Ru as secondary catalyst in the case of the Pt–Ru catalyst [141]. Therefore, the small amount of the oxygen–containing groups presenting on the surface of graphene is also believed to be responsible for better tolerance of the catalyst, and they play a major role to remove those accumulated carbonaceous species like CO formed in the oxidation of methanol and contribute to the low catalyst poisoning in the Pt–Au–graphene catalyst, resulting in the high electrocatalytic activity of the catalyst.

The long–term stability of the Pt–Au–graphene, Pt–Au–XC–72 carbon, and Pt–Au catalysts with the Pt/Au molar ratio of 2:1 was evaluated by chronoamperometric measurement under a constant potential of 0.64 V. The polarization currents at Pt–Au–graphene, Pt–Au–XC–72 carbon, and Pt–Au catalysts decrease rapidly at the initial stage (Figure 28), which is due to the formation of intermediate species during MOR [157]. However, the anodic current of the Pt–Au–graphene/GC electrode (curve a) is still considerably higher than those obtained at Pt–Au–XC–72 carbon/GC (curve b), and Pt–Au/GC electrode (curve c) at a time of 1000 s, indicating that the electrocatalytic stability of the Pt–Au–graphene catalyst for MOR is higher than that of Pt–Au–XC–72 carbon and Pt–Au catalyst. These results are consistent with the voltammetric measurements (Figure 26), further confirming the better tolerance to the intermediate species and higher catalytic activity of Pt–Au–graphene electrocatalyst than Pt–Au–XC–72 carbon and Pt–Au electrocatalysts.

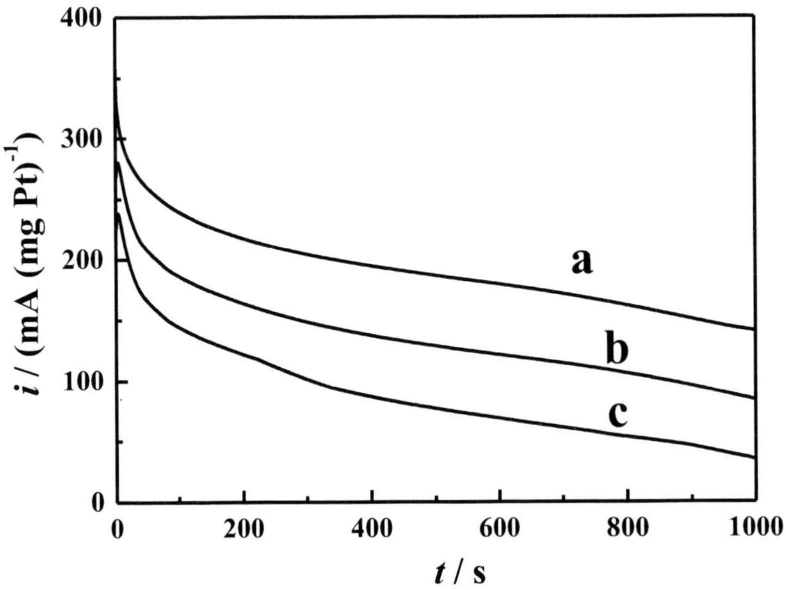

Figure 28. Chronoamperometric response of oxidation of 0.5 M methanol at the Pt–Au–graphene/GC (a), Pt–Au–XC–72 carbon/GC (b), and Pt–Au/GC electrode (c) in 0.5 M H_2SO_4 solution under a constant potential of 0.64 V (versus SCE).

CONCLUSION

Graphene, as a rapidly raised star in nanomaterial science, has attracted great attention all over the world, and graphene is truly exceptional in various aspects including structural, mechanical, electronic, mechanical, thermal and possibly magnetic properties, however, one must establish a viable technique to produce graphene in large quantity and in a controllable fashion.

We predict a bright future for graphene as a metal supports because of its large specific surface area, lack of metallic impurities (which are a major obstacle in electrochemistry research with CNTs), high conductivity and abundance of inexpensive source material (graphene is mainly fabricated from graphite, which is inexpensive). Moreover, an enormous amount of effort has been made in order to understand the electronic and electrochemical behavior of graphene as well as its chemistry. Such rapid progress was possible only because of the accumulation of knowledge and the development of techniques for learning of previous research on graphite and carbon nanotubes. However, much remains need to be done to elucidate the more subtle details of graphene electrochemistry such as the role of the structure of graphene on its electrochemistry, the role of the specific oxygen–containing groups at the edges of graphene sheets, and the mechanism of the oxidation/reduction of the graphene sheets themselves.

ACKNOWLEDGMENTS

The work presented in this chapter is supported by the National Natural Science Foundation of China (20773067, 20833006, and 20905036), Research Fund for the Doctoral Program of Higher Education of China (20103207110004), the Foundation of the Jiangsu Education Committee (09KJA150001, 09KJB150006 and 10KJB150009), and the foundation of Jiangsu Provincial Key Laboratory of Palygorskite Science and Applied Technology (HPK201005).

REFERENCES

[1] H. W. Kroto, J. R. Heath, S. C. O'Brien, R. F. Curl and R. E. Smalley, *Nature,* 318, 162 (1985).
[2] S. Iijima, *Nature,* 354, 56 (1991).
[3] K. S. Novoselov, A. K. Geim, S. V. Morozov, D. Jiang, Y. Zhang, S. V. Dubonos, I. V. Grigorieva and A. A. Firsov, *Science,* 306, 666 (2004).
[4] K. P. Loh, Q. Bao, P. K. Ang and J. Yang, *J. Mater. Chem.* 20, 1 (2010).
[5] N. G. Shang, P. Papakonstantinou, M. McMullan, M. Chu, A. Stamboulis, A. Potenza, S. S. Dhesi and H. Marchetto, *Adv. Funct. Mater.* 18, 3506 (2008).
[6] W. Yang, T. P. Fellinger and M. Antonietti, *J. Am.Chem. Soc.* 133, 206 (2011).
[7] L. J. Cote, R. Cruz-Silva and J. Huang, *J. Am. Chem. Soc.* 131, 11027 (2009).
[8] W. Zhang, J. Cui, C. A. Tao, Y. Wu, Z. Li, L. Ma, Y. Wen and G. Li, *Angew. Chem. Int. Ed.* 48, 5864 (2009).
[9] D. Wei, Y. Liu, H. Zhang, L. Huang, B. Wu, J. Chen and G. Yu, *J. Am. Chem. Soc.* 131, 11147 (2009).
[10] L. D. Landau, *Phys. Z. Sowjetunion,* 11, 26 (1937).
[11] R. E. Peierls, *Ann. Inst. Henri Poincare,* 5, 177 (1935).
[12] J. Stangl, V. Holy and G. Bauer *Rev. Mod. Phys.* 76, 725 (2004).
[13] J. C. Meyer, A. K. Geim, M. I. Katsnelson, K. S. Novoselov, T. J. Booth and S. Roth, *Nature*, 446, 60 (2007).
[14] K. I. Bolotin, K. J. Sikes, Z. Jiang, M. Klima, G. Fudenberg, J. Hone, P. Kim and H. L. Stormer, *Solid State Commun.* 146, 351 (2008).
[15] K. I. Bolotin, K. J. Sikes, J. Hone, H. L. Stormer and P. Kim, *Phys. Rev. Lett.* 101, 096802 (2008).
[16] K. S. Novoselov, D. Jiang, F. Schedin, T. J. Booth, V. V. Khotkevich, S. V. Morozov and A. K. Geim, *Proc. Natl. Acad. Sci. U.S.A.* 102, 10451 (2005).
[17] X. Li, W. F. Zhao and G. H. Chen, *Materials Review,* 22, 48 (2008).
[18] C. Lee, X. Wei, J. W. Kysar and J. Hone, *Science,* 321, 385 (2008).
[19] A. A. Balandin, S. Ghosh, W. Z. Bao, I. Calizo, D. Teweldebrhan, F. Miao and C. N. Lau, *Nano Lett.* 8, 902 (2008).
[20] M. D. Stoller, S. Park, Y. Zhu, J. An and R. S. Ruoff, *Nano Lett.* 8, 3498 (2008).
[21] Y. Zhang, Y. W. Tan, H. L. Stormer and P. Kim, *Nature,* 438, 201 (2005).
[22] C. Xu, X. Wang and J. W. Zhu, *J. Phys. Chem. C,* 112, 19841 (2008).

[23] X. Yang, X. Zhang, Y. Ma, Y. Huang, Y. Wang and Y. Chen, *J. Mater. Chem.* 19, 2710 (2009).
[24] Y. J. Hu, H. Zhang, P. Wu, H. Zhang, B. Zhou and C. X. Cai, *Phys. Chem. Chem. Phys.* 13, 4083 (2011).
[25] E. J. Yoo, J. Kim, E. Hosono, H. S. Zhou, T. Kudo and I. Honma, *Nano Lett.* 8, 2277 (2008).
[26] D. Wang, D. Choi, J. Li, Z. Yang, Z. Nie, R. Kou, D. Hu, C. Wang, L.V. Saraf, J. Zhang, I. A. Aksay and J. Liu, *ACS Nano,* 3, 907 (2009).
[27] S. R. C. Vivekchand, C. S. Rout, K. S. Subrahmanyam, A. Govindaraj and C. N. R. Rao, *J. Chem. Sci.* 120, 9 (2008).
[28] Y. Sui and J. Appenzeller, *Nano Lett.* 9, 2973 (2009).
[29] P. Wu, Q. Shao, Y. J. Hu, J. Jin, Y. J. Yin, H. Zhang and C. X. Cai, *Electrochim. Acta,* 55, 8606 (2010).
[30] X. Wu, Y. Hu, J. Jin, N. Zhou, P. Wu and C. X. Cai, *Anal. Chem.* 82, 3588 (2010).
[31] Y. J. Hu, J. Jin, H. Zhang, P. Wu and C. X. Cai, *Acta Phys. –Chim. Sin.* 26, 2073 (2010).
[32] R. L. McCreery, *Chem. Rev.* 108, 2646 (2008).
[33] C. E. Banks, T. J. Davies, G. G. Wildgoose and R. G. Compton, *Chem. Commun.* 829 (2005).
[34] L. Meng, P. Wu, G. X. Chen, C.X. Cai, Y. M. Sun and Z. H. Yuan, *Biosens. Bioelectron.* 24, 1751(2009).
[35] P. Du, S. N. Liu, P. Wu and C. X. Cai, *Electrochim. Acta,* 53, 1811(2007).
[36] M. S. Raghuveer, A. Kumar, M. J. Frederick, G. P. Louie, P. G. Ganesan and G. Ramanath, *Adv. Mater.* 18, 547 (2006).
[37] X. M. Wu, B. Zhao, P. Wu, H. Zhang and C. X. Cai, *J. Phys. Chem. B,* 113, 13365 (2009).
[38] D. Chen, Q. Wang, J. Jin, P. Wu, H. Wang, S. Yu, H. Zhang and C. X. Cai, *Anal. Chem.* 52, 2448 (2010).
[39] L. N. Wu, X. J. Zhang and H. X. Ju, *Anal. Chem.* 79, 453 (2007).
[40] M. Zhou, J. Ding, L. P. Guo and Q. K. Shang, *Anal. Chem.* 79, 5328 (2007).
[41] M. Zhou, J. D. Guo, L. P. Guo and J. Bai, *Anal. Chem.* 80, 4642 (2008).
[42] Y. E. Seidel, A. Schneider, Z. Jusys, B. Wickman, B. Kasemo and R. J. Behm, *Langmuir,* 26, 3569 (2010).
[43] L. Wang, C. Tian, H. Wang, Y. Ma, B. Wang and H. Fu, *J. Phys. Chem. C,* 114, 8727 (2010).
[44] P. Yu, Q. Qian, Y. Lin and L. Mao, *J. Phys. Chem. C,* 114, 3575 (2010).
[45] C. C. Chen, C. F. Chen, C. M. Chen and F. T. Chuang, *Electrom. Commun.* 9, 159 (2007).
[46] H. Wu, J. Wang, X. Kang, C. Wang, D. Wang, J. Liu, I. A. Aksay, Y. Lin, *Talanta*, 80, 403 (2009).
[47] F. Li, H. Yang, C. Shan, Q. Zhang, D. Han, A. Ivaska and L. Niu, *J. Mater. Chem.* 19, 4022 (2009).
[48] C. Shan, H. Yang, D. Han, Q. Zhang, A. Ivaska and L. Niu, *Biosens. Bioelectron.* 25, 1070 (2010).
[49] F. Li, J. Chai, H. Yang, D. Han and L. Niu, *Talanta*, 81, 1063 (2010).

[50] W. Hong, H. Bai, Y. Xu, Z. Yao, Z. Gu and G. Shi, *J. Phys. Chem. C,* 114, 1822 (2010).
[51] Y. Fang, S. Guo, C. Zhu, Y. Zhai and E. Wang, *Langmuir,* 26, 11277 (2010).
[52] S. Guo, S. Dong and E. Wang, *ACS nano,* 4, 547 (2010).
[53] S. Bong, Y. R. Kim, I. Kim, S. Woo, S. Uhm, J. Lee and H. Kim, *Electrochem. Commun.* 12, 129 (2010).
[54] L. Dong, R. R. S. Gari, Z. Li, M. M. Craig and S. Hou, *Carbon,* 48, 781 (2010).
[55] Y. Li, L. Tang and J. Li, *Electrochem. Commun.* 11, 846 (2009).
[56] E. J. Yoo, T. Okata, T. Akita, M. Kohyama, J. Nakamura and I. Honma, *Nano lett.* 9, 2255 (2009).
[57] Y. Li, W. Gao, L. Ci, C. Wang and P. M. Ajayan, *Carbon,* 48, 1124 (2010).
[58] R. Kou, Y. Shao, D. Wang, M. H. Engelhard, J. H. Kwak, J. Wang, V. V. Viswanathan, C. Wang, Y. Lin, Y. Wang, I. A. Aksay and J. Liu, *Electrochem. Commun.* 11, 954 (2009).
[59] Y. Shao, S. Zhang, C. Wang, Z. Nie, J. Liu, Y. Wang and Y. Lin, *J. Power Sources,* 195, 4600 (2010).
[60] W. S. Hummers, R. E. Offeman, *J. Am. Chem. Soc.* 80, 1339 (1958).
[61] M. Zhou, Y. Wang, Y. Zhai, J. Zhai, W. Ren, F. Wang, S. Dong, *Chem. Eur. J.* 15, 6116 (2009).
[62] M. Zhou, Y. Zhai, S. Dong, *Anal. Chem.* 81, 5603 (2009).
[63] Y. Xu, H. Bai, G. Lu, C. Li, G. Shi, *J. Am. Chem. Soc.* 130, 5856 (2008).
[64] S. Stankovich, D. A. Dikin, R. D. Piner, K. A. Kohlhaas, A. Kleinhammes, Y. Jia, Y. Wu, S. T. Nguyen, R. S. Ruoff, *Carbon,* 45, 1558 (2007).
[65] J. Zhang, H. Yang, G. Shen, P. Cheng, J. Zhang, S. Guo, *Chem. Commun.* 46, 1112 (2010).
[66] M. C. Daniel and D. Astruc, *Chem. Rev.* 104, 293 (2004).
[67] A. C. Templeton, J. J. Pietron, R. W. Murray and P. Mulvaney, *J. Phys. Chem. B,* 104, 564 (2000).
[68] G. Peto, G. L. Molnar, Z. Paszti, O. Geszti, A. Beck and L. Guczi, *Mater. Sci. Eng. C* 19, 95 (2002).
[69] C. Demaille, M. Brust, M. Tsionsky and A. J. Bard, *Anal. Chem.* 69, 2323 (1997).
[70] N. Chandrasekharan and P. V. Kamat, *J. Phys. Chem. B*, 104, 10851 (2000).
[71] M. A. El-Sayed, *Acc. Chem. Res.* 34, 257 (2001).
[72] J. Pendry, *Science,* 285, 1687 (1999).
[73] Y. Tian and T. Tatsuma, *J. Am. Chem. Soc.* 127, 7632 (2005).
[74] K. Naoi, Y. Ohko and T. Tatsuma, *J. Am. Chem. Soc.* 126, 3664 (2004).
[75] Y. Du, X. L. Luo, J. J. Xu and H. Y. Chen, *Bioelectrochemistry,* 70, 342 (2007).
[76] H. Zhang, J. J. Xu, H. Y. Chen, *J. Phys. Chem. C,* 112, 13886 (2008).
[77] B. Scharifker and G. Hills, *Electrochim. Acta,* 28, 879 (1983).
[78] B. Scharifker and J. Mostany, *J. Electroanal. Chem.* 177, 13 (1984).
[79] S. Floate, M. E. Hyde and R. G. Compton, *J. Electroanal. Chem.* 523, 49 (2002).
[80] A. Hamelin, *J. Electroanal. Chem.* 407, 1 (1996).
[81] L. Tang, Y. Wang, Y. Li, H. Feng, J. Lu and J. Li, *Adv. Funct. Mater.* 19, 2782 (2009).
[82] C. M. Sánchez-Sánchez, J. Solla-Gullón, F. J. Vidal-Iglesias, A. Aldaz, V. Montiel and E. Herrero, *J. Am. Chem. Soc.* 132, 5622 (2010).

[83] H. Eriksona, G. Jürmanna, A. Sarapuua, R. J. Potterb and K. Tammeveski, *Electrochim. Acta,* 54, 7483 (2009).
[84] T. Inasaki and S. Kobayashi, *Electrochim. Acta,* 54, 4893 (2009).
[85] M. Bron, *J. Electroanal. Chem.* 624, 64 (2008).
[86] M. S. El-Deab, T. Sotomura and T. Ohsaka, *Electrochim. Acta,* 52, 1792 (2006).
[87] N. Alexeyeva, T. Laaksonen, K. Kontturi, F. Mirkhalaf, D. J. Schiffrin and K. Tammeveski, *Electrochem. Commun.* 8, 1475 (2006).
[88] S. R. Lee, Y. T. Lee, K. Sawada, H. Takao and M. Ishida, *Biosens. Bioelectron.* 24, 410 (2008).
[89] W. Zhao, J. J. Xu, C. G. Shi and H. Y. Chen, *Electrochem. Commun.* 8, 773 (2006).
[90] S. Park, H. Boo and T. D. Chung, *Anal. Chim. Acta,* 556, 46 (2006).
[91] A. Halder, S. Sharma, M. S. Hegde and N. Ravishankar, *J. Phys. Chem. C,* 113, 1466 (2009).
[92] L. Meng, J. Jin, G. X. Yang, T. H. Lu, H. Zhang and C. X. Cai, *Anal. Chem.* 81, 7271 (2009).
[93] M. S. El-Deab, *Electrochim. Acta,* 54, 3720 (2009).
[94] Y. Zhou, L. Z. Fan, H. Z. Zhong, Y. F. Li and S. H. Yang, *Adv. Funct. Mater.* 17, 1537 (2007).
[95] S. F. Zheng, J. S. Hu, L. S. Zhong, L. J. Wan and W. G. Song, *J. Phys. Chem. C,* 111, 11174, (2007).
[96] J. Yuan, Z. Wang, Y. Zhang, Y. F. Shen, D. Han, Q. Zhang and X. Xu, *Thin Solid Films*, 516, 6531 (2008).
[97] L. Ren and Y. Xing, *Electrochim. Acta*, 53, 5563 (2008).
[98] M. C. Gutierrez, M. J. Hortiguela, J. M. Amarilla, R. Jimenez, M. L. Ferrer and F. del-Monte, *J. Phys. Chem. C*, 111, 5557 (2007).
[99] B. C. H. Steele and A. Heinzel, *Nature*, 414, 345 (2001).
[100] H. A. Gasteiger, S. S. Kocha, B. Sompalli and F. T. Wagner, *Appl. Catal. B*, 56, 9 (2005).
[101] E. Antolini, *Mater. Chem. Phys.* 78, 563 (2003).
[102] G. Q. Lu and A. Wieckowski, *Curr. Opin. Coll. Interf. Sci.* 5, 95 (2000).
[103] S. Park, T. D. Chung and H. C. Kim, *Anal. Chem.* 75, 3046 (2003).
[104] J. H. Yuan, K. Wang and X. H. Xia, *Adv. Funct. Mater.* 15, 803 (2005).
[105] Q. Shen, L. Jiang, H. Zhang, Q. Min, W. Hou and J. J. Zhu, *J. Phys. Chem. C,* 112, 16385 (2008).
[106] J. Xu, T. White, P. Li, C. He, J. Yu, W. Y.uan and Y. Han, *J. Am. Chem. Soc.* 132, 10398 (2010).
[107] C. W. Yi, K. Luo, T. Wei and D. W. Goodman, *J. Phys. Chem. B,* 109, 18535 (2005).
[108] T. Wei, J. Wang and D. W. Goodman, *J. Phys. Chem. C,* 111, 8781 (2007).
[109] C. V. Rao and B. Viswanathan, *J. Phys. Chem. C,* 113, 18907 (2009).
[110] Y. Kang, C. B. Murray, *J. Am. Chem. Soc.* 132, 7568 (2010).
[111] J. Kim, Y. Lee and S. Sun, *J. Am. Chem. Soc.* 132, 4996 (2010).
[112] V. Mazumder, M. Chi, K. L. More and S. Sun, *J. Am. Chem. Soc.* 132, 7848 (2010).
[113] C. Wang, D. Vliet, K. C. Chang, H. You, D. Strmcnik, J. A. Schlueter, N. M. Markovic and V. R. Stamenkovic, *J. Phys. Chem. C,* 113, 19365 (2009).

[114] K. Jayasayee, V. A. T. Dam, T. Verhoeven, S. Celebi and F. A. de–Bruijn, *J. Phys. Chem. C,* 113, 20371 (2009).
[115] T. Y. Jeno, S. J. Yoo, Y. H. Cho, K. S. Lee, S. H. Kang and Y. E. Sung, *J. Phys. Chem. C,* 113, 19732 (2009).
[116] J. Wu, J. Zhang, Z. Peng, S. Yang, F. T. Wagner and H. Yang, *J. Am. Chem. Soc.* 132, 4984 (2010).
[117] A. Sarkar and A. Manthiram, *J. Phys. Chem. C,* 114, 4725 (2010).
[118] A. Miura, H. Wang, B. M. Leonard, H. D. Abruna and F. J. Disalvo, *Chem. Mater.* 21, 2661 (2009).
[119] K. K. Tintula, S. Pitchumani, P. Sridhar and A. K. Shukla, *J. Chem. Sci.* 122, 381 (2010).
[120] H. Zhang, Y. J. Yin, Y. J. Hu, C. Y. Li, P. Wu, S. H. Wei and C. X. Cai, *J. Phys. Chem. C,* 114, 11861 (2010).
[121] Q. Yuan, Z. Y. Zhou, J. Zhuang and X. Wang, *Chem. Commun.* 46, 1491 (2010).
[122] W. W. He, X. C. Wu, J. B. Liu, K. Zhang, W. G. Chu, L. L. Feng, X. N. Hu, W. Y. Zhou and S. S. Xie, *J. Phys. Chem. C,* 113, 10505 (2009).
[123] N. Kristian and X. Wang, *Electrochem. Commun.* 10, 12 (2008).
[124] S. Guo, Y. Fang, S. Dong and E. Wang, *J. Phy. Chem. C,* 111, 17104 (2007).
[125] Y. Jin, Y. Shen and S. Dong, *J. Phys. Chem. B,* 108, 8142 (2004).
[126] Y. Zhang, Q. Huang, Z. Zou, J. Yang, W. Vogel and H. Yang, *J. Phys. Chem. C,* 114, 6860 (2010).
[127] J. Jia, L. Cao and Z. Wang, *Langmuir,* 24, 5932 (2008).
[128] J. Anderson, A. Karakoti, D. J. Diaz and S. Seal, *J. Phys. Chem. C,* 114, 4595 (2010).
[129] J. Luo, P. N. Njoki, Y. Lin, D. Mott, L. Wang and C. J. Zhong, *Langmuir,* 22, 2892 (2006).
[130] H. Lang, S. Maldonado, K. J. Stevenson and B. D. Chandler, *J. Am. Chem. Sco.* 126, 12949 (2004).
[131] S. Zhou, K. McIlwrath, G. Jackson and B. Eichhorn, *J. Am. Chem. Soc.* 128, 1780 (2006).
[132] J. Zeng, J. Yang, J. Y. Lee and W. Zhou, *J. Phys. Chem. B*, 110, 24606 (2006).
[133] W. Tang, S. Jayaraman, T. F. Jaramillo, G. D. Stucky and E. W. McFarland, *J. Phys. Chem. C*, 113, 5014 (2009).
[134] X. Ge, R. Wang, P. Liu and Y. Ding, *Chem. Mater.* 19, 5827 (2007).
[135] D. Cameron, R. Holliday and D. Thompson, *J. Power Sources*, 118, 298 (2003).
[136] D. R. Rólison, *Science*, 299, 1698 (2003).
[137] T. F. Jaramillo, S. H. Baeck, B. R. Cuenya and E. W. McFarland, *J. Am. Chem. Soc.* 125, 7148 (2003).
[138] H. Gharibi, K. Kakaei and M. Zhiani, *J. Phys. Chem. C*, 114, 5233 (2010).
[139] S. Liao, K. A. Holmes, H. Tsaprailis and V. I. Briss, *J. Am. Chem. Soc.* 128, 3504 (2006).
[140] H. Tang, J. H. Chen, L. H. Nie, D. Y. Liu, W. Deng, Y. F. Kuang and S. Z. Yao, *J. Colloid Interf. Sci.* 269, 26 (2004).
[141] J. Chen, M. Wang, B. Liu, Z. Fan, K. Cui and Y. Kuang, *J. Phys. Chem. B*, 110, 11775 (2006).
[142] Y. C. Xing, L. Li, C. C. Chusuei and R. V. Hull, *Langmuir*, 21, 4185 (2005).

[143] D. Bom, R. Andrews, D. Jacques, J. Anthony, B. L. Chen, M. S. Meier and J. P. Selegue, *Nano Lett.* 2, 615 (2002).
[144] D. A. Stevens, M. T. Hicks, G. M. Haugen and J. R. Dahn, *J. Electrochem. Soc.* 152, A2309 (2005).
[145] R. Muszynski, B. Seger and P. V. Kamat, *J. Phys. Chem. C*, 112, 5263 (2008).
[146] P. H. Fernandez, S. Rojas, P. Ocon, J. L. G. Fuente, J. S. Fabian, J. Sanza, M. A. Pena, F. J. G. Garcia, P. Terreros and J. L. G. Fierro, *J. Phys. Chem. C* 111, 2913 (2007).
[147] G. C. Bond, *Platinum Metals Rev.* 51, 63 (2007).
[148] R. C. Koffi, C. Coutanceau, E. Garnier, J. M. Leger and C. Lamy, *Electrochim. Acta*, 50, 4117 (2005).
[149] S. S. Kumar and K. L. N. Phani, *J. Power Sources* 187, 19 (2009).
[150] M. Teliska, V. S. Murthi, S. Mukerjee and D. E. Ramaker, *J. Electrochem. Soc.* 152, A2159 (2005).
[151] J. L. Zhang, M. B. Vukmirovic, Y. Xu, M. Mavrikakis and R. R. Adzic, *Angew. Chem. Int. Ed.* 44, 2132 (2005).
[152] J. L. Fernandez, D. A. Walsh and A. J. Bard, *J. Am. Chem. Soc.* 127, 357 (2005).
[153] W. He, J. Y. Liu, Y. J. Qiao, Z. Q. Zou, X. G. Zhang, D. L. Akins and H. Yang, *J. Power Sources*, 195, 1046 (2010).
[154] H. Wang, C. W. Xu, F. L. Cheng, M. Zhang, S. Y. Wang and S. P. Jiang, *Electrochem. Commun.* 10, 1575 (2008).
[155] X. W. Teng, X. Y. Liang, S. Maksimuk and H. Yang, *Small* 2, 249 (2006).
[156] Y. Mu, H. Liang, J. Hu, L. Jiang and L. Wan, *J. Phys. Chem. B* 109, 22212 (2005).
[157] Y. H. Xu, X. Q. Lin, *J. Power Sources*, 170, 13 (2007).

In: Graphene: Properties, Synthesis and Applications
Editor: Zhiping Xu
ISBN: 978-1-61470-949-7

Chapter 10

REDOX PROTEINS–GRAPHENE NANOHYBRIDS: FABRICATION, CHARACTERIZATION, AND BIOSENSING

Ping Wu and Chenxin Cai*
Jiangsu Key Laboratory of New Power Batteries
Jiangsu Key Laboratory of Biofunctional Materials
Laboratory of Electrochemistry
College of Chemistry and Materials Science
Nanjing Normal University, Nanjing, P. R. China

Abstract

Graphene, a flat monolayer of carbon atoms tightly packed into a two–dimensional (2D) honeycomb lattice, holds great promise for application in electrochemistry due to its large surface area, outstanding electrical conductivity, inert electrochemical activity, wide potential window, and so forth. In particular, graphene has high potential application in redox protein immobilization for fabrication of protein–graphene nanohybrid. This chapter describes the preparation and characterization of glucose oxidase/graphene (GOx–graphene) and laccase/graphene (Lac–graphene) nanohybrids. The direct electron transfer characteristics of the GOx on the surface of graphene, the electrocatalytic activities of the GOx–graphene nanohybrid toward the reduction of oxygen, and application in the detection of glucose in human blood serum are presented. The electrocatalytic reduction of oxygen by the Laccase–graphene (Lac–graphene) nanohybrid in the presence of 2,2–azino–bis(3–ethylbenzothiazoline–6–sulfonic acid) (ABTS) as electrochemical mediator is depicted. The detection of the extracellular oxygen released from human erythrocytes with use of the Lac–graphene nanohybrid is also presented in this chapter. This chapter establishes a general route for fabricating redox proteins–based graphene nanohybrids, which has potential utility to bioelectroanalytical chemistry, cellular biology, pathophysiology, and so on.

* E-mail address: cxcai@njnu.edu.cn, caichenxiin@njnu.edu.cn

1. INTRODUCTION

Carbon is a wonder material with several allotropes of which the best known are graphite, diamond, amorphous carbon, carbon nanotubes, fullerenes, and graphene. Among them, graphene is the youngest member but has attractted immense excitements of worldwide reserchers since its discovery by Geim and co–workers at Manchester University in 2004 [1]. This led to an explosion of interest, because graphene is a two–dimentional (2D) sheet of sp^2–hybridized carbon while 2D crystals were thought to be thermodynamically unstable at finite temperatures [2, 3]. Moreover, the discovery of graphene extended the network of carbon family ranging from 0–D, 1–D, 2–D to 3–D. Graphene was regarded as the essential basic building block of graphitic materials of all dimensionalities; it can be wrapped into 0–D fullerenes, rolled into 1–D carbon nanotubes, or stacked into 3–D graphite. Graphene possesses unique electronic, optical, mechanical, thermal and electrochemical properties that are far superior to its counterparts [4]. It is the thinnest known material in the universe and the strongest ever measured [1]. Its charge carriers exhibit giant intrinsic mobility, have zero effective mass, and can travel for micrometers without scattering at room temperature [5, 6]. Graphene can sustain current densities six orders of magnitude higher than that of copper, shows record thermal conductivity and stiffness, is impermeable to gases, and reconciles such conflicting qualities as brittleness and ductility [7]. Electron transport in graphene is described by a Dirac–like equation, which allows the investigation of relativistic quantum phenomena in a benchtop experiment [8–12]. These excellent properitied lead graphene to be promising for application in a wide fields including electronic devices [13, 14], nanocomposite materials [15–18], sustainable energy storage and conversion [19–26], and bioscience/biotechnologies [27–31].

Carbon might be the most widely used electrode material in analytical and industrial electrochemistry due to their low cost, wide potential window, chemical inertness, low residual current, and electrocatalytic activity for a variety of redox reactions. Various allotropes of the carbon, such as graphite [32, 33], carbon black [34, 35], carbon nanofibers [36], highly ordered mesoporous [37, 38], fullerenes [39, 40], and carbon nanotubes [41–50] in particular, have been shown excellent performance in biosensors, biofuel cells, and polymer electrolyte membrane (PEM) fuel cells. These aforementioned fields should be available for graphene, because graphene offers more advanced electrochemical properties than its counterparts. For example, the theoretical surface area, an essential characteristic of an electrode material, that of Graphene is 2630 $m^2\ g^{-1}$, surpassing that of graphite (~10 $m^2\ g^{-1}$), and is two times larger than that of carbon nanotubes (CNTs, 1315 $m^2\ g^{-1}$) [51]. In addition, the electron mobility of graphene is extremely high (ca.200 000 $cm^2\ V^{-1}\ s^{-1}$) [52], meaning that the electron transfer at graphene–based electrodes is fast. Furthermore, the electrical conductivity of graphene has been calculated to be 64 mS cm^{-1}, which is approximately 60 times more than that of single walled carbon nanotubes [53]. More interestingly, it is noteworthy that the presence of oxygen–containing groups at the edges and surface of graphene, which greatly improve the electrochemical performance of graphene [54, 55]. These unusually properties make graphene attractive in electrochemistry and suggest that graphene might outperform its counterparts in the electrochemistry.

The unique properties of graphene may provide insight to apply it for immobilizing redox protein and fabricating protein–graphene nanohybrid for biosensing. This might be greatly attributed to its large surface area, attractive π–π interactions, strong absorptive ability, and the rich oxygen–contaning groups at the edge or surface. These oxygen–containing groups can used as anchoring sites for attachment of proteins [56–58]. These properties can allow the graphene to immobilize protein with high capacity. More important, the proteins are able to retain their native structural conformation and enzymatic activity after the immobilization onto the graphene surface because graphene has good biocompatibility [59]. All these indicate that graphene is a friendly platform to be used as immobilizing matrix to fabricate protein–graphene nanohybrids.

The protein–graphene nanohybrids hold great promise for virtual biosensing. As it used as the electrochemical biosensors, the graphene can act as a nanoconnector for electrical connection with electrode substrates accelerating the electron transfer rates. The high concentration of oxygen–containing groups on the surface of graphene greatly improves the electrocatalytic activity of graphene towards the substrates. Furthermore, unlike most semiconductor systems, its two–dimensional electronic states are not buried deep under the surface, and can be accessed directly by tunnelling and other local probes, indicating that the enzyme catalytic properties of aseembled redox protein can truly be realized at the surface of graphene [60, 61]. Therefore, protein–graphene nanohybrid is regarded as an ideal material for electrochemical biosensing.

Biosensing is significant in biochemical and biophysical science. The biosensing process involves the monitor of substances that are usually apparent within biological systems and concerns electron transfer between redox proteins. Recently, biosensors based on electrochemical transduction have received considerable attention and has become an attractive method for *in vivo* and *ex vivo* biological application, including continuous *in vivo* monitor, measurement of analytes in extremely small volumes, monitor of localized events or biosensing resistive media because of their high sensitivity, low cost, rapid response, compatibility for miniaturization, low manpower requirements, and compatibility with microfabrication technology [62–65].

Glucose is an important component in the blood, determination of its level is crusical in need of diagnosis and management of diabetes mellitus which is a worldwide health problem causing high risks of heart disesase, kidney failure, or blindness. Meanwhile, oxygen is involved in essential cellular functions of all primary animal cells. Evidences have demonstrated that the concentration of oxygen in cells can be regarded as fundamental markers of cell viability and metabolic status and can be used for determining the intensity of cell oxidative processes and intracellular radical generation [66–69]. Moreover, it has also been evidenced that the variations of oxygen concentration throughout the body can significantly impact disease and wound healing processes [70, 71]. Thus, either the determination of the concentration of human blood serum glucose or monitor the cellular oxygen is very significant in biochemistry, clinical chemstry, cellular biology and pathophysiology.

In this chapter, the preparation and characterization of glucose oxidase/graphene (GOx–graphene) and laccase/graphene (Lac–graphene) nanohybrids were discribed. GOx and Lac was chosen as the models for their stability, their ability to be immobilized on the electrode surface, their electrical communication with the electrode surface [72, 73], and their representative for different kinds of oxidoreductase. The fabraication of the GOx–graphene nanohybrids is depicted. The direct electron transfer characteristics of the GOx on the surface of graphene, the electrocatalytic activities of the GOx–graphene nanohybrid toward the reduction of oxygen, and application in the detection of glucose in human blood serum are discussed. The electrocatalytic reduction of oxygen by the Lac–graphene nanohybrid in the presence of ABTS as electrochemical mediator is also presented. The detection of the extracellular oxygen releasing from human erythrocytes with use of the Lac–graphene nanohybrid is presented in this chapter too. The graphene exhibits improved selectivity and sensitivity, offers short response times, displays wider dynamic ranges, as well as exhibits lower detection limits in both the two cases. This chapter establishes a general route for fabricating redox proteins–based graphene nanohybrids, which will generats huge interest in the possible implementation of graphene in bioelectroanalytical chemistry, cellular biology, pathophysiology, and so on.

2. Basic Electrochemistry of Graphene

For the potential application of graphene in electrochemistry, the basic electrochemical behaviors should be first studied to determine several important parameters of graphene electrodes [74–77], for example, electrochemical potential window, electron transfer rate, redox potentials, the appraent electrode area, etc. Graphene exhibits a wide electrochemical potential window of ca.2.5 V in 0.1 M PBS (pH 7.0) [78], which is comparable to that of graphite, glassy carbon (GC), and even boron–doped diamond electrodes. And the charge transfer resistance on graphene as determined from electrochemical impedance spectroscopy (EIS) is much lower than that graphite and GC electrodes, showing graphene electrode has higher electrochemical activity [78]. The electron transfer behavior studies of graphene using cyclic voltammetry (CV) of redox couples, such as $[Fe(CN)6]^{3-/4-}$ and $[Ru(NH_3)_6]^{3+/2+}$, are reported, which exhibit well–defined redox peaks [79–81]. Both anodic and cathodic peak currents in the CVs are linear with the square root of the scan rate, suggestting that the redox process on graphene–based electrodes is a diffusion controlled process [80]. The peak–to–peak potential separations (ΔE_p) in CVs are small indicative of a fast electron transfer [82] on graphene. The apparent surface area (A) of the bare GC and graphene/GC electrode were estimated using cyclic voltammetry. The cyclic voltammetric experiments of 1 mM $Ru(NH_3)_6Cl_3$ in 0.5 M KNO_3 solution were performed at different electrodes and at various scan rates. A straight line of peak currents (i_p) versus scan rates ($v^{1/2}$) could be obtained according to the equation of $i_p = 2.69 \times 10^5 n^{2/3} A D^{1/2} v^{1/2} c^*$, and the values of A for the bare GC and graphene/GC were estimated to be 0.083 ± 0.008 and 0.121 ± 0.026 cm^2 [59], respectively, from the slope of the lines with the use of the diffusion coefficient of $(7.74 \pm 0.23) \times 10^{-6}$ cm^2 s^{-1} for $Ru(NH_3)_6^{3+}$ in 0.5 M KNO_3 solution. The value of the diffusion coefficient of $Ru(NH_3)_6^{3+}$ was estimated by using a Pt microelectrode (5 μm in diameter, BAS, West Lafayette, IN, USA). These results demonstrate that graphene has a high adsorption capacity, and therefore, a large amount of graphene can be assembled.

The graphene on the GC electrode surface is fairly stable because the shape and size of the CV of the graphene/GC electrode remain unchangeable even after the electrode was scanned continuously for a long time in the potential range of interest (ca.100 cycles at a 40 mV s^{-1}). The RSD (relative standard deviation) of background currents of the electrode estimated at a potential of –400 mV is ca.2.5% for 6 different and freshly prepared electrodes, revealing an acceptable reproducibility in the construction of the electrode. The good stability of the electrode is because graphene can form a thin and uniform film on the surface of GC electrode as indicated by SEM image (Figure 1). The uniform nanostructure can significantly increase the effective surface of the electrode for loading of biomolecules and accelerating electron transfer kinetics. The film does not leak from the electrode surface and can stably adhere on the GC surface for a long time. These characteristics are important when the electrode is further used to fabricate a biosensor.

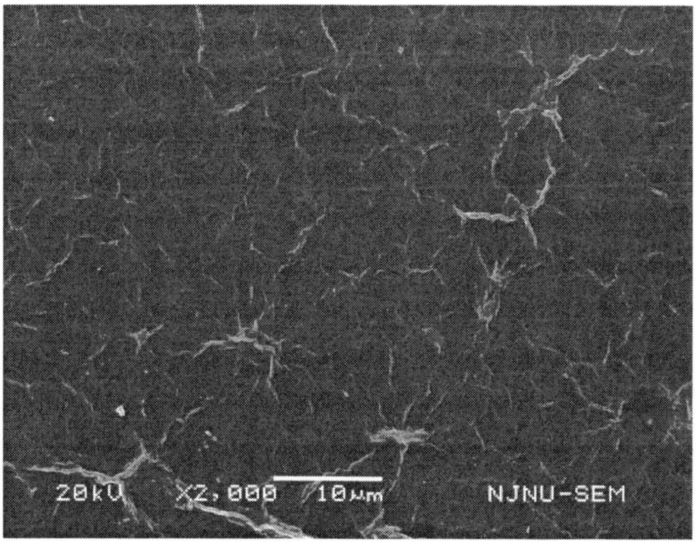

Figure 1. SEM of graphene assemblied on GC electrode.

3. Fabrication of GOx–Graphene Nanohybrids and Application in Biosensing Glucose

A glucose biosensor is a kind of detecting instrument to monitor the content of glucose in human blood. GOx has been widely used in the construction of glucose biosensor due to the high selectivity to glucose and high activity over a broad range of pH. However, the redox center in GOx is usually embedded deeply into the large three dimensional structure of enzyme molecules (Figure 2). Controlling the interactions of enzymes with the substrate to optimize the electron transfer processes plays the impotant role in the biosensing. The unique properties of graphene are benifical for its application in fabrication novel biosensors for virtual applications. The high surface area is helpful in increasing the surface loading of the target enzyme molecules on the surface. The excellent conductivity and small band gap are favorable for conducting electrons from the biomolecule. Here, we describe the fabrication and characterization of a novel nanohybrid (GOx–graphene) in which glucose oxidase (GOx) is directly assembled onto the unmodified graphene surface.

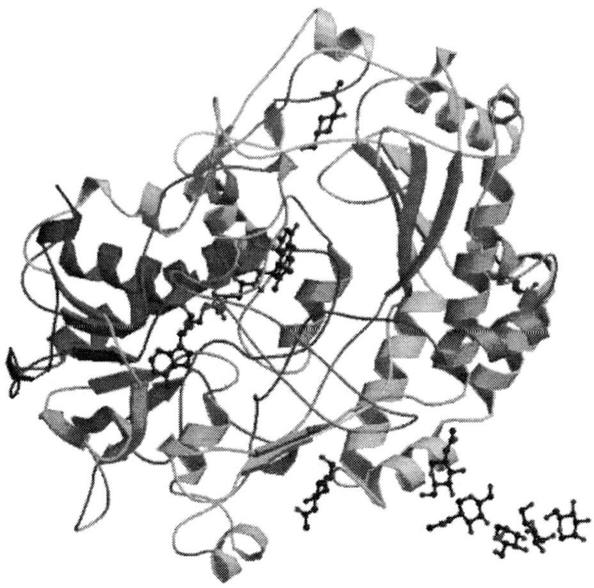

Figure 2. The structure of GOx (adapted from protein data bank).

3.1. Fabrication of GOx–graphene Nanohybrids and Construction of the Glucose Biosensor

Several parameters were needed to be optimized to obtain the nanohybrids with the best voltammetric response. Typically, graphene sheets (2 mg) were dispersed into GOx solution (1 mL, 5 mg mL^{-1} in pH 6.9 PBS) by continuously stirring at 4 °C for 2 h. After that, the mixture was centrifuged and the GOx–graphene nanohybrids were collected by removing the supernatant. The nanohybrids were thoroughly washed with PBS to remove the loosely assembled enzyme molecules. The nanohybrids was characterized by Atom Force Microscopic(AFM), FT–IR and Circular dichroic (CD) spectroscopy.

For constructing the biosensor based on GOx–graphene hybrids, physical adsorption was usually adopted. Typically, 2 mg of graphene was dispersed in 1 mL of water with the aid of ultrasonication forming homogeneous dark suspension (2 mg mL^{-1}). However, the homogeneous suspension cannot stand for a long time, the graphene will be precipitated within ca.2 h. Therefore, the suspension need be ultrasonicated again before next use. The graphene/GC electrode was fabricated by casting 5 μL of graphene suspension (2 mg mL^{-1}) onto the surface of the pretreated GC electrode with a microsyringe. The solvent was allowed to be evaporated at ambient temperature before use. SEM image of graphene on the GC electrode surface (Figure 1) reveals the typical crumpled and wrinkled graphene sheet structure. The experimental results indicate that a thick film of graphene on the electrode surface is not stable and is relatively easy to split from the electrode surface. For example, if the amount of the graphene suspension added to the GC electrode surface is increased to 10 μL, this thick film leaks from the electrode surface within 5–10 scanning cycles. However, fewer amount of graphene suspension (for example 2 μL) cast on the GC electrode surface results to the decrease of the signal of the electrode. Therefore, it is better to choose an appropriate volume of graphene suspension of 5 μL to obtain a good response and stability. GOx–graphene nanohybrids film was fabricated by assembling GOx on graphene via by incubating the graphene/GC electrode in the enzyme solution (5 mg mL^{-1} in PBS) at 4 °C for 1 h. Afterward, the electrode (denoted as GOx–graphene/GC electrode) was thoroughly rinsed with water and stored at 4 °C when not in use.

3.2. Characterization of GOx–graphene Hybrid

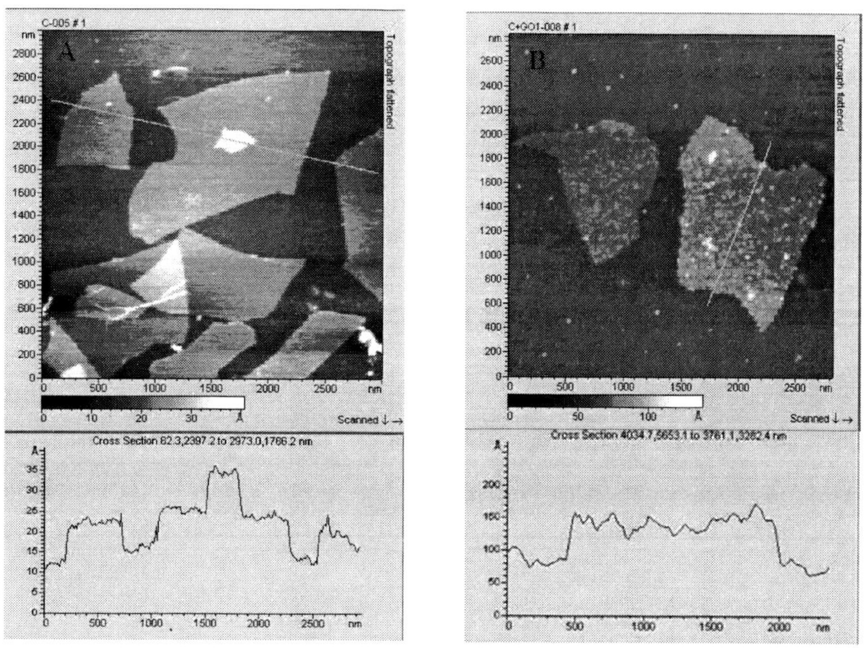

Figure 3. Typical AFM images and the cross–sectional analysis of graphene sheets with (B) and without (A) assembly of GOx.

Figure 3A presents the typical AFM image of graphene. From which, large graphene sheets are observed. The cross–sectional analysis indicates that the thickness of graphene is 0.8–1.0 nm, which matches well with the apparent thickness of graphene sheets reported previously [83], suggesting the single–sheet nature of graphene is obtained in this work. After assembly of GOx, the morphology of graphene changes significantly. Some small and uniformly distributed island–like nanostructures (ca.50 nm in size) are clearly observed (Figure 3B). These nanostructures are the aggregates of GOx molecules since the particle size of GOx aggregate is in the range of 10–100 nm [83, 84]. These results indicate that the enzyme has been effectively assembly on graphene. Moreover, the thickness of graphene has a significant increase with the assembly of GOx (5–6 nm for GOx–graphene), further indicating the formation of GOx–graphene hybrid.

The assembly of GOx was also examined by FTIR spectroscopy. In IR spectrum of the GOx–graphene hybrid (Figure 4a), the amides I and II bands have shapes similar to that of the free GOx (not assembled on the graphene, Figure 4b), except that the bands have a slight shift (1632 to 1633 cm^{-1} and 1543 to 1542 cm^{-1} for amides I and II, respectively), which indicates the interaction between GOx and graphene. Meanwhile, IR spectrum of graphene does not show any observable peaks in the range of amides I and II (Figure 4c). These results further verify that GOx has been successfully assembled on graphene, which is in good agreement with that obtained from AFM image. Moreover, IR results also verify that GOx retains its original structure after assembly on graphene.

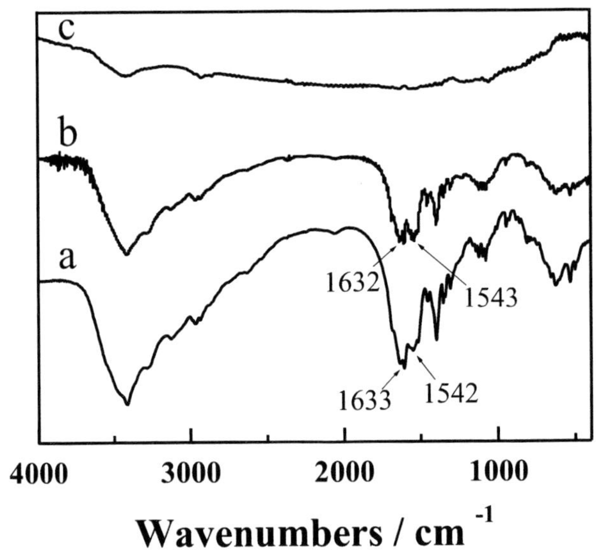

Figure 4. FTIR spectra of the GOx–graphene hybrid (a), free GOx (b), and graphene (c).

The above conclusion is further supported by the CD measurements (Figure 5). The spectra in the far–UV region (200 to 250 nm) correspond to the peptide $n \rightarrow \pi^*$ electronic transition, which is sensitive to the changes of the conformation of the enzyme [85]. The CD spectra of GOx in solution are characterized by two negative peaks at ca.209 and 218 nm (curve a of Figure 5). These peaks almost remain invariable after GOx was immobilized on the surface of graphene (curve b of Figure 5), verifying that the secondary structure and conformation of GOx immobilized on the surface of graphene are essentially the same as that of the native state. Moreover, the CD spectra of GOx–graphene (curve d of Figure 5) in the near–UV region (250 to 300 nm) are also the same as that of GOx in solution (curve c of Figure 5), suggesting that the microenvironment of the active site of GOx immobilized on the surface of graphene also remains unchangeable as that of the native one since there is no signals can be observed in this region for a denatured GOx due to the dissociation of FAD from the enzyme [85].

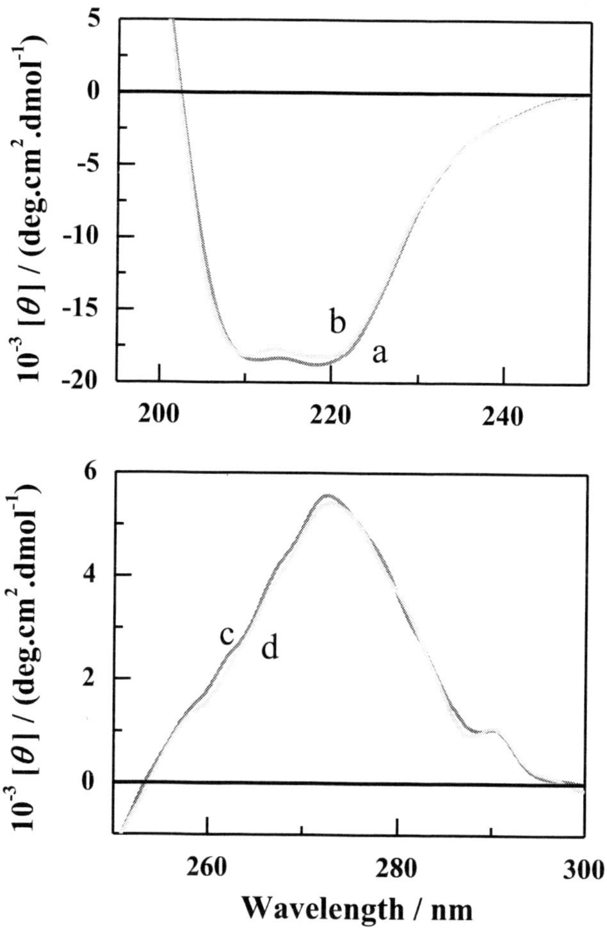

Figure 5. CD spectra of GOx in aqueous solution (a and c) and immobilized on the surface of graphene (b and d) in the far–UV (A) and near–UV (B) region, respectively.

3.3. The Electrochemical Characteristic of GOx–graphene Hybrids

The electrochemical characteristic of GOx–graphene hybrids was studied by cyclic voltammetry. The cyclic voltammograms (CVs) presented in Figure 6 were obtained at the graphene/GC (curve a) and GOx–graphene/GC electrode (curve b) in O_2–free PBS (0.1 M, pH 6.9) at a scan rate of 40 mV s^{-1}. CV of the graphene/GC electrode does not show any observable redox peaks in the potential range of interest (curve a), however, the CV of the GOx–graphene/GC electrode is characterized by a pair of well–defined and nearly symmetrical redox peaks with the cathodic (E_{pc}) and anodic (E_{pa}) peak potential of ca.–468 and –441 mV (vs. SCE), respectively (curve b). The formal potential, $E^{0'}$, calculated by averaging the cathodic and anodic peak potential, is –454.5 mV, which is close to those reported previously for direct electron transfer (DET) of GOx immobilized on the surface of carbon nanotubes (–464 mV) [50], on Au nanoparticles surface (–451 mV) [86], and on boron–doped carbon nanotubes surface (–477 mV) [87]. Therefore, the redox peak depicted in the curve b can be ascribed to the DET reaction of GOx. This conclusion also demonstrates that GOx assembled on graphene surface can still retain its native structure. All of these results indicate that the direct ET reaction of GOx is achieved through GOx integrated with graphene. Graphene plays an active role in prompting DET between the active center of GOx and the electrode due to its extraordinary electron transport property [88, 89].

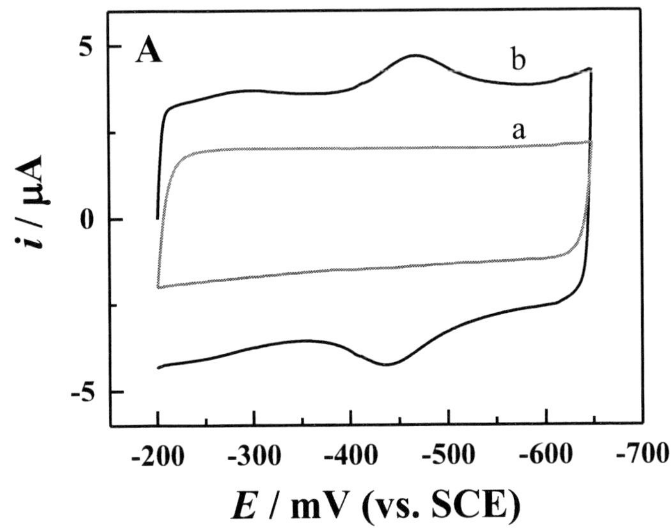

Figure 6. Cyclic voltammograms of graphene/GC (curve a) and GOx–graphene/GC electrode (curve b) in O_2–free PBS (0.1 M, pH 6.9) at a scan rate of 40 mV s^{-1}.

Furthermore, DET reaction of the GOx is fairly stable since the CV features of the GOx–graphene/GC electrode remain unchangeable even after the electrode was scanned continuously for a long time (ca.50 cycles, at 40 mV s^{-1}) in the potential range of interest. In addition, no obvious change is observed after the electrode has been stored in PBS at 4 ºC for one week. These characteristics are important when the electrode is further used as a biosensor for sensing substrates (such as glucose, etc.).

The surface average (Γ, in mol cm^{-2}) of electroactive GOx immobilized on the surface of graphene can be estimated by integrating the anodic or cathodic peak in cyclic voltammograms based on $\Gamma = Q / nFA$, where Q is the charge consumed in the redox reaction in cyclic voltammograms under the background correction, n is the electron transfer number (n = 2 for DET of GOx [50]), F is Faraday constant, and A is the apparent area of the electrode. The value of Γ was calculated to be $(1.3 \pm 0.3) \times 10^{-10}$ mol cm^{-2}, which is higher than those reported form GOx immobilized on CNT/GC electrode (2.35×10^{-11} mol cm^{-2} [90]), on Nafion–CNT/GC electrode (5.93×10^{-11} mol cm^{-2} [91]), on Au nanoparticles modified carbon paste electrode $(9.8 \pm 0.6) \times 10^{-12}$ mol cm^{-2} [86]), on CdS nanoparticles modified pyrolytic graphite electrode $(1.54 \pm 0.06) \times 10^{-11}$ mol cm^{-2} [92]), and on bare GC electrode (2.86×10^{-12} mol cm^{-2} [93]). These results demonstrate that graphene has a high adsorption capacity, and therefore, a large amount of GOx can be assembled.

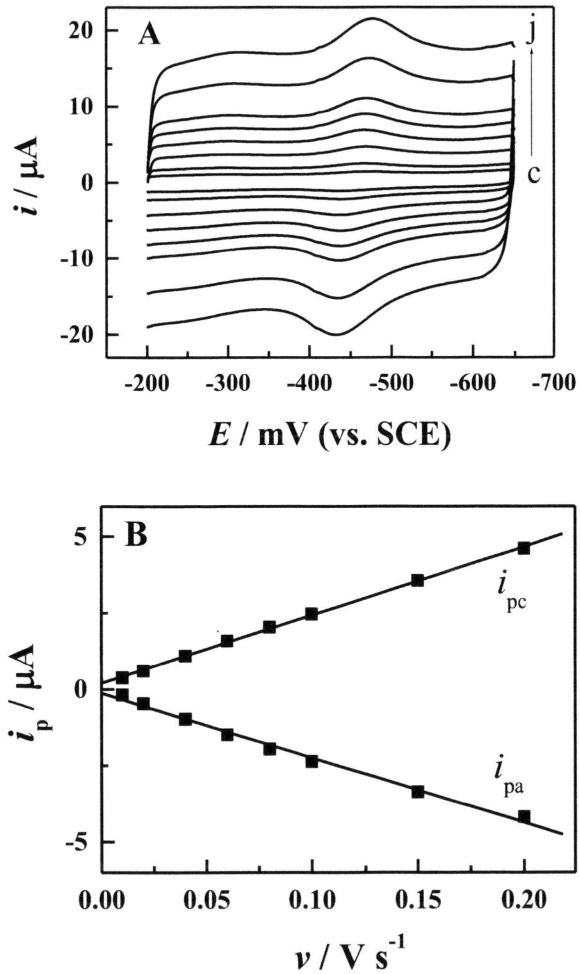

Figure 7. (A) CVs of the GOx–graphene/GC electrode in N_2–saturated PBS (0.1 M, pH 6.9) at different scan rates. Scan rates (from curve c to j) are 10, 20, 40, 60, 80, 100, 150 and 200 mV s^{-1}. (B) Dependence of peak currents on scan rates.

To further investigate the DET characteristics of GOx immobilized on the surface of graphene, the effects of scan rates on the voltammetric response of the GOx–graphene/GC electrode was studied in detail. The results are presented in panel A of Figure 7. It is shows that the anodic and cathodic peak potentials shift slightly to positive and negative direction, respectively, with the increasing of the scan rate (v), the values of E^{0}' (–455 ± 1 mV) is, however, almost independent of v. In addition, almost equal anodic and cathodic peak currents can be observed at each scan rate, and both the anodic and cathodic peak currents increase linearly with scan rates (panel B of Figure 7), indicative of a typical of the surface–confined redox processes. Another important feature of the voltammetric response of the electrode is that the separation of peak potentials (ΔE_p) is small (15 mV at 10 mV s^{-1}, and 30 mV at 200 mV s^{-1}). These results demonstrate a fast DET kinetics of the GOx on the surface of graphene.

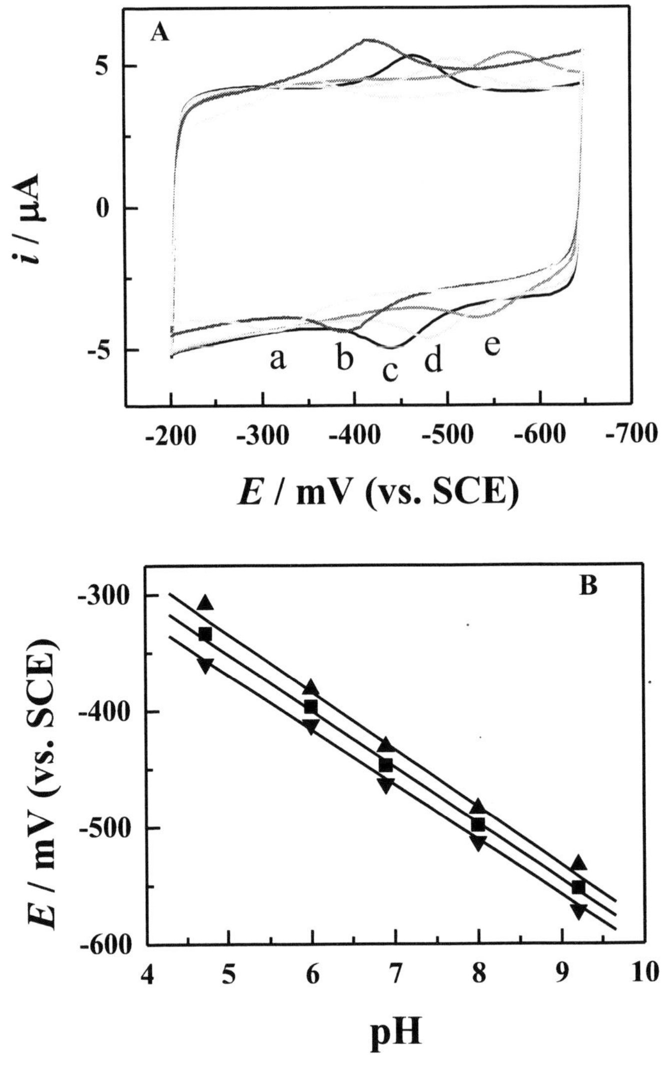

Figure 8. (A) CVs of GOx–graphene/GC electrode in the PBS with different pH value at a scan rate of 40 mV s^{-1}. The pH value (from a to e) is 5, 6, 7, 8 and 9, respectively. (B) Dependence of the anodic (▲), cathodic (▼), and formal porential (■) of the GOx–graphene/GC electrode on solution pH.

The apparent DET rate constant (k_s) is 2.68 s^{-1} estimated from the dependence of ΔE_p on the scan rates based on the model for a surface–controlled electrochemical system developed by Laviron [94]. This value is practically the same as that obtained by Kang et al. [95] at the GOx–graphene–chitosan/GC electrode (2.83 s^{-1}) by considering the experimental error. However, the graphene was dispersed in chitosan solution in Kang's work, thus complicating the electrode fabrication procedures. In the present work, the graphene was directly dispersed in water, the electrode fabrication procedures can be significantly simplified. The k_s value obtained here is higher than those obtained at carbon nanotubes (1.16 s^{-1} [87]) or boron–doped carbon nanotubes (1.56 s^{-1} [87]), implying that graphene can remarkably enhance DET kinetics of GOx. The superior ET kinetics of graphene can be ascribed to its unique electronic structure [96], which are favorable for conducting electrons from the redox center to electrode surface. It may also due to that graphene have more surface defects than carbon nanotubes. These defects result in the high density of electronic states near the Fermi level leading to the enhancement of DET kinetics.

It has been established that DET reaction of GOx is a two–electron coupled with two–proton reaction [86]. Therefore, DET characteristics of GOx would be affected by solution pH. The results demonstrate that the CV characteristics of the GOx–graphene/GC electrode show a considerable dependence on pH of the buffer (Figure 8A). Both the anodic and cathodic peak potentials shift to negative direction with the increasing of the buffer pH, and the value of $E^{0'}$ depends linearly on the pH in the experimental range with a slope of ca.–54.1 mV pH^{-1} (Figure 8B), which is close to the theoretical value (–58.5 mV pH^{-1}, 22 °C) for a reversible two–electron coupled with two–proton reaction process, suggesting the high reversibility of the DET of GOx on the surface of graphene.

3.4. Electrocatalytic Reduction of Oxygen by GOx–graphene and Bisoensing for Glucose

The GOx–graphene hybrid exhibits good electrocatalytic activity toward the reduction of O_2. Upon the presence of O_2 in solution (air–saturated PBS), the voltammetric feature of the GOx–graphene/GC electrode (curve b in part A of Figure 9) changes significantly comparing with that obtained in N_2–saturated PBS (curve a in part A of Figure 9, and the inset in part A for the amplified curve). It is characterized by a large cathodic peak at ca.–470 mV with completely disappearance of the anodic peak. The height of cathodic peak is sensitive to the change of O_2 concentration in solution (not shown here). These characteristics are the typical features of electrocatalytic reactions. The results of part B and C in Figure 9 show that the cathodic current density for the reduction of O_2 (air–saturated PBS) at the graphene/GC and the bare GC electrodes is only ca.1/5 and 1/20, respectively, of that obtained at the GOx–graphene/GC electrode (please note the different current density scale in part A, B, and C). These results confirm that the enhancement of the cathodic peak current in part A (curve b) mainly come from the reduction of oxygen catalyzed by the assembled GOx according to Eqs. (1) and (2):

$$\mathrm{GOx(FAD)} + 2e^- + 2H^+ \rightarrow \mathrm{GOx(FADH_2)}\,, \qquad (1)$$

$$\mathrm{GOx(FADH_2)} + O_2 \rightarrow \mathrm{GOx(FAD)} + H_2O_2\,. \qquad (2)$$

Furthermore, the addition of glucose to air–saturated PBS leads to the decrease of the electrocatalytic current for the reduction of oxygen (part D in Figure 9). This is potentially caused by the enzymatic reaction as Eq. (3):

$$\mathrm{GOx(FAD)} + \mathrm{glucose} \rightarrow \mathrm{gluconolactone} + \mathrm{GOx(FADH_2)}\,. \qquad (3)$$

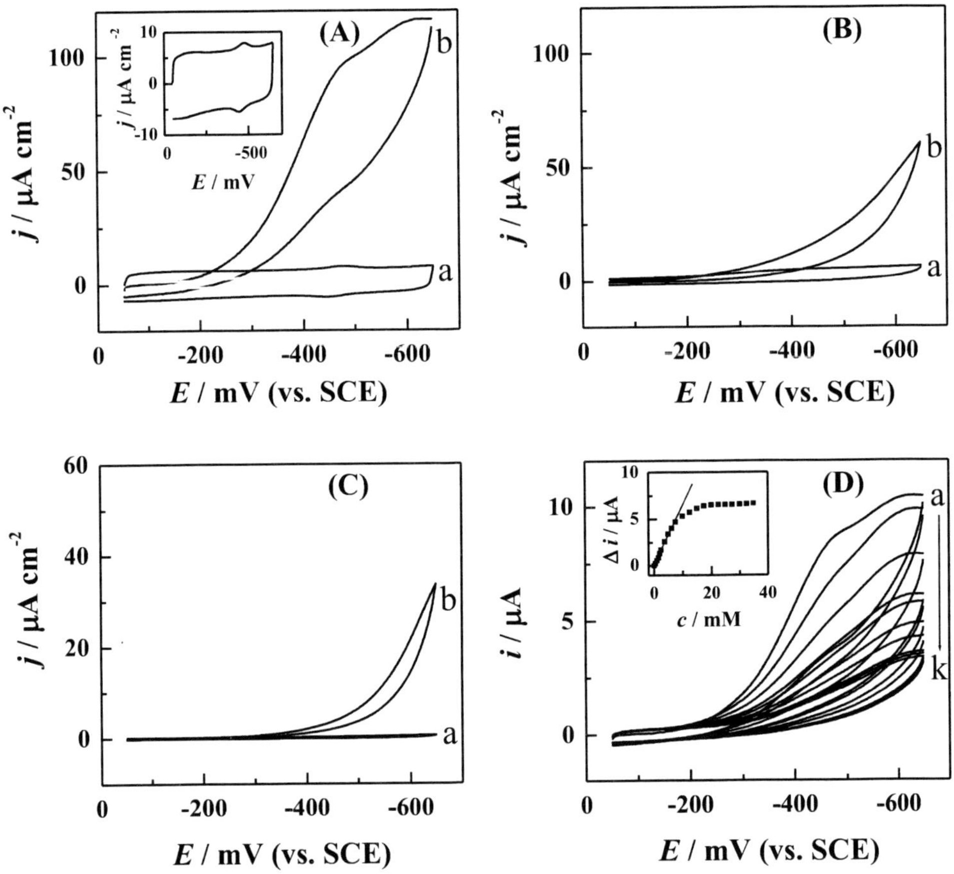

Figure 9. CVs of the GOx–graphene/GC (A), graphene/GC (B), and bare GC (C) in O_2–free (curve a) and air–saturated (curve b) PBS (0.1 M, pH 6.9). Part D shows the effects of addition of glucose into air–saturated PBS on the voltammetric responses of the GOx–graphene/GC electrode. The concentration of glucose is (from curve a to k): 0, 0.5, 1.0, 2.5, 5.0, 7.5, 10.0, 12.5, 15.0, 17.5, and 20.0 mM. Scan rate: 10 mV s^{-1}. Inset: the amplified CV for the GOx–graphene/GC electrode in O_2–free PBS (A), and the dependence of electrocatalytic current at –470 mV on the concentration of glucose (D).

Glucose is the substrate of GOx, whose presence in the air–saturated PBS will result in an enzyme–catalyzed reaction (Eq. 3) and decrease the concentration of GOx(FAD) (the oxidized form of GOx) on electrode surface. Therefore, the electrocatalytic reaction (Eqs. (1) and (2)) is restrained with the addition of glucose, leading to the decrease of reduction current. Based on the decrease, the concentration of glucose can be quantified (the inset in part D). The inset is a calibration curve obtained with GOx–graphene/GC electrode showing relationship between the difference reduction peak current and the concentration of glucose. The difference of oxygen reduction current between that of a glucose free solution and a finite quantity of glucose solution was recorded at a fixed potential of –470 mV (vs. SCE). The electrode linearly responds to glucose at lower concentration and attains a saturation level at a higher concentration (the inset in Part D). The response displays a linear range from 0.1 to 10 mM with the sensitivity of (110 ± 3) μA mM^{-1} cm^{-2} and a detection limit of (10 ± 2) μM at an S/N of 3. The upper–limit of the linear range is far beyond the physiological level (3–8 mM)

[97], suggesting that the electrode developed here would be useful for glucose determination in physiological levels. The linear range obtained here is similar to that obtained for GOx immobilized on graphene–Au nanoparticles–chitosan composite (2–10 mM [98]), and also close to that reported for GOx entrapped in the graphene–chitosan film (0.08–12 mM [95]).

Table 1. Comparison analytical performance of some GOx–based glucose biosensors

Glucose biosensor	Linear range (mM)	Sensitivity ($\mu A\ mM^{-1}\ cm^{-2}$)	Detection limit (μM)	Ref
GOx–graphene	0.1–10	110 ± 3	10 ± 2	This work
GOx–graphene–AuNPs–chitosan[a]	2–10	–	180	[98]
GOx–graphene–chitosan	0.08–12	37.93	20	[95]
Nafion–GOx–GNSs[b]	0.2–1.4	3.4	–	[99]
GOx–MWCNTs–Chitosan[c]	0–7.8	7.36	–	[100]
GOx–colloidal gold	0.08–0.28	8.4	10	[86]
GOx–CNF[d]	0.01–0.35	36.3	2.5	[36]
GOx–nanoPANi[e]	0.01–5.5	97.18	0.3 ± 0.1	[73]
GOx–CNTs[f]	0–30	–	80	[101]
GOx–CNTPE[g]	0–25	11.3 ± 0.1	600	[102]
Nafion–GOx–CNTs–CdTe	0–0.7	14.41	–	[91]
GOx–CNTs–PPF[h]	0.025–2.2	42	6	[103]
GOx–sol–gel–CNTs	0.2–20	0.2	50	[104]
GOx–PAN[i]	0–5	4.2	20	[105]
Nafion–GOx–OMC[j]	0.5–1.5	0.053	156.62	[106]

[a] AuNPs: gold nanoparticles. [b] GNSs: graphite nanosheets. [c] MWCNTs: multi–walled carbon nanotubes. [d] CNF: soluble carbon nanofiber. [e] nanoPANi: highly ordered polyaniline nanotubes. [f] CNTs: carbon nanotubes. [g] CNTPE: carbon nanotubes paste electrodes. [h] PPF: nano–thin plasma–polymerized film. [i] PAN: microporous polyacrylonitrile. [j] OMC: ordered mesoporous carbon.

However, it is much wider than those previously reported for GOx immobilized on GOx–Nafion–graphite nanosheets composite (0.2–1.4 mM [99]), on MWCNTs–chitosan matrix (0–7.8 mM [100]), on colloidal gold modified carbon paste electrode (0.08–0.28 mM [86]), on carbon nanofiber modified GC electrode (0.01–0.35 mM [87]), and on highly ordered polyaniline nanotubes (nanoPANi) (0.01–5.5 mM [73]). The detection limit for glucose at the developed electrode is lower than those reported at the GOx–graphene–chitosan electrode (20 μM [95]), the GOx–graphene–Au nanoparticles–chitosan modified electrode (180 μM [98]), the GOx–CNT nanoelectrode (80 μM [101]), and the GOx–CNT paste electrode (0.6 mM [102]). Moreover, the sensitivity of the developed electrode is much higher than those obtained at the GOx–CNTs–chitosan/GC (7.36 $\mu A\ mM^{-1}\ cm^{-2}$ [100]), the Nafion–CNTs–

CdTe–GOx/GC (14.41 μA mM^{-1} cm^{-2} [101]), the GOx–graphene–chitosan (37.93 μA mM^{-1} cm^{-2} [95]), the GOx–CNTs–PPF/Au (42 μA mM^{-1} cm^{-2} [103]) and the GOx–nanoPANi/Pt (97.18 μA mM^{-1} cm^{-2} [73]). All these results indicate that the developed electrode has advantages over those previous reported glucose biosensors regarding the sensitivity, linear range and detection limit (the detailed comparison of the analytical characteristics of the developed biosensor with those reported previously is summarized in Table 1).

Therefore, a new method for glucose determination has been developed and would be hopeful for glucose detection. The method is different from the usual glucose amperometric sensor on which the determination is based on monitoring either the consumption of oxygen [54] or the production of H_2O_2 [55]. Moreover, this method can efficiently exclude the interference of commonly coexisted electroactive species, such as ascorbic acid (AA), uric acid (UA), and 4–acetamidophenol (AP) due to the use of a low detection potential (–470 mV, vs. SCE). By the way, no obvious decrease in the response to 5 mM glucose was observed when the GOx–graphene/GC electrode was stored in buffer at 4 °C and measured 1–2 days interval, indicating the composite electrode is efficient to keep the bioactivity of GOx and the biosensor had good stability.

3.5. Determination of Glucose in Human Blood Serum

Table 2. Determination of glucose concentration in blood serum samples using the GOx–graphene/GC electrode

No.	Referenced values [a] (mM)	Determined values [b] (mM)	Relative deviation (%)
1	5.02 ± 0.12	4.81 ± 0.14	–4.4
2	4.96 ± 0.11	4.78 ± 0.09	–3.6
3	4.67 ± 0.09	4.86 ± 0.05	4.1
4	6.21 ± 0.06	6.33 ± 0.15	1.9
5	4.46 ± 0.12	4.31 ± 0.11	–3.4

[a] Value obtained at GOx–graphene/GC electrode, it was an average value of the response of 5 independent measurements for each sample (one electrode for per sample). [b] Value obtained by the commercial blood glucose monitor.

The GOx–graphene/GC electrode was used for measurement of glucose in human blood serum. The concentration of glucose in five serum samples obtained from hospital were determined by using five independently prepared GOx–graphene/GC electrodes (one electrode for per sample). The results were compared with that obtained by the commercial blood glucose monitoring system (Table 2). It can be concluded that the values measured by the developed electrode are in good agreement with the data from the commercial monitor, indicating that the developed electrode has great potential for application for determine the concentration of glucose on real clinical sample.

Therefore, the DET of GOx assembled on graphene has been successfully achieved. A novel efficient biosensing platform based on graphene sheets integrated with GOx has also been proposed. The hybrid has been demonstrated to be a good electrocatalyst for the reduction of oxygen. A new approach for detection of glucose based on the bioelectrocatalytic reduction of oxygen has been developed.

4. Fabrication of the Lac–ABTS–Graphene Nanohybrids and Detecting of Extracellular Oxygen Released from Erythrocytes

This section presents a novel electrochemical approach for detection of the extracellular oxygen released from human erythrocytes. The sensing is based on the bioelectrocatalytic system of Lac–ABTS–graphene hybrid toward the reduction of oxygen. ABTS and laccase are assembled on the surface of graphene, which is synthesized by chemistry route, utilizing the π–π and electrostatic interactions of these components. The voltammetric results indicate that ABTS can be used as a redox mediator when it is in immobilized form. The hybrid deposited on the glassy carbon (GC) electrode is demonstrated to be a good bioelectrocatalyst for the reduction of oxygen with inherent enzyme activity, accepted stability, high half–wave potential (ca.670 mV vs NHE), and unimpeded electrical communication to the copper redox sites of laccase. Therefore, this study has not only established a novel approach to detection of extracellular oxygen but also provided a general route for fabricating graphene–based biosensing platform via assembling enzymes/proteins on graphene surface.

4.1. Fabrication of the Lac–ABTS–graphene Nanohybrids

Laccase (the structure is shown in Figure 10), which belongs to the multicopper group of oxidases and drives effectively the four–electron reduction of oxygen to water [107, 108], is one of the most commonly considered enzymes for bioelectrocatalytic cathodes in biofuel cells for the reduction of oxygen [109, 110]. To ensure the effective ET communication between the redox centers of the enzyme and electrode, ABTS is commonly introduced as a mediator [109–112]. ABTS can be easily assembled onto the surface of graphene through π–π electronic interactions forming a new kind of nanocomposites (ABTS–graphene). After the assembly of ABTS, a layer of negative charges is created on the surface of graphene. Therefore, the positively charged laccase (at pH 7.4, the isoelectric point of laccase from *Rhus vernicifera* is ca.8.2 [113]) could be assembled on the surface of ABTS–graphene based on electrostatic interaction with the formation of the Lac–ABTS–graphene hybrid.

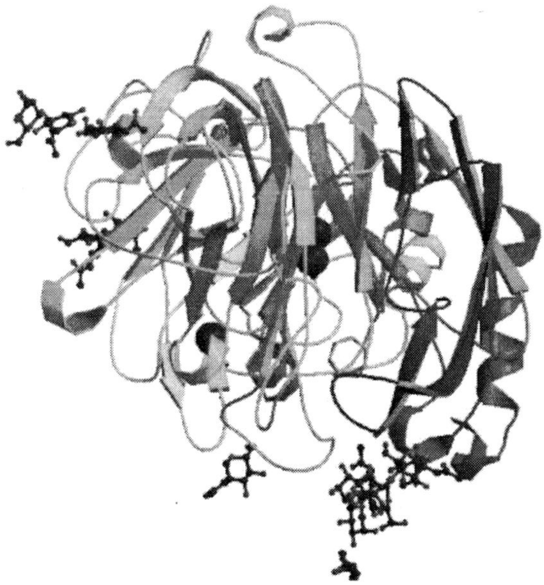

Figure 10. The structure of Lac (adapted from protein data bank).

To fabricate the modified electrode, the Lac–ABTS–graphene hybrid (2 mg) was dispersed into PBS (1 mL, pH 7.4) to form a homogeneous suspension (2 mg mL^{-1}). Then, the suspension (5 μL) was cast onto the surface of the pretreated GC electrode [114] (3 mm in diameter, CH Instruments) with a microsyringe, and solvent was allowed to be evaporated at ambient temperature before use. The electrode (denoted as the Lac–ABTS–graphene/GC electrode) was stored at 4 °C when not in use.

4.2. Characterization of the Lac–ABTS–graphene Nanohybrids

Figure 11 shows a typical TEM image of graphene. Large graphene sheets with length up to ca.4 μm and width of ca.3 μm are observed. The graphene sheets are rippled and resemble crumpled silk veil waves. However, when its size is down to ca.1 × 1 μm, the surface is flat as shown by AFM image in Figure 12a. The cross–sectional analysis indicates that the thickness of graphene is ca.0.7 nm, which matches well with the reported apparent thickness of graphene sheets [115, 116], suggesting the single–sheet nature of graphene is obtained in this work. The AFM image reveals that no sheets either thicker or thinner than 0.7 nm is observed, demonstrating that complete exfoliation of graphene oxide down to individual graphene sheet is indeed achieved under our experimental conditions.

The image in Figure 12a also shows that graphene sheets distribute separately on the surface of the electrode without forming aggregates or entangling each other. The uniform nanostructure can significantly increase the effective surface of the electrode for loading of biomolecules and accelerating ET kinetics. After the assembly of laccase, the morphology of graphene changes significantly. Some small and uniformly distributed island–like nanostructures appear (Figure 12b). These nanostructures are the aggregates of laccase molecules, indicating that the enzyme has been effectively loaded on the graphene. Moreover, the thickness of graphene has a significant increase with the assembly of laccase (5–6 nm for laccase–assembled graphene, please refer to the corresponding cross–sectional analysis), further indicating the formation of Lac–ABTS–graphene hybrid. The morphology of the hybrid on the electrode surface processes the similar structure to that of graphene. Such a structure is expected to be very attractive for the detection of substrates because each of the hybrids is fully and easily accessible to substrates, and therefore can be used as an electrochemical sensing unit, yielding a high ratio of signal–to–noise for electrochemical determination.

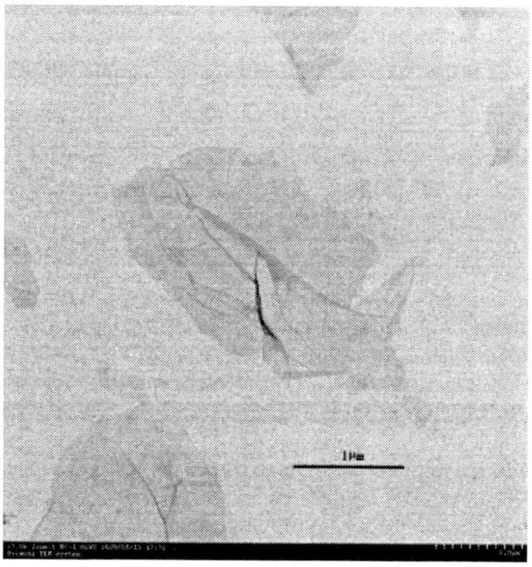

Figure 11. Typical TEM image of graphene sheet.

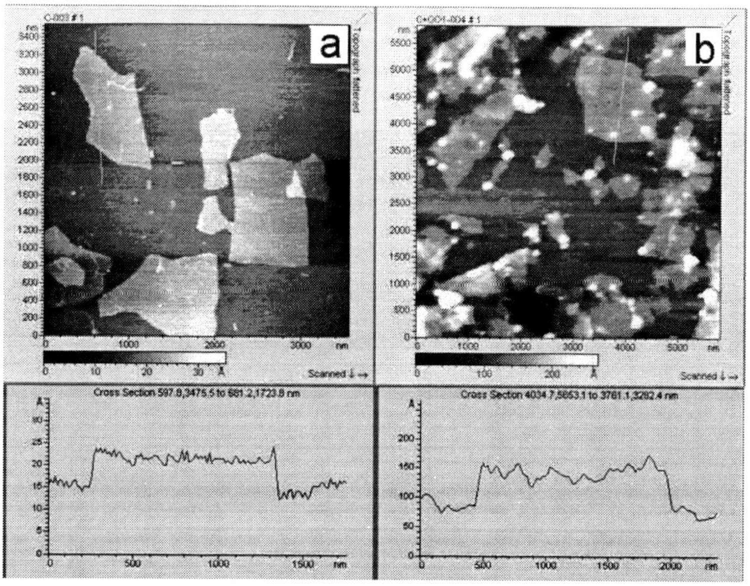

Figure 12. Typical tapping mode AFM images and the cross–sectional analysis of graphene sheets with (b) and without (a) assembly of ABTS and laccase.

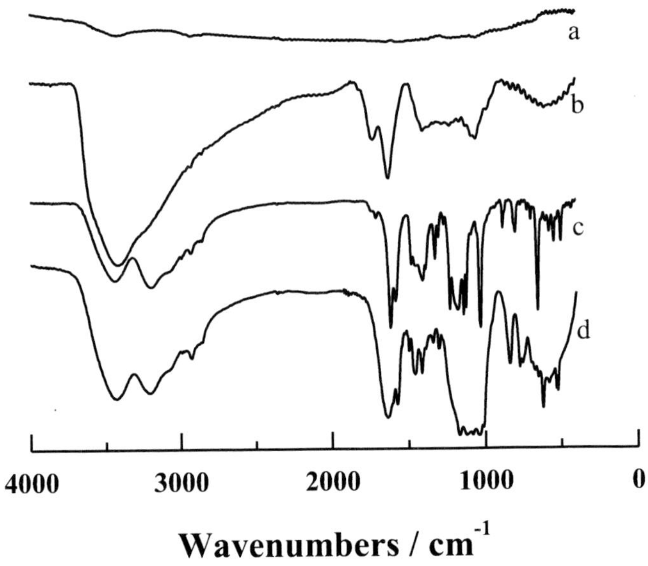

Figure 13. FTIR spectra of the graphene (a), graphite oxide (b), ABTS–graphene (c), and Lac–ABTS–graphene hybrid (d).

The assembly of ABTS and laccase were also examined by FTIR spectroscopy. In IR spectrum of graphene oxide (Figure 13b), the characteristic vibration modes of the O–H groups (~3410 cm^{-1}), C=O groups (~1730 cm^{-1}), the deformation peak of O–H groups (1410 cm^{-1}), the stretching peak of C–OH (1225 cm^{-1}), and the stretching peak of C–O (1050 cm^{-1}) are obvious, demonstrating the successful oxidation of graphite [117]. The peak at 1620 cm^{-1} is assigned to the vibrations of the adsorbed water molecules and the skeletal vibrations of unoxidized graphitic domains [117]. After reduction with hydrazine, the residual oxygen functionalities are still present on the graphene surface, however, the IR intensities of those oxygen–containing groups decrease significantly (Figure 13a), indicating that graphene oxide has been well deoxygened and the graphene is successfully prepared. The assembly of ABTS on graphene results in the characteristic vibration of $-CH_3$ and $-CH_2-$ (~2852 and ~2927 cm^{-1}) groups in the IR spectrum (Figure 13c). The peaks at ~1180 and 1020 cm^{-1} is assigned to $-SO_3$ group antisymmetric and symmetric vibration adsorption, respectively [118]. Peaks at ~1120 and 1010 cm^{-1} can be assigned to the in–plane skeleton vibration and in–plane bending vibration of benzene ring, respectively. These results indicate that ABTS has been effectively assembled on the surface of graphene. The demonstration of amide I (1642 cm^{-1}) and amide II (1540 cm^{-1}) [119] in the IR spectrum of Lac–ABTS–graphene hybrid (Figure 13d) indicates that laccase has been successfully assembled on graphene via ABTS. This result is in good agreement with that obtained from AFM image.

4.3. Voltammetric Characteristics of the Lac–ABTS–graphene Hybrid

The use of laccase combined with ABTS as a diffusion mediator for the reduction of oxygen in a biofuel cell has already been well–established [106]. More recently, attempts have been made to immobilize ABTS together with laccase on the electrode surface [111, 120, 121]. In the present work, the cyclic voltammetric responses of ABTS–graphene and Lac–ABTS–graphene hybrid deposited on GC electrode (denoted as the ABTS–graphene/GC and Lac–ABTS–graphene/GC electrode, respectively) are studied. For comparison, cyclic voltammogram of ABTS adsorbed on the bare GC electrode (ABTS/GC) is also presented.

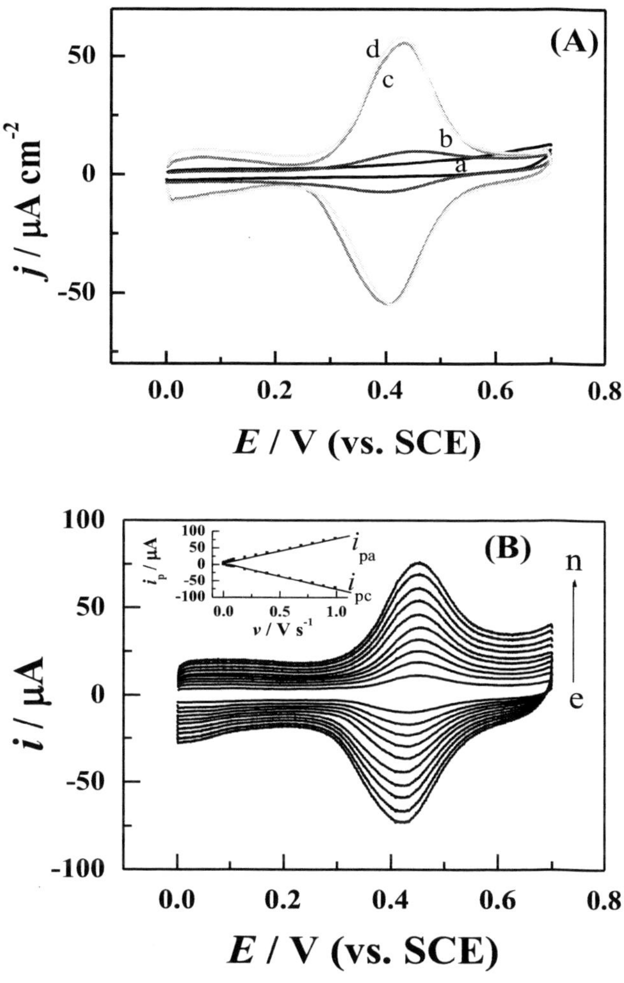

Figure 14. (Part A) Voltammetric responses of the graphene/GC (a), ABTS/GC (b), ABTS–graphene/GC (c), and Lac–ABTS–graphene/GC electrode (d) at a scan rate of 10 mV s^{-1} in O_2–free PBS (0.1 M, pH 7.4). (Part B) Cyclic voltammograms of the Lac–ABTS–graphene/GC electrode at scan rates of (from curve e to n) 100, 200, 300, 400, 500, 600, 700, 800, 900, and 1000 mV s^{-1} in oxygen–free PBS (0.1 M, pH 7.4). The inset of part B shows dependence of peak currents on scan rates for the Lac–ABTS–graphene/GC electrode.

The cyclic voltammogram of the graphene/GC electrode does not show any observable redox peaks (Figure 14a, 0.1 M PBS, pH 7.4) in the potential range of interest, indicating that graphene cannot undergo the redox reaction in the experimental potential ranges. However, the cyclic voltammetric response of the ABTS–graphene/GC electrode is characterized by a pair of well–defined redox peaks with the cathodic (E_{pc}) and anodic (E_{pa}) peak potential of ca.415 and 425 mV (at 10 mV s^{-1}), respectively (Figure 14c). These peak potentials are similar to those for the oxidation and reduction of $ABTS^{2-}/ABTS^{\bullet-}$ couple adsorbed on the surface of carbon nanotubes [121]. This result indicates the formation of ABTS–graphene hybrid, which is in agreement with that obtained from the spectroscopic results. The cathodic and anodic peak currents are almost equal at all scan rates (v) and increase linearly with v up to at least 1000 mV s^{-1} (part A, Figure 14), indicating the presence of surface–confined redox processes. Although E_{pc} and E_{pa} shift slightly with v increasing up to 1000 mV s^{-1}, the value of $E^{0\prime}$, defined as the average value of the E_{pc} and E_{pa}, is independent of v, demonstrating the fast ET kinetics of the $ABTS^{2-}/ABTS^{\bullet-}$ couple on the surface of graphene, implying ABTS can be utilized as a redox mediator when it is in immobilized form. Another important feature of the cyclic voltammetric response of ABTS–graphene is that the redox peaks are symmetrical with small separation (ΔE_p) between the cathodic and anodic peak (10 mV at 10 mV s^{-1}, and 30 mV at 1000 mV s^{-1}), although the value of ΔE_p is not zero. The nonzero separation parameter most likely reflects differences in the nature and relative strengths of adsorption of the oxidized and reduced forms of ABTS on the graphene surface. The full width at half–maximum of the peaks for ABTS oxidation is equal to ca.130 mV (Figure 14c), and it is somewhat higher than the theoretical value of 90 mV expected for an ideal one–electron surface–confined voltammetric peak [122]. The above observation may originate from strong repulsive interactions between the oxidized ($ABTS^{\bullet-}$) or reduced ($ABTS^{2-}$) states of the mediator, or both [121].

To illustrate the special effects of graphene in assembling ABTS, the cyclic voltammetric responses the ABTS/GC electrode were also recorded, and compared with that of the ABTS–graphene/GC electrode. As shown in Figure 14b, ABTS at the bare GC electrode also produces a well–defined redox peak with E_{pc} and E_{pa} of 390 and 450 mV, respectively. While the formal potential (420 mV) is comparable to that of the ABTS–graphene/GC electrode, a higher value of ΔE_p (60 mV) is observed at the ABTS/GC electrode. Moreover, the peak current density of the electrode is only ca.1/7 of that obtained at the ABTS–graphene/GC electrode. Please note that the cyclic voltammograms in part A of Figure 14 are plotted as current density (j, in μA cm^{-2}) versus potential in order to obtain more information on the relative ET characteristic itself, compare the results better, and eliminate the effects of the apparent surface area (A) of the bare GC and graphene/GC electrodes on the peak current. The above results demonstrate that graphene can remarkably enhance ET kinetics and peak current of the $ABTS^{2-}/ABTS^{\bullet-}$ couple with respect to the bare GC electrode, similarly to what has been reported in carbon nanotube [121, 123–125]. The superior ET kinetics of graphene can be ascribed to its unique electronic structure [126]. It may also due to that graphene has more surface defects than polished GC electrode because the graphene sheets were synthesized via a chemical method. These defects results to the high density of electronic states near the Fermi level leading to the enhancement of ET kinetics. Moreover, the high current obtained at the graphene/GC electrode may also be due to that graphene has a higher adsorption capacity and can interact much stronger with ABTS than that of the bare GC, and therefore, a large amount of ABTS can be assembled. The value of surface concentration (Γ, in mol cm^{-2}) corresponding to the saturated adsorption of ABTS on the graphene is estimated to be ca.$(4.3 \pm 0.2) \times 10^{-10}$ mol cm^{-2} using $\Gamma = Q/nAF$ (where Q is the charge consumed in coulombs, obtained from integrating the anodic or cathodic peak area in cyclic voltammograms under the background correction, n is the number of electron transfer for the reaction of ABTS (n = 1), F is the Faraday constant, and A is the apparent area of the electrode). The value obtained at bare GC electrode is only ca.$(1.8 \pm 0.3) \times 10^{-11}$ mol cm^{-2}.

Cyclic voltammogram of the Lac–ABTS–graphene/GC electrode (Figure 14d) also shows a pair of redox peak with the peak potential and currents almost identical with those presented in Figure 14c, demonstrating that assembly of the enzyme does not affect the electrochemical characteristics of ABTS. This feature is important for ABTS being utilized as useful mediator after it was immobilized on the electrode surface with an enzyme.

The ABTS–graphene and Lac–ABTS–graphene on the GC electrode surface are fairly stable because the voltammetric characteristics of the ABTS–graphene/GC electrode or Lac–ABTS–graphene/GC electrode remain unchangeable even after the electrode was scanned continuously for a long time in the potential range of interest (ca.100 cycles at a 10 mV s^{-1}). This result is consistent with the persistence of the attachment of ABTS and laccase on graphene. Therefore, the hybrid is expected to act as a stable mediating system of potential utility to bioelectrocatalytic applications.

4.4. Biosensing Oxygen

Since ABTS dissolved in solution or immobilized on suitable substrates [120, 121] is a good mediator for the reduction of oxygen catalyzed by laccase, Lac–ABTS–graphene is also expected to be a good catalyst for the reduction of oxygen. The electrocatalytic reduction of oxygen at the Lac–ABTS–graphene/GC (part A in Figure 15), ABTS–graphene/GC (part B in Figure 15), graphene/GC (part C in Figure 15), Lac–graphene/GC (part D in Figure 15), ABTS/GC (part E in Figure 15), and Lac–ABTS/GC electrode (part F in Figure 15) were studied in PBS (pH 7.4), and their electrocatalytic characteristics were compared. Upon the presence of O_2 in solution, the voltammetric features of the Lac–ABTS–graphene/GC electrode change significantly characterizing with a large cathodic peak at ca.330 mV and completely disappearance of the anodic peak (part A, Figure 15). The height of cathodic peak is sensitive to the change of the O_2 concentration in solution (not show here). The reduction currents in air–saturated solution are less than 1/5 of that obtained in oxygen–saturated one. These characteristics are the typical features of electrocatalytic reactions. The reduction of O_2 starts at a very positive potential of ca.580 mV, and the observed reduction current increases quite rapidly during negative potential scanning to reach the current density of more than 120 μA cm^{-2} at ca.330 mV. The onset potential is 250 mV more positive than the reduction peak actually appears. The half–wave potential of the catalytic wave is ca.430 mV (vs SCE; i.e., ca.670 mV vs NHE). Such positive reduction potentials are usually observed only for ABTS mediator dissolved in the solution, in gel phase, or on the surface of carbon nanotubes [120,121]. The above results clearly imply that the ABTS–graphene hybrid functions as an effective mediating system in the laccase–catalyzed electroreduction of O_2. It is reasonable to depict the electrocatalytic processes as that electrons are transferred from the electrode surface along the conductive graphene to reduce $ABTS^{\bullet-}$ to $ABTS^{2-}$ at ca.430 mV (as shown in Figure 14c). $ABTS^{2-}$ should donate effectively four electrons to four Cu(II) ions in the active site of laccase. This step also results in regeneration of $ABTS^{\bullet-}$. Finally, O_2 molecules are activated at two of four Cu(I) ions, and they are subsequently reduced to water at 670 mV (vs NHE) [103].

The results of part B, C, D, and E in Figure 15 show that the presence of O_2 in PBS does not affect the voltammetric features of the ABTS–graphene/GC, graphene/GC, Lac–graphene/GC, and ABTS/GC electrode, demonstrating that graphene, ABTS–graphene, Lac–graphene, or ABTS does not have any appreciable electrocatalytic activity toward the reduction of O_2.

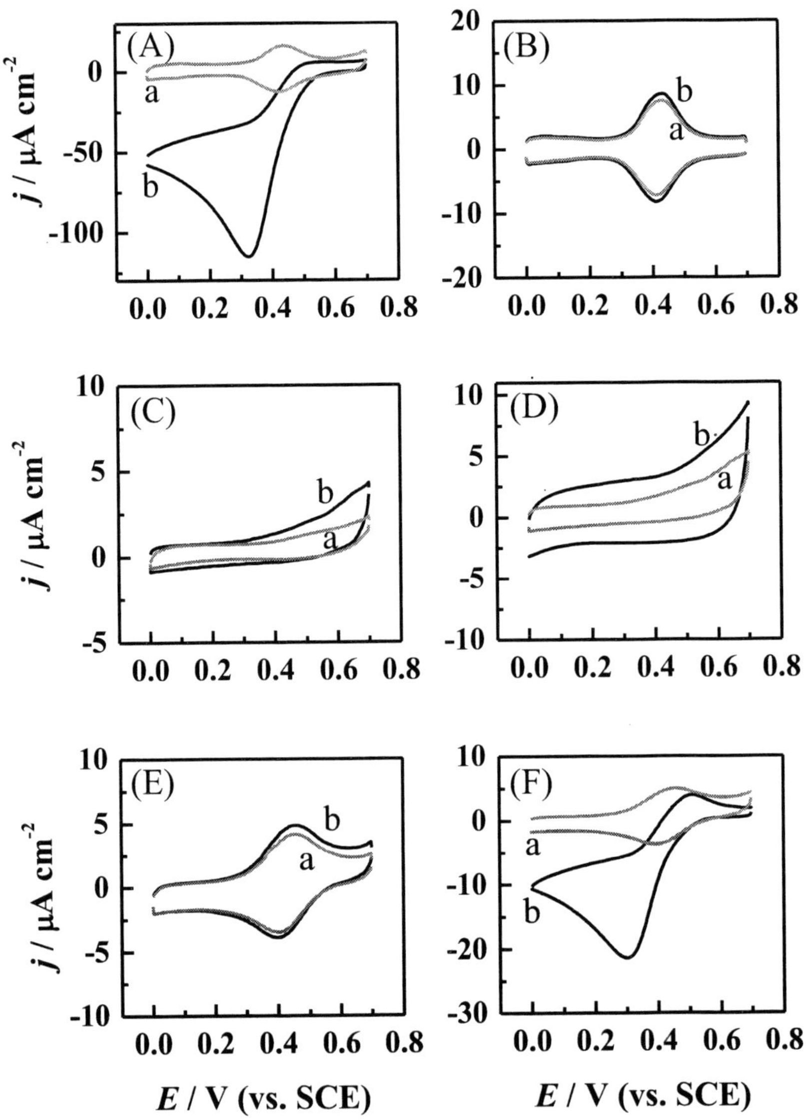

Figure 15. Cyclic voltammograms of the Lac–ABTS–graphene/GC (part A), ABTS–graphene/GC (part B), graphene/GC (part C), Lac–graphene/GC (part D), ABTS/GC (part E), and Lac–ABTS/GC electrode (part F) in oxygen–free (curve a) and oxygen–saturated (curve b) PBS (0.1 M, pH 7.4). Scan rate: 10 mV s^{-1}.

The electrocatalytic reduction of O_2 by Lac–ABTS deposited on the GC electrode (denoted as the Lac–ABTS/GC electrode) were also studied and compared with those obtained at the Lac–ABTS–graphene/GC electrode. The cyclic voltammograms (part F of Figure 15) show that the electrocatalytic features of the Lac–ABTS/GC electrode are similar to those of the Lac–ABTS–graphene/GC electrode in terms of the onset and cathodic peak potentials. However, by carefully analyzing the curve b in part A and part F, it is obvious that the electrocatalytic current density at the Lac–ABTS–graphene/GC electrode is more than 6 times higher than that obtained at the Lac–ABTS/GC electrode (please note the different current density scale in part A and part F). Moreover, the Lac–ABTS/GC electrode loses its electrocatalytic activity quickly due to the enzyme denaturation. However, it should be stressed that the electrocatalytic characteristics of Lac–ABTS–graphene hybrid are fairly stable since the features of the cyclic voltammograms remain almost invariable after the electrode was scanned continuously for a long time (ca.50 cycles, at 10 mV s^{-1}) in O_2–saturated PBS. In addition, no obvious change is detected after the electrode has been stored in PBS at 4 °C for one week. Therefore, the hybrid has acceptable stability in the potential range where it is used as a bioelectrocatalyst for the reduction of oxygen.

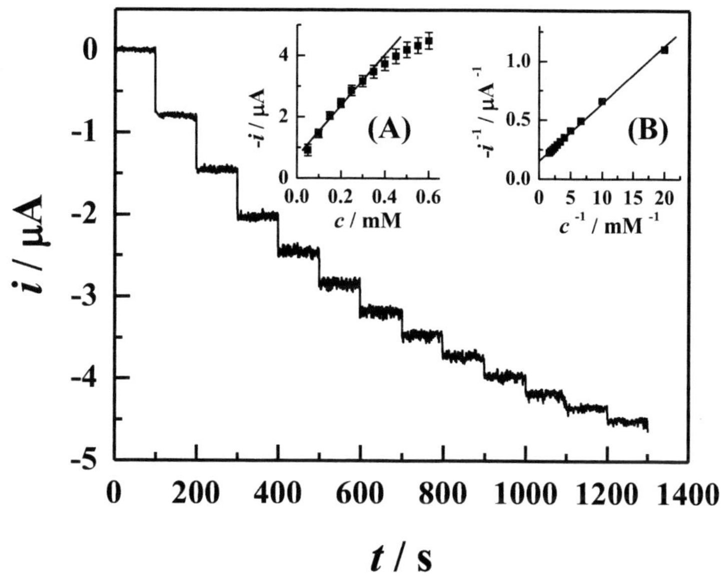

Figure 16. Typical steady state current response of the Lac–ABTS–graphene/GC electrode on successive increase of O_2 concentration in solution (PBS, pH 7.4) at an applied potential of 0.3 V. Insets show the calibration curve of the electrocatalytic current on the concentration of O_2 (inset A) and the Lineweaver–Burk plot (inset B).

Here, the developed Lac–ABTS–graphene/GC electrode can also be used to detect O_2 quantitatively. The concentration of O_2 in a saturated supporting electrolyte was 1.1 mM [123], the solution with different concentration of O_2 was prepared by diluting the O_2–saturated solution with the deaerated buffer. Figure 16 displays the typical steady state amperometric response of the electrode to stepwise increasing of O_2 concentration (0.05 mM for each addition) at an optimum condition. Immediately after the addition, the response increases and reaches 95% of the steady state value within 5 s, suggesting that the electrode responds rapidly to the change of the O_2 concentration. The electrode linearly responds to O_2 concentration at lower concentrations and attains saturation levels at higher concentrations (inset A, Figure 16). The response displays a good linear range from 0.05 to 0.4 mM with a correlation coefficient of 0.995 and a slope of (10.55 ± 0.35) μA mM^{-1}. Therefore, the sensitivity is evaluated to be ca.90 μA mM^{-1} cm^{-2}. The detection limit is estimated to be ca.(10 ± 2) μM (at a signal/noise of 3), which is comparable to that obtained at the carbon nanotubes–chitosan–laccase modified electrode (7.8μM) [127]. The apparent Michaelis–Menten constant (K_M^{app}) is estimated to be (0.32 ± 0.05) mM from the Lineweaver–Burk plot (the inset B of Figure 16). The K_M^{app} and the sensitivity are superior to those reported at monoolein cubic phases–laccase (5.5 mM and 0.23 μA mM^{-1} cm^{-2}, data was estimated from the given plot) [128] and at carbon nanotubes–chitosan–laccase modified electrode (3.22 mM and 27.31 μA mM^{-1} cm^{-2}) [127]. The sensitivity is also higher than that obtained at carbon nanotubes–laccase modified boron–doped diamond electrode (15–37 μA mM^{-1} cm^{-2}, the value was calculated from the reported data) [129]. The linear range is much wide than that obtained at poly(nile blue) modified electrode (1.3–26 μM) [130]. These results suggest that graphene can be used as a biocompatible platform for development of laccase–based amperometric biosensors for O_2 determination.

4.5. Detection of Extracellular Oxygen Released from Human Erythrocytes

The erythrocyte, lacking protein synthesis machinery and living in the presence of permanently high concentrations of oxygen and haemoglobin iron, is a good model to investigate the processes of O_2 release and consumption [131]. Figure 17 depicts amperometric response obtained at the Lac–ABTS–graphene/GC electrode located near the cell in PBS (pH 7.4) at a detection potential of 0.3 V (vs SCE). After a steady state background was reached, 2 mM $NaNO_2$ is injected into PBS. A reduction current of ca.(3 ± 0.3) μA (this value is an average of five independent measurements) is observed within ca.10–20 s after the addition of $NaNO_2$ (Figure 17b), which has been reported to induce the release of oxygen from human erythrocytes [132]. This measurement was repeated up to five times to establish the observation, Figure 17b represents a typical response curve. However, no response is attained at the ABTS–graphene/GC electrode (Figure 17a). Therefore, the observed cathodic current is ascribed to the electrocatalytic reduction of the released O_2.

Due to a relative high detection potential (0.3 V vs SCE) being used in this work, some coexisting oxidizable species in cells, such as ascorbic acid (AA) and uric acid (UA), etc., might be oxidized at the electrode and their responses inevitably affect the electrocatalytic responses of the reduction of O_2 provided these substances can also be released from the cells after addition of $NaNO_2$. To further verify the response depicted in Figure 17b is only due to the electrocatalytic reduction of the released O_2 from erythrocyte, a commercial O_2 meter (CyberScan DO600, Eutech Cyberscan) was connected to the system and used for monitoring the amount of O_2 release induced by $NaNO_2$. The measurements were also repeated up to five times, and the average O_2 concentration released from the cells was ca.(0.31 ± 0.05) mM. This value is in good agreement with that obtained at the developed electrode (0.28 ± 0.08 mM), which is calculated based on the average response currents (3 ± 0.3 μA) and the calibration curve of the electrocatalytic current on the concentration of O_2 (inset A in Figure

16). These results show that there is a good concordance between the responses of the developed sensor and the O_2 meter, suggesting the addition of $NaNO_2$ can only induce the release of O_2 from human erythrocytes and cannot lead to the release of the other oxidizable species including AA and UA etc. These results also demonstrate that the observed signal at the Lac–ABTS–graphene/GC electrode (Figure 17b) is actually due to the electrocatalytic reduction of O_2 released from the cells. Therefore, the determination of O_2 release by the developed method is not interfered by the oxidizable species in the cells.

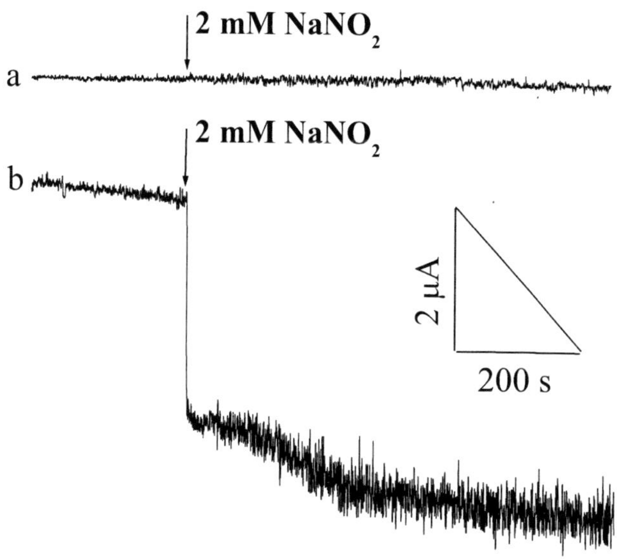

Figure 17. Typical Amperometric responses obtained at the ABTS–graphene/GC (a) and Lac–ABTS–graphene/GC electrode (b) for the electrocatalytic reduction of oxygen released from human erythrocytes.

Though it is difficult to estimate precisely the concentration of O_2 in erythrocyte using the developed method since a little of O_2 is inevitably released before $NaNO_2$ was injected into PBS (this can be evidenced from that steady state background presented in Figure 17b is not level (slightly inclined) before addition of $NaNO_2$), the above observation substantially demonstrates that the developed Lac–ABTS–graphenc/GC electrode represents a new biosensing platform for determination of extracellular O_2 released from cell. Therefore, the method should be extremely helpful for studying the kinetics of O_2 release from cell and could be potentially useful for further physiological and pathological studies. This study has not only established a general route for fabricating graphene–based hybrid via assembling enzymes/proteins on graphene but also expanded the scope of graphene applications to the field of bioelectroanalytical chemistry and cellular biology, which may open up a new challenge and approach to explore the electrochemical features of graphene or its hybrid materials for the potential utilizations.

Conclusion

In summary, the novel efficient biosensing platforms based on graphene sheets integrated with glucose oxidase and laccase has been proposed in this chapter. The graphene has been demonstrated to be appropriate immobilization matrix for assembly redox protein to fabrication redox–protein nanohybrids. Moreover, the hybrids have been demonstrated to be a good electrocatalyst for the reduction of oxygen with not only inherent enzyme activity, accepted stability, and high selectivity but also unimpeded charge transport and electrical communication to the redox sites of proteins. An electrochemical approach for detection of glucose based on GOx–graphene nanohybrids and determination the glucose concentration in human blood serum were discussed; An electrochemical approach for monitor the extracellular oxygen released from human erythrocytes based on the bioelectrocatalytic reduction of oxygen by Lac–ABTS–graphene has also been depicted. The study in this chapter has not only established a general route for fabricating graphene–based hybrid via assembling enzymes/proteins on graphene but also expanded the scope of graphene applications to the field of bioelectroanalytical chemistry and cellular biology, which may open up a new challenge and approach to explore the electrochemical features of graphene or its hybrid materials for the potential utilizations.

Acknowledgments

The work presented in this chapter is supported by the National Natural Science Foundation of China (20773067, 20833006, and 20905036), Research Fund for the Doctoral Program of Higher Education of China (20103207110004), the Foundation of the Jiangsu Education Committee (09KJA150001, 09KJB150006 and 10KJB150009), and the foundation of Jiangsu Provincial Key Laboratory of Palygorskite Science and Applied Technology (HPK201005)..

References

[1] K. S. Novoselov, A. K. Geim, S. V. Morozov, D. Jiang, Y. Zhang, S. V. Dubonos, I. V. Grigorieva and A. A. Firsov, *Science*, 306, 666 (2004).
[2] L. D. Landau, *Phys. Z. Sowjetunion* 11, 26 (1937).
[3] R. E. Peierls, *Ann. Inst. Henri Poincare* 5, 177 (1935).
[4] M. Pumera, *Chem. Rec.* 9, 211 (2009).
[5] J. H. Chen, C. Jang, S. D. Xiao, M. Ishigami and M. S. Fuhrer, *Nat Nanotechnol.* 3, 206 (2008).
[6] R. F. Service, *Science*, 324, 875 (2009).
[7] A. K. Geim, *Science* 324, 1530 (2009).
[8] K. S. Novoselov, E. McCann, S. V. Morozov, V. I. Fal'ko, M. I. Katsnelson, U. Zeitler, D. Jiang, F. Schedin and A. K. Geim, *Nat. Phys.* 2, 177 (2006).
[9] Z. Jiang, Y. Zhang, Y. W. Tan, H. L. Stormer and P. Kim, *Solid State Commun.* 143, 14 (2007).

[10] Z. Jiang, Y. Zhang, H. L. Stormer and P. Kim, *Phys. Rev. Lett.* 99, 106802 (2007).
[11] Y. B. Zhang, Y. W. Tan, H. L. Stormer, P. Kim, *Nature*, 438, 201 (2005).
[12] K. S. Novoselov, Z. Jiang, Y. Zhang, S. V. Morozov, H. L. Stormer, U. Zeitler, J. C. Maan, G. S. Boebinger, P. Kim and A. K. Geim, *Science*, 315, 1379 (2007).
[13] J. Hass, W. A. de Heer and E. H. Conrad, *J. Phys.: Condens. Matter.* 20, 323202 (2008).
[14] F. Chen, J. L. Xia, D. K. Ferry and N. J. Tao, *Nano Lett.* 9, 2571 (2009).
[15] T. Ramanathan, A. A. Abdala, S. Stankovich, D. A. Dikin, M. Herrera–Alonso, R. D. Piner, D. H. Adamson, H. C. Schniepp, X. Chen, R. S. Ruoff, S. T. Nguyen, I. A. Aksay, R. K. Prud'homme and L. C. Brinson, *Nat. Nanotechnol.* 3, 327 (2008).
[16] S. Stankovich, D.A. Dikin, G. H. B. Dommett, K. M. Kohlhaas, E. J. Zimney, E.A. Stach, R. D. Piner, S. T. Nguyen and R. S. Ruoff, *Nature*, 442, 282 (2006).
[17] G. Williarris, B. Seger and P. V. Kamat, *ACS Nano*, 2, 1487 (2008).
[18] Y. X. Xu, H. Bai, G. W. Lu, C. Li and G. Q. Shi, *J. Am. Chem. Soc.*130, 5856 (2008).
[19] M. D. Stoller, S. J. Park, Y. W. Zhu, J. H. An and R. S. Ruoff, *Nano Lett.* 8, 3498 (2008).
[20] D. W. Wang, F. Li, J. P. Zhao, W. C. Ren, Z. G. Chen, J. Tan, Z. S. Wu, I. Gentle, G. Q. Lu and H. M. Cheng, *ACS Nano*, 3, 1745 (2009).
[21] Y. Wang, Z. Q. Shi, Y. Huang, Y. F. Ma, C. Y. Wang, M. M. Chen and Y. S. Chen, *J. Phys. Chem. C*, 113, 13103 (2009).
[22] B. Seger and P. V. Kamat, *J. Phys. Chem. C*, 113, 7990 (2009).
[23] E. Yoo, J. Kim, E. Hosono, H. Zhou, T. Kudo and I. Honma, *Nano Lett.* 8, 2277 (2008).
[24] X. Wang, L. J. Zhi, N. Tsao, Z. Tomovic, J. L. Li and K. Mullen, *Angew. Chem. Int. Ed.* 47, 2990 (2008).
[25] E. Yoo, T. Okata, T. Akita, M. Kohyama, J. Nakamura and I. Honma, *Nano Lett.* 9, 2255 (2009).
[26] H. Chen, M. B. Muller, K. J. Gilmore, G. G. Wallace and D. Li, *Adv. Mater.* 20, 3557 (2008).
[27] Y. C. Si and E. T. Samulski, *Chem. Mater.* 20, 6792 (2008).
[28] Z. Liu, J. T. Robinson, X. M. Sun and H. J. Dai, *J. Am. Chem. Soc.* 130, 10876 (2008).
[29] Z. J. Wang, X. Z. Zhou, J. Zhang, F. Boey and H. Zhang, *J. Phys. Chem. C*, 113, 14071 (2009).
[30] C. S. Shan, H. F. Yang, J. F. Song, D. X. Han, A. Ivaska and L. Niu, *Anal. Chem.* 81, 2378 (2009).
[31] L. Tang, Y. Wang, Y. Li, H. Feng, J. Lu and J. Li, *Adv. Funct. Mater.* 19, 2782 (2009).
[32] R. M. Ianniello, T. J. Lindsay and A. M. Yacynych, *Anal. Chem.* 54, 1098 (1982).
[33] P. Ramesh and S. Sampath, *Anal. Chem.* 75, 6949 (2003).
[34] H. Iwata, T. Nakanoya, H. Morohashi, J. Chen, T. Yamauchi and N. Tsubokawa, *Sens. Actuators B: Chemical*, 113, 875 (2006).
[35] X. L. Wang, H. M. Zhang, J. L. Zhang, H. F. Xu, Z. Q. Tian, J. Chen, H. X. Zhong, Y. M. Liang and B. L. Yi, *Electrochim. Acta*, 51, 4909 (2006).
[36] L. N. Wu, X. J. Zhang and H. X. Ju, *Biosens. Bioelectron.* 23, 479 (2007).
[37] V. Vamvakaki, K. Tsagaraki and N. Chaniotakis, *Anal. Chem.* 78, 5538 (2006).
[38] M. Zhou, J. Ding, L. P. Guo and Q. K. Shang, *Anal. Chem.* 79, 5328 (2007).
[39] E. Y. Zhang and C. R. Wan, *Curr. Opin. Colloid Interface Sci.* 14, 148 (2009).

[40] J. J. Yin, F. Lao, P. P. Fu, W. G. Wamer, Y. Zhao, P. C. Wang, Y. Qiu, B. Sun, G. Xing, J. Dong, X. J. Liang and C. Chen, *Biomaterials*, 30, 611 (2009).
[41] S. N. Kim, J. F. Rusling and F. Papadimitrakopoulos, *Adv. Mater.* 19, 3214 (2007).
[42] K. Balasubramanian and M. Burghard, *Anal. Bioanal. Chem.* 385, 452 (2006).
[43] Q. Zhao, Z. H. Gan and Q. K. Zhuang, *Electroanalysis*, 14, 1609 (2002).
[44] J. Wang, M. Musameh and Y. H. Lin, *J. Am. Chem. Soc.* 125, 2408 (2003).
[45] M. E. G. Lyons and G. P. Keeley, *Chem. Commun.* 22, 2529 (2008).
[46] J. J. Gooding, R. Wibowo, J. Q. Liu, W. R. Yang, D. Losic, S. Orbons, F. J. Mearns, J. G. Shapter and D. B. Hibbert, *J. Am. Chem. Soc.* 125, 9006 (2003).
[47] G. Wang, J. J. Xu and H.Y. Chen, *Electrochem. Commun.* 4, 506 (2002).
[48] D. X. Chen, Q. Wang, J. Jin, P. Wu, H. Wang, S. Q. Yu, H. Zhang and C. X.Cai, *Anal. Chem.* 82, 2248 (2010).
[49] C. X. Cai and J. Chen, *Anal. Biochem.* 325, 285 (2004).
[50] C. X. Cai and J. Chen, *Anal. Biochem.* 332, 75 (2004).
[51] D. A. C. Brownson and C. E. Banks, *Anayst*, 135, 2768 (2010).
[52] K. I. Bolotin, K. J. Sikes, Z. Jiang, M. Klima, G. Fudenberg, J. Hone, P. Kim and H. L. Stormer, *Solid State Commun.* 146, 351 (2008).
[53] X. Wang, L. Zhi and K. Mullen, *Nano Lett.* 8, 323 (2008).
[54] C. E. Banks, T. J. Davies, G. G. Wildgoose and R. G. Compton, *Chem. Commun.* 7, 829 (2005).
[55] G. G. Wallace, J. Chen, D. Li, S. E. Moulton and J. M. Razal, *J. Mater. Chem.* 20, 3553 (2010).
[56] W. J. Lin, C. S. Liao, J. H. Jhang and Y. C. Tsai, *Electrochem. Commun.* 11, 2153 (2009).
[57] C. Liu, S. Alwarappan, Z. Chen, X. Kong and C. Z. Li, *Biosens. Bioelectron.* 25, 1829 (2010).
[58] G. G. Wildgoose, C. E. Banks, H. C. Leventis and R. G. Compton, *Microchim. Acta,* 152, 187 (2006).
[59] P. Wu, Q. Shao, Y. J. Hu, J. Jin, Y. J. Yin, H. Zhang and C. X. Cai, *Electrochim. Acta*, 55, 8606 (2010).
[60] A. K. Geim and K. S. Novoselov, *Nature Mater.* 6, 183 (2007).
[61] D. A. C. Brown and C. E. Banks, *Analyst*, 135, 2768 (2010).
[62] R. Kahl, A. Weimann, S. Weinke and A. G. Hildebrandt, *Arch. Toxicol.* 60, 158 (1987).
[63] Y. Morel and R. Barouki, *Biochem. J.* 342, 481 (1999).
[64] J. Malda, T. J. Klein and Z. Upton, *Tissue Eng.* 13, 2153 (2007).
[65] X. M. Wu, Y. J. Hu, J. Jin, N. L. Zhou, P. Wu, H. Zhang and C. X. Cai, *Anal. Chem.* 82, 3588 (2010).
[66] P. C. Thomas, M. Halter, A. Tona, S. R. Raghavan, A. L. Plant and S. P. Forry, *Anal. Chem.* 81, 9239 (2009).
[67] I. K. Dremza1, E. A. Lapshina1, J. Kujawa and I. B. Zavodnik, *Redox Report*, 11, 185 (2006).
[68] N. P. Rodrigues, Y. Sakai and T. Fujii, *Sens. Actuators B:chemical*, 132, 608 (2008).
[69] G. S. Wilson and R. Gifford, *Biosens. Bioelectron.* 20, 2388 (2005).
[70] Y. Luo, H. Liu, Q. Rui and Y. Tian, *Anal. Chem.* 81, 3035 (2009).
[71] P. Rowiński, R. Bilewicz, M.–J. Stébé and R. Rogalska, *Anal. Chem.* 76, 283 (2004).

[72] X. Q. Liu, L. H. Shi, W. X. Niu, H. J. Li and G. B. Xu, *Biosens. Bioelectron.* 23, 1887 (2008).
[73] Z. Y. Wang, S. N. Liu, P. Wu and C.X. Cai, *Anal. Chem.* 81,1638 (2009).
[74] Y. Y. Shao, J. Wang, H. Wu, J. Liu, I. A. Aksay and Y. H. Lin, *Electroanalysis*, 22, 1027 (2010).
[75] J. B. Jia, D. Kato, R. Kurita, Y. Sato, K. Maruyama, K. Suzuki, S. Hirono, T. Ando and O. Niwa, *Anal. Chem.* 79, 98 (2007).
[76] O. Niwa, J. Jia, Y. Sato, D. Kato, R. Kurita, K. Maruyama, K. Suzuki and S. Hirono, *J. Am. Chem. Soc.* 128, 7144 (2006).
[77] R. L. McCreery, *Chem. Rev.* 108, 2646 (2008).
[78] M. Zhou, Y. M. Zhai and S. J. Dong, *Anal. Chem.* 81, 5603 (2009).
[79] S. L. Yang, D. Y. Guo, L. Su, P. Yu, D. Li, J. S. Ye and L. Q. Mao, *Electrochem. Commun.* 11, 1912 (2009).
[80] W. J. Lin, C. S. Liao, J. H. Jhang, Y. C. Tsai, *Electrochem. Commun.* 11, 2153 (2009).
[81] L. H. Tang, Y. Wang, Y. M. Li, H. B. Feng, J. Lu and J. H. Li, *Adv. Funct. Mater.* 19, 2782 (2009).
[82] N. G. Shang, P. Papakonstantinou, M. McMullan, M. Chu, A. Stamboulis, A. Potenza, S. S. Dhesi and H. Marchetto, *Adv. Funct. Mater.* 18, 3506 (2008).
[83] D. Li, M. B. Müller, S. Gilje, R. B. Kaner and G. G. Wallace, *Nat. Nanotechnol.* 3, 101 (2008).
[84] D. Shan, Y. He, S. Wang, H. Xue and H. Zheng, *Anal. Biochem.* 356, 215 (2006).
[85] G. Zoldák, A. Zubrik, A. Musatov, M. Stupák and E. Sedlák, *Biol. Chem.* 279, 47601 (2004).
[86] S. Q. Liu and H. X. Ju, *Biosens. Bioelectron.* 19, 177 (2003).
[87] C. Deng, J. Chen, X. Chen, C. Xiao, L. Nie and S. Yao, *Biosens. Bioelectron.* 23, 1272 (2008).
[88] Y. B. Zhang, Y. W. Tan, H. L. Stormer and P. Kim, *Nature*, 438, 201 (2005).
[89] X. L. Li, G. Y. Zhang, X. D. Bai, X. M. Sun, X. R. Wang, E. Wang and H. J. Dai, *Nat. Nanotechnol.* 3, 538 (2008).
[90] J. D. Zhang, M. L. Feng and H. Yachikawa, *Biosens. Bioelectron.* 22, 3036 (2007).
[91] Q. Liu, X. Lu, J. Li, X. Yao and J. H. Li, *Biosens. Bioelectron.* 22, 3203 (2007).
[92] Y. X. Huang, W. J. Zhang, H. Xiao and G. X. Li, *Biosens. Bioelectrons.* 21, 817 (2005).
[93] J. Z. Xu, J. J. Zhu, Q. Wu, Z. Hu and H.Y. Chen, *Chin. J. Chem.* 21, 1088 (2003).
[94] E. Laviron, *J. Electroanal. Chem.* 101, 19 (1979).
[95] X. H. Kang, J. Wang, H. Wu, I. A. Aksay, J. Liu and Y. H. Lin, *Biosens. Bioelectron.* 25, 901 (2009).
[96] N. G. Shang, P. Papakonstantinou, M. McMullan, M. Chu, A. Stamboulis, A. Potenza, S. S. Dhesi and H. Marchetto, *Adv. Funct. Mater.* 18, 3506 (2008).
[97] Y. Li, Y. Y. Song, C. Yang and X. H. Xia, *Electrochem. Commun.* 9, 981, (2007).
[98] C. S. Shan, H. F. Yang, D. X. Han, Q. X. Zhang, A. Ivaska and L. Niu, *Biosens. Bioelectron.* 25, 1070 (2010).
[99] C. L. Fu, W. S. Yang, X. Chen and D. G. Evans, *Electrochem. Commun.* 11, 997 (2009).
[100] Y. Liu, M. K. Wang, F. Zhao, Z. A. Xu and S. J. Dong, *Biosens. Bioelectron.* 21, 984 (2005).
[101] Y. H. Lin, F. Lu, Y. Tu, and Z. F. Ren, *Nano. Lett.* 4, 191 (2004).

[102] M. D. Rubianes and G. A. Ravies, *Electrochem. Commun.* 5, 689 (2003).
[103] H. Muguruma, Y. Shibayama and Y. Matsui, *Biosens. Bioelectron.* 23, 827 (2008).
[104] A. Salimi, R.G. Compton and R. Hallaj, *Anal. Biochem.* 333, 49 (2004).
[105] H. Zheng, H. Xue, Y. Zhang, Z. Shen, *Biosens. Bioelectron.* 17, 541 (2002).
[106] M. Zhou, L. Shang, B. Li, L. Huang, S.J. Dong, *Biosens. Bioelectron.* 24, 442 (2008).
[107] E. I. Solomon, U. M. Sundaram and T. E. Machonkin, *Chem. Rev.* 96, 2563 (1996).
[108] S. Shleev, J. Tkac, A. Christenson, T. Ruzgas, A. I. Yaropolov, J. W. Whittaker and L. Gorton, *Biosens. Bioelectron.* 20, 2517 (2005).
[109] G. T. Palmore and H.–H. Kim, *J. Electroanal. Chem.* 464, 110 (1999).
[110] S. C. Barton, J. Gallaway and P. Atanassov, *Chem. Rev.* 104, 4867 (2004).
[111] J. Fei, H.–K. Song and G. T. Palmore, *Chem. Mater.* 19, 1565 (2007).
[112] J. H. Thomas, J. M. Drake, J. R. Paddock, S. Conklin, J. Johnson, C. J. Seliskar, H. B. W. Halsall and R. Heineman, *Electroanalysis*, 16, 547 (2004).
[113] Y. Y. Wan, Y. M. Du, F. X. Yang, Y. Xu, R. Z. Chen and J. F. Kennedy, *Int. J. Biol. Macromol.* 38, 232 (2006).
[114] X. M. Wu, B. Zhao, P. Wu, H. Zhang and C. X. Cai, *J. Phys. Chem. B*, 113, 13365 (2009).
[115] M. Zhou, Y. L. Wang, Y. M. Zhai, J. F. Zhai, W. Ren, F. Wang and S. J. Dong, *Chem. Eur. J.* 15, 6116 (2009).
[116] D. Li, M. B. Müller, S. Gilje, R. B. Kaner and G. G. Wallace, *Nat. Nanotechnol.* 3, 101 (2008).
[117] J. C. Yang, M. J. Jablonsky and J. W. Mays, *Polymer*, 43, 5125 (2002).
[118] K. Niwa, M. Furukawa and K. Niki, *J. Electroanal. Chem.* 245, 275 (1988).
[119] G. Irace, E. Bismuto, F. Savy and G. Colonna, *Arch. Biochem. Biophys.* 244, 459 (1986).
[120] W. Nogala, E. Rozniecka, I. Zawisza, J. Rogalski and M. Opallo, *Electrochem. Commun.* 8, 1850 (2006).
[121] K. Karnicka, K. Miecznikowski, B. Kowalewska, M. Skunik, M. Opallo, J. Rogalski, W. Schuhmann and P. J. Kulesza, *Anal. Chem.* 80, 7643 (2008).
[122] A. J. Bard and L. R. Faulkner, *Electrochemical Methods*, 2nd ed.; John Wiley & Sons, New York (2001).
[123] L. Meng, P. Wu, G. X. Chen, C. X. Cai, Y. M. Sun and Z. H. Yuan, *Biosens. Bioelectron.* 24, 1751 (2009).
[124] L. Meng, P. Wu, G. X. Chen and C. X. Cai, *J. Electrochem. Soc.* 155, F231 (2008).
[125] L. Meng, J. Jin, L. G. Yang, T. H. Lu, H. Zhang and C. X. Cai, *Anal. Chem.* 81, 7271 (2009).
[126] N. G. Shang, P. Papakonstantinou, M. McMullan, M. Chu, A. Stamboulis, A. Potenza, S. S. Dhesi and H. Marchetto, *Adv. Funct. Mater.* 18, 3506 (2008).
[127] Y. Liu, X. H. Qu, H. W. Guo, H. J. Chen, B. F. Liu and S. J. Dong, *Biosens. Bioelectron.* 21, 2195 (2006).
[128] P. Rowiński, R. Bilewicz, M.–J. Stébé and R. Rogalska, *Anal. Chem.* 76, 283 (2004).
[129] K. Stolarczyk, E. Nazaruk, J. Rogalski and R. Bilewicz, *Electrochim. Acta*, 53, 3983 (2008).
[130] H. X. Ju and C. Z. Shen, *Electroanalysis*, 13, 789 (2001).

[131] J. van der Zee, T. M. Dubbelman and J. Van Stereninck, *Biochim. Biophys. Acta* 818, 38 (1995).
[132] I. K. Dremza, E. A. Lapshina, J. Kujawa and I. B. Zavodnik, *Redox Report*, 11, 185 (2006).

In: Graphene: Properties, Synthesis and Applications
Editor: Zhiping Xu
ISBN: 978-1-61470-949-7

Chapter 11

GRAPHENE OXIDE: CHARACTERIZATION BY FLOW CYTOMETRY AND BIOMEDICAL APPLICATIONS

Sunil K. Singh, Paresh P. Kulkarni and Debabrata Dash*
Department of Biochemistry, Institute of Medical Sciences,
Banaras Hindu University,Varanasi, India

Abstract

Graphene, a true 2-dimensional nanomaterial, has received increasing attention due to its unique physicochemical properties like large surface area, excellent conductivity, high mechanical strength and ease of bio-functionalization. Based on its unique optical, physical and electrical properties, graphene has found applications in a wide spectrum of biomedical utilities like imaging, sensing and therapeutics. Attempts are also being made to synthesize highly stable sheets of graphene, which can be easily ingested or injected into the body. Graphene can be characterized with various characterization tools like transmission electron microscopy (TEM) and atomic force microscopy (AFM). In a recent report, we have characterized graphene oxide (GO) sheets by flow cytometry, which stands out as a remarkable and versatile means for study of physical properties of graphene and graphene-cell interactions. Flow cytometry also provides rapid access to high-dimension statistical information on individual GO sheets that is not achievable with existing characterization tools. This review discusses latest advancements in graphene research highlighting its unique characteristics, potential biomedical applications, future prospective, as well as toxicity considerations.

* E-mail address: ddass@satyam.net.in

1. INTRODUCTION

Carbon is one of the most intriguing elements in the Periodic Table. Among all the elements, carbon has the unique characteristics that allow formation of complicated and extensive networks of its own atoms [1]. Carbon is fundamental to organic chemistry and the basis for the existence of life, at least in its known forms. Diamond, graphite, and amorphous (noncrystalline) structures were the only allotropes of pure carbon that were known until 1985. Carbon nanostructures have been at the center stage of science for at least 25 years, since the discovery of a novel form of carbon called fullerene in 1985 by Kroto and co-workers [2], which has been recognized with Nobel Prize for Chemistry in 1996. Since then a large family of new carbon materials has been discovered. The most well-known form of fullerene is the buckyball (C60), which is made of 20 hexagons and 12 pentagons of carbon atoms. Almost six years later, carbon nanotubes (CNTs), which are considered a new form of fullerenes, were discovered by Sumio Iijima [3]. Thus, only three dimensional (diamond, graphite), one-dimensional (nanotubes), and zero-dimensional (fullerenes) allotropes of carbon were known. Although the nanocarbon community had grown rapidly due to the entry of fullerene-based researchers, and later CNTs-based researchers, graphene *per se* did not attract much of a following until the 2004 publication of Geim and Novoselov [4], where the authors have reported a simple easily reproducible technique for the preparation of high-quality monolayer and few-layer graphene flakes. They also described some highly innovative experiments that not only attracted a great deal of attention from the physicists but also provided exciting challenges and opportunities for chemists, biologists, engineers, and material scientists. Graphene is the name given to a flat monolayer of carbon atoms tightly packed into a two-dimensional (2D) honeycomb lattice. Graphene is the basic building block of other important allotropes; it can be stacked to form 3D graphite, rolled to form 1D nanotubes, and wrapped up to form 0D fullerenes. Being a gapless semiconductor with a linear energy spectrum, single-layer graphene realizes two-dimensional, massless relativistic particles that are of crucial importance for understanding unusual electronic properties. Besides the high mobility and minimum conductivity, graphene shows very interesting behavior in the presence of a magnetic field. Measurements have shown that graphene has a breaking strength 200 times greater than steel, making it the strongest material ever tested.

Based on its unique optical, physical and electrical properties, graphene has found applications in a wide spectrum of biomedical areas like imaging, sensing and therapeutics. As a robust yet flexible membrane, graphene provides essentially infinite possibilities for the modification or functionalization of its carbon backbone [5]. Their remarkable features like high conductivity, electron mobility, stability and optical properties are function of thickness of the sheets, which is equivalent to the number of constituent layers of carbon atoms in individual graphene sheets [6]. Therefore, method of their synthesis and characterization are one of the primary thrust areas of research. Graphene can be characterized by utilizing various characterization tools including TEM and AFM. In a recent report, we have successfully characterized graphene oxide sheets using flow cytometry [7], which is a versatile tool to generate diverse information on graphene population not obtainable by the existing characterization methods. Attempts are also being made to synthesize highly stable graphene, which can be easily ingested or injected into the body. In this chapter, we present the recent advances in this emerging field and discuss its future perspectives.

The first section of this review provides an overview of different synthesis procedures for biocompatible graphene nanosheets, their functionalization and characterization. This section also highlights the important properties of nanosheets, which can be uniquely exploited for biomedical applications. In the subsequent section we have discussed prominent biological applications of graphene and their derivatives under subsections like imaging, biosensors, diagnostics and therapeutics. In the concluding section we have focused on the issue of toxicity that needs to be fully and decisively addressed before realization of biological potential of the nanomaterials.

1.1. Synthesis and Characterization

Graphene, as a building block of graphite, has been theoretically investigated since the 1940s [8]. But Lang *et al* [9] for the first time synthesized mono- and multi-layered graphite by thermal decomposition of carbon on single crystal platinum substrates. However, Novoselov *et al.* [4] first showed repeatable synthesis of graphene through exfoliation and the same is still the most followed technique for graphene synthesis, although efforts are being made to develop new processing routes for efficient large-scale synthesis of graphene. So far, the original top-down approach of mechanical exfoliation is the most straightforward way to produce the highest quality samples, but the method is neither high-throughput nor high-yield. Generally the exfoliation can be performed either mechanically or chemically [10], or by combination of both the techniques, the most successful method being mechanical exfoliation of graphite onto oxidized silicon wafers [4]. Alternatives to mechanical exfoliation include primarily three general approaches: chemical efforts to exfoliate and stabilize individual sheets in solution [11, 12], bottom-up methods to grow graphene directly from organic precursors [13, 14] and attempts to catalyze growth *in situ* on a substrate [15-17]. The primary advantage of chemical exfoliation over the mechanical approach lies in its high-yield and scalability. As the present review focuses on biomedical applications of graphene, it is beyond its scope to describe all the existing methods for preparation of graphene sheets. This section will be restricted to synthesis of GO by chemical exfoliation. Subsequently, we will also discuss the advances being made to enhance the biocompatibility of graphene oxide, which can then be easily administered under *in vivo* conditions. In general, GO is synthesized by the Brodie [18], Staudenmaier [19] or Hummers method [20], or some variation of these methods. All three methods involve chemical exfoliation of graphite powders by oxidation of graphite to various levels. Brodie and Staudenmaier used a combination of potassium chlorate with nitric acid to oxidize graphite, while the Hummers method involves treatment of graphite with strong oxidants such as potassium permanganate dissolved in concentrated sulfuric acid. In a recent report, Morcono *et al.* [21] have modified Hummer method to increase the number of hydrophilic oxidized graphene. During reaction and processing, the graphene sheets in the particles are not only functionalized with oxygen-containing groups [22, 23] but also torn up into smaller pieces. As a result, the lateral sizes of the as-synthesized GO sheets are usually polydisperse, ranging from a few nanometers to tens of micrometers [24, 25], which may even vary with different synthesis procedures.

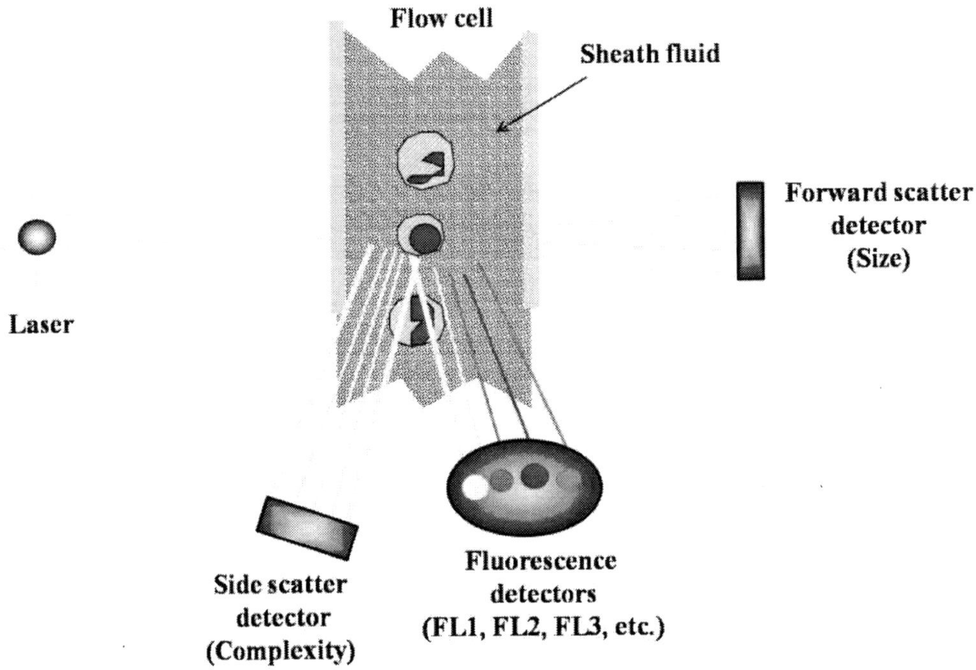

Figure1. Schematic diagram illustrating working principle of flow cytometer.

For biomedical applications of graphene, it is extremely important for it to be compatible with biological systems, and it is desirable to be water dispersible and non-toxic. This can be achievable by employing synthesis procedures which use non-toxic materials, and by appropriate functionalization of graphene. Further, there is a growing need to develop biologically benign graphene sheets using nature-friendly synthesis procedures to address the possible environmental concerns with widespread use of graphene. Both non-covalent and covalent functionalizations of graphene have been proposed to obtain sheets, which are well dispersed in range of polar solvents, particularly in water [26, 27]. Non-covalent approaches usually rely on the stabilization effect of surfactants that adsorb at the surface of graphene to produce colloidal suspensions of graphene sheets, while covalent ones involve chemical modification of GO [28-30]. Due to the higher density of charges originating from the ionized –COOH groups on the edges, smaller GO sheets become more hydrophilic, which suggests that the colloidal stability of GO is size-dependent. The presence of oxygen-containing suspending groups in GO provides a means to functionalize graphene sheets. Further, to make GO colloidal solutions more biocompatible and easier to administer under *in vivo* conditions, polyethylene glycol (PEG) was covalently grafted on hydroxyl and carboxylic groups of GO with the help of 1,1-carbonyldiimidazole [31]. Dai and coworkers [31] demonstrated that branched PEG could be grafted onto the carboxyl groups of GO sheets. This conjugate exhibited inherent photoluminescence in the near-infrared (NIR) range, thus providing for a NIR probe for cellular imaging. They also explored the use of GO as a carrier for the delivery of water-insoluble aromatic drugs via non-covalent interactions. Recently, Peng *et al.* [32] further illustrated the advantages of PEG-conjugated graphene. They showed that PEG can serve as a 'bridge' between fluorescein and graphene sheets, preventing fluorescence quenching of fluorophore by graphene sheets. As a typical polymeric stabilizer, PEGylation increases the stability and dispersity of graphene sheets in aqueous solvents, particularly in cell culture medium, and enhances cellular uptake of nanomaterials. Hence, PEGylation of GO could prove to be a valuable tool in improving biocompatibility of graphene.

As the properties of graphene depend critically on the number of layers and size, finding reliable and easy-to-use methods to characterize graphene in terms of its thickness and size is crucial. Techniques like electron microscopy (TEM and SEM) and AFM are well established for direct atomic resolution imaging of GO sheets. However, these methods are relatively more expensive and time consuming. Besides this, TEM requires a complicated sample preparation procedure and AFM suffers from the fact that there is always an offset in the measurement [33]. In this context, Ferrari *et al.* [34] demonstrated the capability of using 2D band in Raman spectra to study the number of layers of Bernal or AB stacking graphene and thereafter Raman spectroscopy/imaging has become a general method in graphene thickness determination. But each technique has its own limitations. Data on size distribution emanating from these methods will rely on observation of fewer GO sheets and could easily be influenced by observer bias.

Table 1. Comparison of flow cytometer with conventional characterization tools for GO sheets

	Flow cytometry	TEM/AFM
Measurement of size	Yes (Relative)	Yes (Absolute)
Measurement of thickness	No	Yes
Size distribution analysis	Yes (Based on tens of thousands of observations)	Yes (Based on a maximum of few hundred observations)
Measurement of Fluorescence	Yes	No
Sorting	Yes	No
Easy Handling	Yes	No
Rapidity of characterization	Yes	No

Flow cytometry, on the other hand, is a technique that integrates light scattering and fluorescence signals emanating from a group of cells or particles in the path of laser beam (Figure 1), thereby generating extensive statistical data on size, shape and internal characteristics of individual particles within the population. The method has found wide applications in biological research in view of its predilection for multi-parameter analysis at the single-cell level. Flow cytometry usually characterizes cells or particles with typical size range between 0.5 and 70 μm. In a recent report [7], we have projected flow cytometry as an alternative characterization tool having unique advantages, like simultaneous measurement of size distribution and fluorescence properties of individual GO sheets in a population and the possibility of analyzing thousands of (50,000 or more events) sheets in a matter of few seconds. Because of its unique ability to acquire scatter as well as multi-channel fluorescence emission signals emanated from individual GO sheets, this approach would add new dimension to conventional graphene characterization methods. Moreover, a flow cytometer with sorting capability can be employed to sort and retrieve different subpopulations of a GO preparation based on parameters like size, physical characteristics or fluorescence yield, thus opening up immense application possibilities. Thus, flow cytometry is more versatile than TEM, AFM and other available techniques in terms of rapidity, high sensitivity, ability to simultaneously evaluate size distribution and fluorescence properties, sorting different populations of GO and monitoring GO interaction with cells (Table 1). Thus, flow cytometry and microscopy-based characterization tools can complement each other to generate data unique to the particular method.

1.2. Properties

Graphene has not only attracted a great deal of attention from physicists but has also provided exciting challenges and opportunities for chemists, biologists, engineers and material scientists. Keeping in view the scientific interest generated by graphene and its future potential in electronics and sensing applications, lot of research effort are devoted in understanding the structure and properties of graphene. A single-layer graphene is defined as a single two-dimensional hexagonal sheet of carbon atoms [35]. Bi-layer and few-layer graphene has 2 and 3 to 10 layers of such two-dimensional sheets, respectively. Graphene structures consisting more than 10 such layers are considered as thick graphene sheets and are of less scientific interest. In the subsequent sections we have discussed the properties mostly associated with single-layer, bi-layer and few-layer graphene. Detailed discussion of properties of graphene is beyond the scope of this review. This section aims to introduce the basic properties of graphene, so that the biological applications are well understood.

1.2.1. Electrical properties

Graphene is unique in electronic structure among the carbon allotropes, as it shows band-overlap in two conical points K and K' (Direc points) at the corners of the first Brillouin zone. This gives rise to a number of unusual physical properties that the graphene is expected to show as compared with metals, semiconductors and semi-metals. Due to its distinctive structure, graphene exhibits several peculiar electronic properties like quantum Hall effect with electrons on its surface obeying a linear dispersion relation, behaving like zero mass relativistic particles [36, 37]. Few-layer graphenes and nanographite particles exhibit semiconducting or insulating behavior with their resistance showing little change in the range 100–300 K. The resistivity increases sharply below 50 K and decreases markedly if the graphene is heated to high temperatures. Thus, graphene obtained from exfoliation of graphite demonstrates semiconducting properties. Graphene sheets prepared from graphite oxide display well-behaved field-effect transistor properties [38]. Due to these remarkable electrical properties, graphene reveals outstanding electron transfer promoting ability for oxido-reductase enzymes and excellent catalytic behavior towards small biomolecules such as H_2O_2

and NADH [39-46], thus making it extremely attractive for use in enzyme-based biosensors. We shall discuss this in detail under the section on biomedical applications of graphene.

1.2.2. Mechanical properties

Graphite, diamond and carbon nanotubes have each set their own record in terms of mechanical robustness, be it hardness or Young's modulus. The Young's modulus of graphene was experimentally investigated with force-displacement measurements by AFM. The force-displacement behavior obtained from the nanoindentation experiments yielded Young's modulus of 1.0 TPa and intrinsic strength of 130 GPa for monolayer graphene sheets, which establish graphene as the strongest material ever measured. The high Young's modulus, extremely low mass, large surface area, high sustainable tension in a single sheet, combined with relatively low cost make this nanomaterial an ideal candidate for mechanical reinforcement and mechanical devices best suited for sensing mass, force, and charges [47-49].

1.2.3. Thermal properties

Since the carrier density of non-doped graphene is relatively low, the electronic contribution to thermal conductivity (Wiedemann-Franz law) is negligible. The thermal conductivity (κ) of graphene is thus dominated by phonon transport, namely diffusive conduction at high temperature and ballistic conduction at sufficiently low temperature [50]. A study by Berber and colleagues predicted that suspended monolayer graphene at room temperature showed thermal conductivity of about 6000 Wm^{-1} K^{-1}, which was stated to be much higher than that of graphitic carbon [51]. In a recent optical measurement using the shift in the Raman G band, thermal conductivity value of about 5000 Wm^{-1} K^{-1} was obtained for suspended monolayer graphene 'flake', produced by using mechanical exfoliation [52].

1.2.4. Magnetic properties

In some reports, magnetic properties of nanographene sheets have also been discussed [53, 54]. The edge states and/or defects in the graphene sheets are the key players behind these magnetic properties. But, magnetism in carbon-based materials with networks of sp^2 hybridized carbon atoms has been controversial due to the possibility of contamination with magnetic impurities [55].

1.2.5. Optical properties

GO (yellowish brown color) exhibited optical absorption in the visible and near-infrared range. The maximum absorbance at 230 nm could originate from $\pi - \pi^*$ transition of aromatic sp^2 domains [4, 7, 56], whereas a minor peak at ~320 nm might be attributed to $n - \pi^*$ transitions of $C = O$ [4, 7, 10]. The inherent optical properties and stable nature of GO suspensions could facilitate biomedical application in areas of cellular imaging and drug delivery. We shall discuss in detail about this under the next section on biomedical applications.

1.3. Biomedical applications

Among various nanostructures developed, carbon-based nanomaterials (e.g., fullerene and carbon nanotubes) have received much attention due to their remarkable physical, chemical, and biological properties. The distinct structural properties of carbon nanomaterials, in particular their high aspect ratio, propensity to functional modification, unique electronic and optical properties as well as their potential biocompatibility, make them extremely attractive candidates for biomedical applications like biosensor development, imaging and drug delivery (Figure 2).

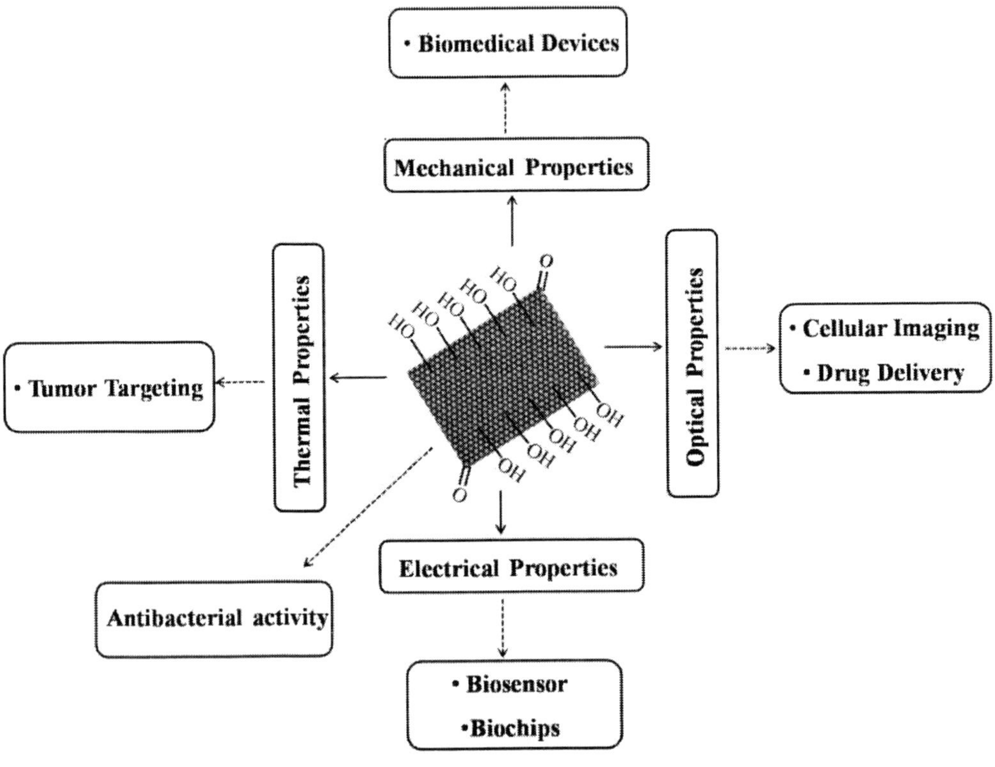

Figure 2. Schematic diagramme showing biomedical applications of GO.

1.3.1 Biosensors

Electrochemical biosensors utilizing nanomaterials have recently attracted considerable attention [57-59]. Carbon nanotubes [60, 61], gold nanoparticles [62, 63], metal oxides [64] and semiconductors [65] have been successfully employed in biosensors with applications in fields of medicine and food quality control. Biosensors are based on the principle of electrochemical analysis and cyclic voltammetry, which measure potential and/or current between the electrodes in an electrochemical cell containing the analyte. Potential is applied between the reference electrode and the working electrode and current is measured between the working electrode and the counter electrode within the cell. Voltammetry involves measurement of cell's current, while the electrical potential between the electrodes is actively altered. In cyclic voltammetry the electrode potential is varied linearly with time, which is measured as scan rate (V/s). The data are then plotted as current vs potential to give the cyclic voltammogram trace. Current will increase as potential reaches the reduction potential of the analyte but subsequently falls off as the concentration of the analyte is depleted close to the electrode surface. Electrolyte is usually added to the test solution to ensure sufficient conductivity. Electrodes are static and sit in unstirred solutions. Common materials used for working electrodes include glassy carbon or platinum. Coating of working electrodes with nanomaterials enhances their electron transfer rates and thus improves sensitivity of the device. Coating is simply carried out by dropping a uniformly dispersed solution of the nanomaterial on the electrode and allowing it to dry. The counter electrode, also known as the *auxiliary* or *second electrode*, can be made of any material (usually a platinum wire), which conducts easily and will not react with the bulk solution. The reference electrode is usually an Ag/AgCl electrode. Presence of biomolecules in the analyte can be evaluated by cyclic voltammetry. Binding of biomolecule to *working electrode*'s surface will lead to a progressive increase in electron transfer resistance that can be measured from the radius of the cyclic voltammetric curve.

In view of its distinctive electronic properties described in earlier sections, graphene is extremely suitable for application in biosensors. Several studies have established the utility of graphene-based biosensors using different redox probes including H_2O_2, NADH and ascorbic acid (39-46). Graphene-based biosensors have been employed for sensing several important biomolecules including glucose, DNA, dopamine, and alcohol (39-43). In a recent study cytochrome c was immobilized on graphene-coated electrode, which was then utilized in the sensing of nitric oxide. This illustrates how graphene-coated electrode with an antibody or enzyme immobilized on its surface can be utilized in sensing corresponding antigen or substrate. This substantially widens the possible array of biomolecules that can be detected by graphene-based biosensors. Further, graphene's modifiable chemistry, atomic thickness and molecularly gatable structure make it an excellent candidate for mammalian and microbial detection and diagnosis [66].

With growing application of graphene in biosensors, efforts are also being directed towards functionalization of graphene to further improve its utility. Functionalization of GO with methylene green during its chemical reduction was found to significantly improve the water dispersibility of GO. The chemically functionalized graphene can readily mix with polymers in solution to form stable dispersion and yield novel electrically conductive nanocomposites [67-69]. In another study, covalent functionalization of GO with poly L-lysine was shown to provide a compatible environment for bioconjugation, such as attaching bioactive molecules like enzymes. A recent study has also established that electrochemically reduced GO has faster electron transfer rates than pristine GO utilizing different redox probes. Thus, with the advances in graphene functionalization and chemical modification techniques the array of applications for graphene in biosensors is bound to increase dramatically. Apart from electrochemical analyzers, biosensors based on the principle of fluorescence resonance energy transfer (FRET) have also utilized GO to detect thrombin, DNA-DNA hybridization and tumor markers [70-72].

1.3.2 Imaging

A recent study by Peng *et al.* [73] demonstrated that fluorescein-functionalized PEGylated GO exhibits excellent fluorescent properties, and that it can be efficiently taken up by cells and serve as a fluorescent nanoprobe for cellular imaging. A previous study by Sun *et al.* [74] had already found that GO sheets are photoluminescent in the visible and infrared regions, and that this intrinsic photoluminescence can be used for live cell imaging in the near-infrared range with little background.

1.3.3 Therapeutics

Owing to its small size, large specific surface area and useful non-covalent interactions with drug molecules GO is a promising tool in drug delivery. Antibody-conjugated PEGylated GO has been shown to be effective in targeted delivery of anti-cancer drugs to tumor cells [74]. More importantly, a recent study has shown that potent, water insoluble anti-cancer drugs could be rendered therapeutically useful by complexing with PEGylated GO. The resulting complex exhibited excellent water solubility while maintaining its high cancer cell killing potency similar to that of the free drug molecules in organic solvents [75]. Apart from drug delivery, graphene has also been shown to have potential for application in thermal ablation therapy of tumors, where GO targeted to tumor cells were heated up by near-infrared laser leading to destruction of tumor cells [76]. Further, recent reports have implicated GO nanowalls and nanosheets with antibacterial properties [77, 78]. In the former report, GO nanowalls deposited on stainless steel substrates were shown to have significant antibacterial activity against both Gram positive and Gram negative bacteria, which was found to be mediated by cell membrane damage. They also found that reduced GO had a higher antibacterial activity and attributed it to better charge transfer to the bacterial cell [77]. The second report [78] also described significant bactericidal and bacteriostatic activity of GO nanosheets against E coli through disruption of cell integrity. However, contrary to the previous report these investigators observed slightly lower antibacterial toxicity of reduced GO compared to GO [78]. Thus, the antibacterial activity of GO needs to be evaluated further both in terms of relative efficacy towards different bacteria and the influence of different graphene synthesis, modification or functionalization procedures. It can be speculated that, the next generation antimicrobials would be composed of nanosheets of graphene with a market-driven thrust comparable to that observed with nanosilver.

1.4. Toxicity assessment

Concern over harmful effects of nanomaterials has stimulated the advent of nanotoxicology as a significant research discipline. Literature is stuffed with reports that carbon nanomaterials can cause *in vitro* and *in vivo* toxicity. Most *in vivo* toxicity data have been generated from studies using carbon nanotubes. Animal studies have shown that exposure to nanotubes can cause inflammation, fibrosis, epithelioid granuloma formation in lungs, [79-81] cellular cytotoxicity and cardiopulmonary and vascular irregularities [82-84]. Carbon nanotubes are also known to induce oxidative stress in human keratinocytes via nuclear factor- κB [85]. Nanodiamond is known to be biologically inert with relatively low cytotoxicity [86], supporting its suitability for biomedical applications. But there have been reports to suggest that, nanodiamond induces marked elevation in IL-8 expression in human airway epithelial cells through a post-transcriptional mechanism that involves stabilization of IL-8 transcripts [87]. Recently Yuan *et al.* [88] investigated the accumulation and translocation of diamond nanoparticles in mouse liver, lung and spleen. In other recent reports, carbon nanomaterials were found to adversely affect the platelets, an important component of circulatory system responsible for blood clotting and haemostasis [84, 89, 90]. Nanocarbon material like SWNTs (single walled nanotubes), MWNTs (multiwalled nanotubes), C60 fullerenes (C60CS) and mixed carbon nanoparticles (MCN) were found to aggravate platelet aggregation potentially [84] ,which may lead to deadly consequences like stroke, myocardial infarction and deep vein thrombosis [91-93]. Enhanced aggregation of platelets by different nanocarbons has been suggested to be the outcome of extracellular calcium influx into the cell [90].

However, till date, very few studies have been carried out regarding the toxicity of graphene towards cells or live organisms [94-97]. In vitro study showed that GO induces slight decrease of the cell viability at higher dose and can elicit concentration-dependent cytotoxicity in A549 cells and PC12 cells [94, 95], which is mediated by oxidative stress and apoptosis. The shape of graphene sheets or their agglomeration may be important factors behind graphene toxicity. Animal studies have indicated that GO could induce severe cytotoxicity in dose- and time dependent manner, and can enter into cytoplasm and nucleus, decreasing cell adhesion, inducing cell floating and apoptosis [96]. GO can enter into lung tissues, induce inflammation and subsequent granuloma formation. Further comparative study suggests that cells respond differently to reduced GO and carbon nanotubes, which may be attributed to their distinct nanotopographic features [97]. The reduced GO film is biocompatible with all tested cells, implying its potential application in biology. Recently we have demonstrated that GO can potentially induce integrin mediated platelet aggregation both *in vitro* and *in vivo* [98]. Src and Syk tyrosine kinases were found to play key role in GO-induced platelet activation. From comparative studies with reduced GO, we inferred that charge distribution on GO surface regulates its capability to interact with cells and activate and aggregate platelets. Surface characteristics of nanomaterials, like shape, aggregate formation, coating and functionalization, may be deciding factors for their adverse toxic effects. Toxicity depends upon the surface area/mass ratio and retention time of the particles – particles having larger surface area and retention time have greater contact with the cell membrane, and hence have greater capacity for adsorption and transport into cells. Compared to carbon nanotubes, whose toxicity may be attributed to their water insolubility and susceptibility to form aggregates, more water dispersible graphene can serve as a less toxic material for applications in physiological systems. But it would be advisable to critically assess the adverse prothrombotic and cytotoxic effects of GO before exploiting its biomedical potential as therapeutic and diagnostic tool.

References

[1] L. Pauling, *The Nature of the Chemical Bond*, Cornell University Press, Ithaca, NY, (1960).

[2] S. Iijima, *Nature* 354, 56 (1991).

[3] H. W. Kroto, J. R. Heath, S. C. O'Brien, R. F. Curl and R. E. Smalley, *Nature* 318, 162 (1985).

[4] K. S. Novoselov, A. K. Geim, S. V. Morozov, D. Jiang, Y. Zhang, S. V. Dubonos, I. V. Grigorieva, and A. A. Firsov, *Science* 306, 666 (2004).

[5] D. W. Boukhvalov and M. I. Katsnelson, *J. Phys. Condens. Matter* 21, 344205 (2009).

[6] N. R. Wilson, P. A. Pandey, R. Beanland, R. J. Young, I. A. Kinloch, L. Gong, Z. Liu , K. Suenaga, J. P. Rourke , S. J. York and J. Sloan, *ACS Nano 3,* 2547 (2009).

[7] S. K. Singh, M. K. Singh, M. K. Nayak, S. Kumari, J. J. A. Gracio and D. Dash, C*arbon* 49, 684 (2011).

[8] P. R. Wallace, *Phys. Rev.* 71, 622 (1947).

A. Lang, *Surface Sci.* 53, 317 (1975).

[9] S. J. Park and R. S. Ruoff, *Nat. Nanotechnol.* 4, 217 (2009).

[10] S. Stankovich, D. A. Dikin, R. D. Piner, K. A. Kohlhaas, A. Kleinhammes, Y. Jia, Y. Wu, S. T. Nguyen and R. S. Ruoff, *Carbon* 45, 1558 (2007).

[11] Y. Hernandez, V. Nicolosi, M. Lotya, F. M. Blighe, Z. Y. Sun, S. De, I. T. McGovern, B. Holland, M. Byrne, Y. K. Gun'ko, J. J. Boland, P. Niraj, G. Duesberg, S. Krishnamurthy, R. Goodhue, J. Hutchison, V. Scardaci, A. C. Ferrari and J. N. Coleman, *Nat. Nanotechnol.* 3, 563 (2008).

[12] N.Tyutyulkov, G. Madjarova, F. Dietz and K. Mullen, *J. Phys. Chem. B* 102, 10183 (1998).

[13] X. Y. Yang, X. Dou, A. Rouhanipour, L. J. Zhi, H. J. Rader and K. Mullen, *J. Am. Chem. Soc.* 130, 4216 (2008).

[14] C. Berger, Z. M. Song, T. B. Li, X. B. Li, A. Y. Ogbazghi, R. Feng, Z. T. Dai, A. N. Marchenkov, E. H. Conrad, P. N. First and W. A. de Heer, *J. Phys. Chem. B* 108, 19912 (2004).

[15] K. S. Kim, *Nature* 457, 706 (2009).

[16] P. W. Sutter, J. I. Flege and E. A. Sutter, *Nat. Mater.* 7, 406 (2008).

[17] B. C. Brodie, *Ann. Chim. Phys.* 59, 466 (1860).

[18] L. Staudenmaier, *Ber. Deut. Chem. Ges.* 31, 1481 (1898).

[19] W. S. Hummers and R. E. Offeman, *J. Am. Chem. Soc.* 80, 1339 (1958).

[20] D. C. Marcano, D. V. Kosynkin, J. M. Berlin, A. Sinitskii, Z. Sun, A. Slesarev, L. B. Alemany, W. Lu and J. M. Tour, *ACS Nano* 4, 4806 (2010).

[21] Lerf, H. Y. He, M. Forster and J. Klinowski, *J. Phys. Chem. B* 102, 4477 (1998).

[22] K. Erickson, R. Erni, Z. Lee, N. Alem, W. Gannett and A. Zettl, *AdV. Mater.* 22, 4467 (2010).

[23] M. J. Allen, V. C. Tung and R. B. Kaner, *Chem. ReV.* 110, 132 (2009).

[24] O. C. Compton and S. T. Nguyen, *Small* 6, 711 (2010).

[25] J. Patil, J. L. Vickery, T. B. Scott and S. Mann, *Adv. Mater.* 21, 1 (2009).

[26] S. Zu and B. Han, *J. Phys. Chem. C* 113, 13651 (2009).

[27] N. Mohanty and V. Berry, *Nano Lett.* 8, 4469 (2008).

[28] G. Wang, X. Shen, B. Wang, J. Yao and J. Parka, *Carbon,* 47, 1359 (2009).

[29] C. Shan, H. Yang, J. Song, D. Han, A. Ivaska and L. Niu, *Anal. Chem.* 21, 2378 (2009).

[30] X. Sun, Z. Liu, K. Welsher, J. T. Robinson, A. Goodwin, S. Zaric and H. Dai, *Nano Res.* 1, 203 (2008).
[31] C. Peng, W. Hu, Y. Zhou, C. Fan, and Q. Huang, *Small* 6, 1686 (2010).
[32] Gupta, G. Chen, P. Joshi, S. Tadigadapa, and P. C. Eklund, *Nano Lett.* 6, 2667 (2006).
[33] Ferrari, J. C. Meyer, V. Scardaci, C. Casiraghi, M. Lazzeri, F. Mauri, S. Piscanec, D. Jiang, K. S. Novoselov, S. Roth, and A. K. Geim, *Phys. Rev. Lett.* 97, 187401 (2006).
[34] J. Hass, W. A. de Heer, and E. H. Conrad, *J. Phys. Condens. Matter* 20, 323202 (2008).
[35] C. N. R. Rao, K. Biawas, K. S. Subramanyam, and A. Govindaraj, *J. Mater. Chem.* 19, 2457 (2009).
[36] Y.B. Zhang, Y. Tan, H.L. Stormer and P. Kim, *Nature* 438, 201(2005).
[37] S. Gilje, S. Han, M. Wang, K. L.Wang and R. B. Kaner, *Nano Lett.* 7, 3394 (2007).
[38] Y. Shao, J. Wang, H. Wu, J. Liu, I. A. Aksay and Y. Lin, *Electroanalysis*, 22, 1027 (2010)
[39] Z. Wang, S.N. Liu, P. Wu and C. X. Cai, *Anal. Chem.* 81, 1638 (2009).
[40] S. Alwarappan, A. Erdem, C. Liu and C. Z. Li, *J. Phys. Chem. C*, 113, 8853 (2009).
[41] Y. Wang, Y. M. Li, L. H. Tang, J. Lu and J. H. Li, *Electrochem.Commun.* 11, 889 (2009).
[42] C.S. Shan, H.F. Yang, J. F. Song, D.X. Han, A. Ivaska and L. Niu, Anal. *Chem.* 81 2378 (2009).
[43] M. Zhou, Y. M. Zhai and S. J. Dong, *Anal. Chem.* 81, 5603 (2009).
[44] L. H. Tang, Y. Wang, Y. M. Li, H. B. Feng, J. Lu and J. H. Li, *Adv. Funct. Mater.* 19, 2782 (2009).
[45] H. Liu, J. Gao, M. Q. Xue, N. Zhu, M. N. Zhang and T. B. Cao, *Langmuir* 25, 12006 (2009).
[46] C. D. Reddy, S. Rajendran and K. M. Liew, *Nanotechnology* 17, 864 (2006).
[47] C. Lee, X. D. Wei, J. W. Kysar and J. Hone, *Science* 321, 385 (2008).
[48] J. W. Suk, R. D. Piner, J. An and R. S. Ruoff, *ACS Nano* 4, 6557 (2010).
[49] C. H. Yu , L. Shi , Z. Yao , D. Y. Li and A. Majumdar, *Nano Lett.* 5, 1842 (2005).
[50] S. Berber, Y. K. Kwon and D. Tomanek, *Phys. Rev. Lett.* 84, 4613 (2000).
[51] S. Ghosh , I. Calizo , D. Teweldebrhan , E. P. Pokatilov , D. L. Nika , A. A. Balandin , W. Bao, F. Miao and C. N. Lau , *Appl. Phys. Lett.* 92, 151911 (2008).
[52] T. Enoki and Y. Kobayashi, *J. Mater. Chem.* 15, 3999 (2005).
[53] T. Enoki, Y. Kobayashi and K. Fukui, *Int. Rev. Phys. Chem.* 26, 609 (2007).
[54] C. N. R. Rao, A. K. Sood, K. S. Subrahmanyam, and A. Govindaraj, *Angew. Chem.* 48, 7752 (2009).
[55] K. P. Loh, Q. Bao, G. Eda and M. Chhowalla *Nature chemistry* 2, 1015 (2010).
[56] Willner, B. Willner and E. Katz, *Bioelectrochemistry* 70, 2 (2007).
[57] P. Pandey, M. Datta and B. D. Malhotra, *Anal. Lett.* 41, 159 (2008).
[58] F. Valentini and G. Palleschi, *Anal. Lett.* 41, 479 (2008).
[59] B. R. Azamian, J. J. Davis, K. S. Coleman, C. B. Bagshaw and M. L. H. Green, *J. Am. Chem. Soc.* 124, 12664 (2002).

[60] G. A. Rivas, M. D. Rubianes, M. C.Rodriguez, N. E. Ferreyra, G. L. Luque, M. L. Pedano, S. A. Miscoria and C. Parrado, *Talanta*, 74, 291 (2007).
[61] B. K. Jena and C. R. Raj, *Anal. Chem.* 78, 6332 (2006).
[62] J. M. Pingarron, P. Yanez-Sedeno and A. Gonzalez-Cortes, *Electrochim. Acta*, 53, 5848 (2008)
[63] M. Hsing, Y. Xu and W. T. Zhao, *Electroanalysis*, 19, 755 (2007).
[64] W. Vastarella and R. Nicastri, *Talanta*, 66, 627 (2005).
[65] N. Mohanty and V. Berry, *Nano Lett.* 8, 4469 (2008).
[66] D. Li and R. B. Kaner, *Science*, 320 1170 (2008).
[67] X. L. Li, X. R. Wang, L. Zhang, S. Lee and H. J. Dai, *Science* 319, 1229 (2008).
[68] C. Xu, X. Wang and J. W. Zhu, *J. Phys. Chem. C* 112, 19841 (2008).
[69] F. Liu, J. Y. Choi and T. S. Seo, *Biosensors and Bioelectronics* 25, 2361 (2010).
[70] H. Chang, L. Tang, Y. Wang, J. Jiang, and J. Li, *Anal. Chem.* 82, 2341 (2010).
[71] R. S. Swathi and K. L. Sebastiana, *J. Chem. Phys.* 129, 054703 (2008).
[72] C. Peng, W. Hu, Y. Zhou, C. Fan and Q. Huang, *Small* 6, 1686 (2010).
[73] Sun, X. Z. Liu, K. Welsher, J. T. Robinson, A. Goodwin, S. Zaric and H. Dai, *Nano Res.* 1, 203 (2008).
[74] Z. Liu, J. T. Robinson, X. Sun, and H. Dai, *J. Am. Chem. Soc.* 130, 10876 (2008).
[75] K.Yang, S. Zhang, G. Zhang, X. Sun, S. -T. Lee, and Z. Liu, *Nano lett.* 10, 3318 (2010).
[76] O. Akhavan and E. Ghaderi, *ACS Nano* 4, 5731 (2010).
[77] W. Hu, C. Peng, W. Luo, M. Lv, X. Li, D. Li, Q. Huang and C. Fan, *ACS Nano* 4, 4317 (2010).
[78] J. B. Mangum, E. A. Turpin, A. Antao-Menezes, M. F. Cesta, E. Bermudez and J. C. Bonner, *Part Fibre Toxicol.* 3, 15 (2006).
A. Shvedova, J. P. Fabisiak, E. R. Kisin, A. R. Murray, J. R. Roberts, Y. Y.Tyurina, J. M. Antonini, W. H. Feng, C. Kommineni, J. Reynolds, A. Barchowsky, V. Castranova and V. E. Kagan, *Am. J. Respir. Cell Mol. Biol.* 38, 579 (2008).
[79] D. B. Warheit, B. R. Laurence, K. L. Reed, D. H. Roach, G. A. Reynolds and T. R. Webb, *Toxicol. Sci.* 77, 117 (2004).
[80] G. Jia, H. Wang, L. Yan, X. Wang, R. Pei, T. Yan, Y. Zhao and X. Guo, *Environ. Sci. Technol.*, 39, 1378 (2005).
[81] J. Muller, F. Huaux, N. Moreau, P. Misson, J. F. Heilier and M. Delos, *Toxicol. Appl. Pharmacol.* 207, 221 (2005).
[82] Radomski, P. Jurasz, D. Alonso-Escolano, M. Drews, M. Morandi, T. Malinski and M. W. Radomski, *Br. J. Pharmacol.* 146, 882–893 (2005).
[83] S. K. Manna, S. Sarkar, J. Barr, K. Wise, E. V. Barrera, O. Jejelowo, A. C. Rice-Ficht and G. T. Ramesh, *Nano Lett.* 5, 1676 (2005).
[84] S. J. Yu, M. W. Kang, H. C. Chang, K. M. Chen and Y. C. Yu, *J. Am. Chem. Soc.* 127, 17604 (2005).
[85] R. Silbajoris, J. M. Huang, W-Y. Cheng, L. Dailey, T. L. Tal, I. Jaspers, A. J. Ghio, P. A. Bromberg and J. M. Samet, *Nanotoxicology* 3,152 (2009)
[86] Y. Yuan, Y. Chen, J-H. Liu, H. Wang and Y. Liu, *Diamond & Related Materials* 18, 95 (2009).

[87] P. Bihari, M. Holzer, M. Praetner, J. Fent, M. Lerchenberger, C. A. Reichel, M. Rehberg, S. Lakatos and F. Krombach, *Toxicology* 269, 148 (2010).

[88] J. Semberova, S.H.D.P. Lacerda, O. Simakova, K. Holada, M. P. Gelderman and J. Simak, *Nano Lett.* 9, 3312 (2009).

[89] F. Saller, M.Schapira and A. Angelillo-Scherrer, *Curr. Signal Transd. T.* 3, 22 (2008).

[90] Badruddin and Philip B. Gorelick. *Curr. Treat. Options in Neurol.* 11,452 (2009).

[91] Y. A. Talavera, I. M. Hernandez and C.V. Portilla, *Revista Ecuatoriana De Neurologia* 16, 127 (2007).

[92] Y. Chang, S-T. Yanga, J-H. Liua, E. Dong, Y. Wang, A. Cao, Y. Liu and H. Wang, *Toxicol.Lett.* 200, 201 (2011).

[93] Y. Zhang, S. F. Ali, E. Dervishi, Y. Xu, Z. Li, D. Casciano and A. S. Biris, *ACS Nano* 4, 3181 (2010).

[94] K.Wang, J. Ruan, H. Song, J. Zhang, Y. Wo, S. Guo and D. Cui, *Nanoscale Res. Lett*, 6, 8 (2011).

[95] S. Agarwal, X. Zhou, F.Ye, Q. He, G. C. K. Chen, J. Soo, F. Beoy, H. Zhang and P. Chen, Langmuir 26, 2244 (2010).

[96] S. K. Singh, M. K. Singh , M. K. Nayak, S. Kumari, S. Shrivastava, J. J. A. Grácio and D. Dash, ACS Nano 5, 4987 (2011)

INDEX

C

D

E

F

G

H

M

N

O

P

V

W

X

Y

Z